AF182940

Sorption and Degradation of Pesticides and Organic Chemicals in Soil

Sorption and Degradation of Pesticides and Organic Chemicals in Soil

Proceedings of a symposium sponsored by Divisions S-3, S-1, S-2, and A-5 of the Soil Science Society of America and American Society of Agronomy in Denver, Colorado, 30 Oct. 1991.

Editorial Committee
D. M. Linn (chair)
T. H. Carski
M. L. Brusseau
F. -H. Chang

Organizing Committee
D. M. Linn
T. H. Carski

Editor-in-Chief SSSA
R. J. Luxmoore

Editor-in-Chief ASA
G. A. Peterson

Managing Editor
S. H. Mickelson

SSSA Special Publication Number 32

Soil Science Society of America, Inc.
American Society of Agronomy, Inc.
Madison, Wisconsin, USA

1993

Cover Design: Patricia Scullion

Cover Artwork: Scanning electron micrograph of a bacterial microcolony in soil, bar = 1 μm. (Courtesy of D.M. Linn.)

Soil Science Society of America, Inc.
American Society of Agronomy, Inc.
677 South Segoe Road, Madison, WI 53711 USA

Library of Congress Cataloging-in-Publication Data

Sorption and degradation of pesticides and organic chemicals in soil : proceedings of a symposium sponsored by Divisions S-3, S-1, S-2, and A-5 of the Soil Science Society of America and American Society of Agronomy in Denver, Colorado. 20 Oct. 1991 / editorial committee, D.M. Linn (chair) ... [et al.].
p. cm. — (SSSA special publication ; no. 32)
ISBN 0-89118-803-7
1. Soils—Pesticide content—Congresses. 2. Soils—Organic compound content—Congresses. 3. Pesticides—Biodegradation—Congresses. 4. Organic compounds—Biodegradation—Congresses. I. Linn, D.M. (Daniel M.) II. Soil Science Society of America. Division S-3. III. Series.
S592.6.P43S67 1993
631.4′1—dc20 93-25573

Printed in the United States of America

CONTENTS

FOREWORD

Man and synthetic and naturally occurring organic compounds ultimately interact with soil. It is essential that we have detailed knowledge about the movement, sorption, degradation, and dissipation that organic compounds undergo in soil if the contamination of natural resources is to be minimized. This special publication contains the papers presented at a symposium entitled "Sorption and Degradation of Agricultural Chemicals in Soils" that was held at the 1991 Annual Meetings of SSSA and ASA. The papers present state-of-the-art information regarding the relationships between sorption and degradation of anthropogenic compounds in soils. This needed information is critical to the development of models predicting the fate of organic chemicals in soils. The SSSA and ASA Executive Committees commend the organizers of the symposium and the editors of the special publication for identifying an important issue and producing a high quality symposium and publication.

DARREL W. NELSON, *president,*
Soil Science Society of America

DENNIS R. KEENEY, *president*
American Society of Agronomy

PREFACE

Few, if any, processes that occur in soil are isolated from the effects of their surrounding environment. The degradation of man made or naturally occurring organic compounds is no exception. Over the past several decades considerable time and energy has been spent on defining the conditions that influence the breakdown of organic compounds in soil. For example, the effects of water content, temperature, nutrient and oxygen availability, pH, and the types of microorganisms involved have all been examined. However, until recently the idea that an organic compound's physical or chemical behavior in soil could have an impact on its rate of degradation received far less attention.

One area in which interest is rapidly growing concerns the role that adsorption may play in affecting the rate and extent of an organic compound's degradation in soil. Bioavailability of sorbed compounds, effects of sorption on degradation and technologies required to measure degradation of sorbed versus non-sorbed compounds are some of the questions currently being addressed. Ultimately, answers to these questions will allow us to couple the dynamic processes of sorption and degradation in a way that will improve our understanding and ability to predict the fate of organic compounds in the soil environment.

As a result of the interest in this area, the symposium "Sorption and Degradation of Agricultural Chemicals in Soils" was conceived, organized and held at the annual SSSA meetings in Denver, CO during October, 1991. The symposium was sponsored by Division S-3 and cosponsored by Divisions S-1 and S-2 of the SSSA and Division A-5 of the ASA. This Special Publication represents the proceedings of that symposium. The chapters that follow attempt to bring together current thinking regarding the issues and constraints involved in coupling the processes of sorption and degradation. It is hoped that these proceedings will be of value not only to seasoned soil and environmental scientists, but to students of these disciplines as well.

I would like to extend my thanks to the authors involved in this publication for their valuable contributions, to the editorial committee and numerous other scientists for their help in reviewing manuscripts and to the SSSA staff for their patience, help, and professional guidance during the publication process.

DANIEL M. LINN, *chair and editor*
DuPont Agricultural Products
Stine-Haskell Research Center
Newark, Delaware

CONTRIBUTORS

D. D. Adrian Professor, Department of Civil Engineering, Louisiana State University, Baton Rouge, LA 70803-7511

Calvin C. Ainsworth Senior Research Scientist, Geosciences Laboratory, Pacific Northwest Laboratory, Richland, WA 99352

K. Balu Senior Group Leader, Residue Chemistry Department, CIBA-GEIGY Corporation, Greensboro, NC 27419

C. A. Bellin Research Chemist, DuPont Agricultural Products, Stine-Haskell Research Center, Newark, DE 19714. Formerly Graduate Research Assistant, Soil and Water Sciences Department, University of Florida, Gainesville, FL 32611

Jack A. Best Regional Manager, Product Development, Sandoz AGRO, Inc., Des Plaines, IL 60018

Stephen A. Boyd Professor, Department of Crop and Soil Sciences, Michigan State University, East Lansing, MI 48824-1325

M. L. Brusseau Assistant Professor of Environmental Chemistry and Subsurface Hydrology, Soil and Water Science Department, University of Arizona, Tucson, AZ 85721

G. A. Buyanovsky Professor of Soil Science, Department of Soil and Atmospheric Science, University of Missouri, Columbia, MO 65211

Jennifer J. Crawford Graduate Research Assistant, Environmental Science Program, Ohio State University, Columbus, OH 43210-1292

R. D. DeLaune Professor, Wetland Biogeochemistry Institute, Louisiana State University, Baton Rouge, LA 70803-7511

D. W. Dickson Professor of Nematology, Entomology and Nematology Department, University of Florida, Gainesville, FL 32611

Kathryn C. Dowling Post-Doctoral Researcher, USDA-ARS, Department of Soil and Environmental Sciences, University of California, Riverside, CA 92521. Formerly graduate student, Graduate field of Environmental Toxicology, Cornell University, Ithaca, NY 14853.

Allan S. Felsot Associate Professor, Crop and Soil Sciences, Washington State University-Tri Cities, Food and Environmental Quality Laboratory, Richland, WA 99352. Formerly Associate Professor, Pesticide Chemistry and Toxicology, Illinois Natural History Survey, Champaign, IL 61820

James K. Fredrickson Senior Research Scientist, Environmental Sciences Department, Pacific Northwest Laboratory, Richland, WA 99352

Jeffry J. Fuhrmann Associate Professor, Department of Plant and Soil Sciences, University of Delaware, Newark, DE 19717-1303

A. M. Gajda Research Associate, University of Missouri-Columbia, Columbia, MO 65211

A. P. Gamerdinger Assistant Professor of Environmental Chemistry, Department of Natural Resources Science, University of Rhode Island, Kingston, RI 02881

Robert T. Gavenda Post-Doctoral Fellow, Department of Agronomy and Soil Science, University of Hawaii-Manoa, Honolulu, HI 96822

Jane U. Gonese Graduate Assistant, North Carolina State Univesity, Raleigh, NC 27695

Richard E. Green Professor of Soil Science, University of Hawaii, 1910 East-West Road, Honolulu, HI 96822

William F. Guerin Research Associate, Department of Crop and Soil Science, Michigan State University, East Lansing, MI 48824-1325

Richard M. Johnson Research Soil Chemist (Formerly Post-Doctoral Research Fellow), USDA-ARS-SRRC, 1100 Robert E. Lee Blvd., New Orleans, LA 70179. Formerly Post-doctoral Research Fellow, Department of Plant and Soil Science, University of Delaware, Newark, DE 19717-1303.

R. J. Kremer Microbiologist, USDA-ARS, University of Missouri, Columbia, MO 65211

A. T. Lemley Professor of Environmental Toxicology, College of Human Ecology, Cornell University, Ithaca, NY 14853

R. A. McLaughlin Associate Professor, Soil Science Department, North Carolina State University, Raleigh, NC 27695. Formerly Project Scientist, CIBA-GEIGY Corporation, Greensboro, NC 27419

Carl J. Miles Chemical Services Manager, Toxikon Environmental Sciences, 106 Coastal Way, Jupiter, FL 33477

Kye-Heon Oh Assistant Professor of Genetic Engineering, Soonchunhyang University, 53-1, Eupnae-ri, Shinchang-myon, Asan-gun, Chung-Nam 337-880, South Korea. Formerly The Ohio State University, Columbus, OH 43210.

L.-T. Ou Associate Research Scientist, Soil and Water Science Department, University of Florida, Gainesville, FL 32611

J. H. Pardue Assistant Professor, Wetland Biogeochemistry Institute, Louisiana State University, Baton Rouge, LA 70803-7511

W. H. Patrick, Jr. Boyd Professor and Director, Wetland Biogeochemistry Institute, Louisiana State University, Baton Rouge, LA 70803-7511

G. J. Pieczonka Staff Scientist, Department of Agricultural Chemistry, Hazleton, Madison, WI 53707-7545. Formerly Research Associate, Environmental Trace Substances Research Center, University of Missouri, Columbia, MO.

Mark Radosevich Graduate Research Associate, Department of Agronomy, Ohio State University, Columbus, OH 43210

P. S. C. Rao Professor, Soil and Water Science Department, University of Florida, Gainesville, FL 32611

R. H. Ross Manager, Residue Chemistry Department, CIBA-GEIGY Corporation, Agricultural Division, Greensboro, NC 27419

R. C. Schneider Junior Researcher, Department of Agronomy and Soil Science, University of Hawaii, 1910 East-West Road, Honolulu, HI 96822

Kate M. Scow Assistant Professor, Department of Land, Air and Water Resources, University of California, Davis, CA 95616

Daniel R. Shelton USDA-ARS, Pesticide Degradation Laboratory, Bldg. 050 BARC-West, Beltsville, MD 20705-2350

Steven C. Smith Senior Technical Specialist, Pacific Northwest Laboratory, Richland, WA 99352

J. E. Thomas Chemist, Department of Soil and Water Science, University of Florida, Gainesville, FL 32611

Samuel J. Traina Associate Professor, Department of Agronomy, Ohio State University, Columbus, OH 43210

Olli H. Tuovinen Professor of Microbiology, Department of Microbiology, Ohio State University, Columbus, OH 43210-1292

Jerome B. Weber Professor of Crop Science and Soil Science, North Carolina State University, Raleigh, NC 27695

Conversion Factors for SI and non-SI Units

Conversion Factors for SI and non-SI Units

To convert Column 1 into Column 2, multiply by	Column 1 SI Unit	Column 2 non-SI Unit	To convert Column 2 into Column 1, multiply by
	Length		
0.621	kilometer, km (10^3 m)	mile, mi	1.609
1.094	meter, m	yard, yd	0.914
3.28	meter, m	foot, ft	0.304
1.0	micrometer, μm (10^{-6} m)	micron, μ	1.0
3.94×10^{-2}	millimeter, mm (10^{-3} m)	inch, in	25.4
10	nanometer, nm (10^{-9} m)	Angstrom, Å	0.1
	Area		
2.47	hectare, ha	acre	0.405
247	square kilometer, km^2 $(10^3\ m)^2$	acre	4.05×10^{-3}
0.386	square kilometer, km^2 $(10^3\ m)^2$	square mile, mi^2	2.590
2.47×10^{-4}	square meter, m^2	acre	4.05×10^{3}
10.76	square meter, m^2	square foot, ft^2	9.29×10^{-2}
1.55×10^{-3}	square millimeter, mm^2 $(10^{-3}\ m)^2$	square inch, in^2	645
	Volume		
9.73×10^{-3}	cubic meter, m^3	acre-inch	102.8
35.3	cubic meter, m^3	cubic foot, ft^3	2.83×10^{-2}
6.10×10^{4}	cubic meter, m^3	cubic inch, in^3	1.64×10^{-5}
2.84×10^{-2}	liter, L (10^{-3} m^3)	bushel, bu	35.24
1.057	liter, L (10^{-3} m^3)	quart (liquid), qt	0.946
3.53×10^{-2}	liter, L (10^{-3} m^3)	cubic foot, ft^3	28.3
0.265	liter, L (10^{-3} m^3)	gallon	3.78
33.78	liter, L (10^{-3} m^3)	ounce (fluid), oz	2.96×10^{-2}
2.11	liter, L (10^{-3} m^3)	pint (fluid), pt	0.473

	Mass		
2.20×10^{-3}	gram, g (10^{-3} kg)	pound, lb	454
3.52×10^{-2}	gram, g (10^{-3} kg)	ounce (avdp), oz	28.4
2.205	kilogram, kg	pound, lb	0.454
0.01	kilogram, kg	quintal (metric), q	100
1.10×10^{-3}	kilogram, kg	ton (2000 lb), ton	907
1.102	megagram, Mg (tonne)	ton (U.S.), ton	0.907
1.102	tonne, t	ton (U.S.), ton	0.907
	Yield and Rate		
0.893	kilogram per hectare, kg ha^{-1}	pound per acre, lb $acre^{-1}$	1.12
7.77×10^{-2}	kilogram per cubic meter, kg m^{-3}	pound per bushel, lb bu^{-1}	12.87
1.49×10^{-2}	kilogram per hectare, kg ha^{-1}	bushel per acre, 60 lb	67.19
1.59×10^{-2}	kilogram per hectare, kg ha^{-1}	bushel per acre, 56 lb	62.71
1.86×10^{-2}	kilogram per hectare, kg ha^{-1}	bushel per acre, 48 lb	53.75
0.107	liter per hectare, L ha^{-1}	gallon per acre	9.35
893	tonnes per hectare, t ha^{-1}	pound per acre, lb $acre^{-1}$	1.12×10^{-3}
893	megagram per hectare, Mg ha^{-1}	pound per acre, lb $acre^{-1}$	1.12×10^{-3}
0.446	megagram per hectare, Mg ha^{-1}	ton (2000 lb) per acre, ton $acre^{-1}$	2.24
2.24	meter per second, m s^{-1}	mile per hour	0.447
	Specific Surface		
10	square meter per kilogram, m^2 kg^{-1}	square centimeter per gram, cm^2 g^{-1}	0.1
1000	square meter per kilogram, m^2 kg^{-1}	square millimeter per gram, mm^2 g^{-1}	0.001
	Pressure		
9.90	megapascal, MPa (10^6 Pa)	atmosphere	0.101
10	megapascal, MPa (10^6 Pa)	bar	0.1
1.00	megagram per cubic meter, Mg m^{-3}	gram per cubic centimeter, g cm^{-3}	1.00
2.09×10^{-2}	pascal, Pa	pound per square foot, lb ft^{-2}	47.9
1.45×10^{-4}	pascal, Pa	pound per square inch, lb in^{-2}	6.90×10^3

(continued on next page)

Conversion Factors for SI and non-SI Units

To convert Column 1 into Column 2, multiply by	Column 1 SI Unit	Column 2 non-SI Unit	To convert Column 2 into Column 1, multiply by
	Temperature		
1.00 (K − 273)	Kelvin, K	Celsius, °C	1.00 (°C + 273)
(9/5 °C) + 32	Celsius, °C	Fahrenheit, °F	5/9 (°F − 32)
	Energy, Work, Quantity of Heat		
9.52×10^{-4}	joule, J	British thermal unit, Btu	1.05×10^{3}
0.239	joule, J	calorie, cal	4.19
10^{7}	joule, J	erg	10^{-7}
0.735	joule, J	foot-pound	1.36
2.387×10^{-5}	joule per square meter, J m^{-2}	calorie per square centimeter (langley)	4.19×10^{4}
10^{5}	newton, N	dyne	10^{-5}
1.43×10^{-3}	watt per square meter, W m^{-2}	calorie per square centimeter minute (irradiance), cal cm^{-2} min^{-1}	698
	Transpiration and Photosynthesis		
3.60×10^{-2}	milligram per square meter second, mg m^{-2} s^{-1}	gram per square decimeter hour, g dm^{-2} h^{-1}	27.8
5.56×10^{-3}	milligram (H_2O) per square meter second, mg m^{-2} s^{-1}	micromole (H_2O) per square centimeter second, μmol cm^{-2} s^{-1}	180
10^{-4}	milligram per square meter second, mg m^{-2} s^{-1}	milligram per square centimeter second, mg cm^{-2} s^{-1}	10^{4}
35.97	milligram per square meter second, mg m^{-2} s^{-1}	milligram per square decimeter hour, mg dm^{-2} h^{-1}	2.78×10^{-2}
	Plane Angle		
57.3	radian, rad	degrees (angle), °	1.75×10^{-2}

	Electrical Conductivity, Electricity, and Magnetism		
10	siemen per meter, S m^{-1}	millimho per centimeter, mmho cm^{-1}	0.1
10^4	tesla, T	gauss, G	10^{-4}
	Water Measurement		
9.73×10^{-3}	cubic meter, m^3	acre-inches, acre-in	102.8
9.81×10^{-3}	cubic meter per hour, $m^3\ h^{-1}$	cubic feet per second, $ft^3\ s^{-1}$	101.9
4.40	cubic meter per hour, $m^3\ h^{-1}$	U.S. gallons per minute, gal min^{-1}	0.227
8.11	hectare-meters, ha-m	acre-feet, acre-ft	0.123
97.28	hectare-meters, ha-m	acre-inches, acre-in	1.03×10^{-2}
8.1×10^{-2}	hectare-centimeters, ha-cm	acre-feet, acre-ft	12.33
	Concentrations		
1	centimole per kilogram, cmol kg^{-1} (ion exchange capacity)	milliequivalents per 100 grams, meq 100 g^{-1}	1
0.1	gram per kilogram, g kg^{-1}	percent, %	10
1	milligram per kilogram, mg kg^{-1}	parts per million, ppm	1
	Radioactivity		
2.7×10^{-11}	becquerel, Bq	curie, Ci	3.7×10^{10}
2.7×10^{-2}	becquerel per kilogram, Bq kg^{-1}	picocurie per gram, pCi g^{-1}	37
100	gray, Gy (absorbed dose)	rad, rd	0.01
100	sievert, Sv (equivalent dose)	rem (roentgen equivalent man)	0.01
	Plant Nutrient Conversion		
	Elemental	***Oxide***	
2.29	P	P_2O_5	0.437
1.20	K	K_2O	0.830
1.39	Ca	CaO	0.715
1.66	Mg	MgO	0.602

1 Coupling Biodegradation of Organic Chemicals to Sorption and Transport in Soils and Aquifers: Paradigms and Paradoxes

P. S. C. Rao and C.A. Bellin

University of Florida
Gainesville, Florida

M. L. Brusseau

University of Arizona
Tucson, Arizona

Most laboratory-scale experiments, and some field-scale studies, are designed for investigating environmental processes in isolation; at least attempts are made to control most variables except the one whose impact upon the system is being investigated. In real-world scenarios, even in the simplest of laboratory experiments, however, the rates and magnitudes of a reaction or process are often controlled by one or more other processes, each of which may have its own set of unique control variables at different spatial and temporal scales. This is indeed the case for laboratory experiments and field studies on fate and transport of organic chemicals in soils and aquifers. An explicit understanding of the coupling and feedback among simultaneous processes is essential in explaining experimental observations and for developing predictive models.

Two simple examples will illustrate this point. Let us first consider the total amount of pesticide degradation occurring before a surface-applied pulse of pesticide leaches beyond the crop root zone. The residence time for the pesticide pulse within the root zone and the degradation rate are the two key variables. Residence time depends upon the extent of sorption and dynamics of water flow and solute transport, while the pesticide degradation rate depends on microbial activity, bioavailability and the recalcitrance of the chemical. These factors, in turn, are influenced by several environmental variables and molecular properties of the chemical. The effect of shorter residence time due to rapid leaching can be muted by faster degradation rates.

 Sorption and Degradation of Pesticides and Organic Chemicals in Soil. SSSA Special Publication no. 32.

With such offsetting effects, it is possible that in two different soils, or in the same soil under different conditions, the measured pesticide loss via degradation may be essentially equal, although for entirely different reasons.

Another example of interest is the possible linkage between microbial degradation and sorption of pesticides and other organic chemicals. Let us consider two likely scenarios—one in which sorption renders the solute unavailable for degradation, while in the other sorption has no measurable effect on biodegradation. In the first case, the observed degradation rates among several soils might vary due to differences in the magnitude of sorption even if microbial communities and their activities were similar. In the second case, the observed differences in pesticide degradation rates are due to variations in microbial activities (or populations), regardless of the sorption capacity of the soils. Of course, it is also possible that a combination of both effects needs to be considered. Thus, in order to explain the observed differences in pesticide degradation rates among several soils, it is necessary to make several additional measurements to evaluate the coupling between sorption and degradation.

Understanding sorption, biodegradation, and transport of organic chemicals in soils and aquifers is essential for predicting their fate and transport in the environment. Currently, the success of bioremediation of contaminated soils and groundwater is limited due to a failure to achieve the extent and rate of biodegradation under field conditions similar to those anticipated from laboratory studies. Understanding the physical and chemical constraints to biodegradation in soils and aquifers may improve the designs of bioremediation programs and provide an understanding of the reasons for chemical persistence. Therefore, information is needed regarding microbial transformation of organic chemicals in soil-water systems, as affected by the interaction of chemical, physical, and biological processes.

Lack of consideration of the interactions between physical-chemical and biological proceses results in discrepancies between model predictions and experimental observations. Investigating organic chemical behavior in natural systems and developing chemical transport models that account for the interactive effects of biodegradation and sorption are necessary to adequately describe the environmental behavior of chemicals. Coupled process models could then be used to gain insight into the dominant interactions among the processes that affect the fate of chemicals in the environment for: (i) prescribing management strategies that prevent or minimize groundwater contamination, and (ii) designing effective remediation procedures for contaminated soils and aquifers.

The scientific and practical issues discussed in the foregiong paragraphs provide the focus for this chapter, with a specific emphasis on biodegradation of organic chemicals in soils and aquifers. We will examine: (i) the need for understanding coupling of certain key processes governing environmental behavior of pesticides and other organic chemicals in soils and aquifers; (ii) the reasons for decoupling of processes in explaining experimental observations; and (iii) the difficulties and successes encountered in developing and validating coupled-processes models and estimating the required model parameters. No attempt is made in this chapter to provide an exhaustive literature review; rather, certain important issues are identified by discussing some key papers.

EXPERIMENTAL OBSERVATIONS OF BIODEGRADATION

Biodegradation is a dominant mechanism of organic chemical transformations in soils and aquifers. One of the prevailing paradigms is that microbial degradation of most small organic compounds (molecular mass <600) occurs intracellularly. Thus, the rate of biodegradation is limited by the following factors:

1. Any physical-chemical process (e.g., solubility, sorption, and hydrodynamic dispersion) that lowers solute concentrations in the solution phase.
2. Any soil or environmental factor (e.g., temperature, soil-water content, and electron acceptors) that limits physiological activity of the appropriate microbial consortia.
3. Any microbial factor that limits substrate uptake by the microorganisms (e.g., cell membrane permeation, and hydrophobicity).
4. Any intracellular genetic or biochemical factor (e.g., presence of appropriate enzyme systems, and presence or expression of genes) that limits utilization of the compound.

The recalcitrance of different organic chemicals in a specific soil, or the variations in degradation rates of a specific compound in several soils, may be explained to a large extent by understanding the effects of aforementioned factors.

Bioaugmentation of soils and aquifers with microorganisms capable of readily degrading specific organic chemicals may not result in biodegradation of the target compounds due to various environmental stresses not present under laboratory conditions. For example, biodegradation rates may be limited by the following factors (Madsen, 1985; Goldstein et al., 1985; Alexander et al., 1991):

1. Low substrate concentrations not supporting microbial growth.
2. Microorganisms encountering toxins or predators.
3. Microorganisms using more-readily-available C sources.
4. Introduced microorganisms not reaching the contaminated site.

A summary of these and other factors influencing biodegradation is shown in Table 1-1.

Bioenhancement, either on site or in situ, of biodegradation is an alternative to bioaugmentation for removing organic chemicals in soils and aquifers. Using indigenous microorganisms is preferable to "inoculation" or injection because the microorganisms are already present at the site and adapted to the local environment. In the subsurface environment, however, the complex interaction between microorganisms, substrates, and surfaces may

Table 1-1. Interactions of substrates, organisms, and surfaces (Modified from Madsen, 1985; and van Loosdrecht et al., 1990).

Observation	Some probable causes
	I. Substrate/Organism interactions
Limited degradation	Substrate resists attack by existing enzymes
	Substrate unable to penetrate the cell wall or cell membrane
	Site on substrate not recognized by extracellular enzymes due to intramolecular folding
	Complex substrate structure, not all enzymes available for complete degradation
	Limited energy or C produced for microbial growth
	Substrate or product is toxic to organisms
	Substrate is insoluble
	Substrate concentration too low for enzyme induction
	II. Substrate/Surface interactions
Limited reversibility of sorption	Sites/domains with different sorption mechanism
	Polymerization with surface
	Chemical bonding—cation exchange
	Selectivity of substrate for surface
	Nonspecific partitioning (Hydrophobic)
Nonequilibrium sorption	Diffusion-limited sorption site/domain
	Intraorganic matter diffusion (IOMD)
	Intraparticle diffusion (IPD)
	Intramineral diffusion (IMD)
	III. Organism/Surface interactions
Increased growth rate	Increased substrate concentration at the interface
	More efficient use of proton motive force
	Detoxification of substrate
	Detoxification of inhibitors
	pH buffering by ion exchange
Decreased growth rate	Less cell surface available for substrate uptake
	Higher maintenance coefficient
	Substrate transport limitation
	Growth inhibition due to accumulation of toxic metabolites
Increased assimilation and decreased respiration rates	No explanation ?
Decreased assimilation	No explanation ?
Increased respiration	Change in membrane process
	pH buffering by ion exchange
Increased adhesion of active cells	Increased cell hydrophobicity
Higher activity of attached cells	No explanation
Decreased substrate utilization	Desorption limitation
	Diffusion limitation
	Lower substrate concentration
Lower substrate affinity	Diffusion limitation
Change in pH optimum	Proton concentration at surface different from the bulk
Difference in fermentation pattern	Surface is electron acceptor
Increase in productivity	Immobilization of biomass
Decreased mortality	Decreased phagocytosis
No effect ?	No explanation ?

alter this process. Biodegradation rates may be limited by chemical properties of substrates, interactions of the substrate with surfaces, or simply by the lack of necessary enzymes (Madsen, 1985). Bioavailability of slightly soluble substrates may be controlled by the dissolution rate (Stucki & Alexander, 1987; Miller & Bartha, 1989; Huang & Chou, 1990), or low aqueous concentrations that may not induce the necessary enzymes for biodegradation (Madsen, 1985). Similarly, sorption of the substrate by soil may reduce substrate concentrations in solution below levels necessary for enzyme induction.

Sorption of substrates might also enhance biodegradation rates by decreasing the substrate concentration to levels that are not toxic to microorganisms responsible for degradation (van Loosdrecht et al., 1990). However, sorption more likely reduces or inhibits biodegradation rates in soils (Stotzky & Rem, 1966; Madsen, 1985; van Loosdrecht et al., 1990). For example, sorption was found to decrease substrate availability, hence the biodegradation rates of several compounds, including: diquat (1,1′-ethylene-2,2′-bipyridylium) (Weber & Coble, 1968); benzylamine (Subba-Rao & Alexander, 1982; Miller & Alexander, 1991); alkylamines (Wszolek & Alexander, 1979); glucose (Gordon & Millero, 1985); amino acids (Dashman & Stotzky, 1986); toluene (Robinson et al., 1990); benzidine (Weber, 1991); quinoline (Smith et al., 1992); and flumetsulam {*N*(2,6-difluorophenyl)-5methyl[1,2,4]triazolo [1,5*a*]pyrimidine-2-sulfonamide (DE-498)} (Lehman et al., 1992). Degradation was adequately described by a second-order rate equation with the assumption that only solution-phase chlorproham (isopropyl m-chlorocarbanilate) and dibutyl phthalate are biodegraded in the presence of sediments (Steen et al., 1980). Ogram et al. (1985) demonstrated that sorbed 2,4-D (2,4-dichlorophenoxyacetic acid) was protected from biodegradation and only solution-phase 2,4-D was degraded by bacteria in both the sorbed and solution phases.

Biodegradation may also be limited by mass transfer [intraparticle diffusion (IPD) and intraorganic matter diffusion (IOMD)] from the interior of the sorbent to the exterior solution. Intra-aggregate pores are too small to be accessible to bacteria (Steinberg et al., 1987; Scow & Alexander, 1992) and the substrate sorbed within organic matter is accessible only after desorption. For several hydrophobic organic compounds (HOCs) mass transfer has been shown to limit sorption/desorption rates (Wu & Gschwend, 1986; Brusseau & Rao, 1989b; Brusseau et al., 1991), and biodegradation of HOCs (Rijnaarts et al., 1990; Robinson et al., 1990). Myrold and Tiedje (1985) studied mass transfer constraints for denitrification. Mass transfer from solid to solution was shown to influence the biodegradation rate of naphthalene in suspensions of larger particles, but not in suspensions of smaller particles (Mihelcic, 1988; Mihelcic & Luthy, 1988a,b). Quinoline, a *N*-heterocyclic organic base, forms a strong cation exchange complex on soil clays that desorb slowly and limits biodegradation (Smith et al., 1992). The foregoing discussion suggests that models coupling the sorption, biodegradation, and transport in soils and aquifers are needed to elucidate the rate-limiting steps of organic chemical biodegradation.

SIMULATION MODELS FOR COUPLED PROCESSES

Several simulation models, with varying levels of sophistication and intended applications, have been proposed since the 1970s for predicting the biodegradation of organic chemicals during transport through porous media such as soils and aquifers. Brusseau et al. (1992a) reviewed coupled-process models for chemical transport.

The governing differential equation that serves as the basis for most coupled-processes models is:

$$[\partial(\theta C)/\partial t] = [\nabla \cdot D\theta\ \nabla C] - [\nabla \cdot qC] - [\partial(\rho S)/\partial t] \pm \Sigma\ \phi_i \qquad [1]$$

where C = solution-phase concentration (M/L^3); S = sorbed-phase concentration (M/M); t = time (T); ρ = soil bulk density (M/L^3); θ = fractional volumetric water content (dimensionless); D = local-scale hydrodynamic dispersion coefficient (L^2/T); q = Darcy flux for water flow (L/T); ϕ_i = rates (M/L^3T) of loss or gain via various sinks and sources; and ∇ is the Laplacian differential operator. In Eq. [1], multidimensional, advective-dispersive solute transport in a heterogeneous porous medium under transient water flow condition (first two terms on the r.h.s.) is coupled to sorption dynamics (third term on r.h.s.) and biodegradation kinetics (last term on r.h.s.). Differences in published models arise from the specific manner in which sorption and degradation kinetics are modeled, whether transient or steady flow is considered, and if one- or multidimensional transport is of interest.

For multidimensional, steady, unsaturated water flow conditions in a homogeneous medium, Eq. [1] can be restated as:

$$[\theta(x,y,z)(\partial C/\partial t)] = [\nabla\quad D\theta\ \nabla C] - [\nabla \cdot qC] - [\rho\ (\partial S/\partial t)] \pm \Sigma\ \phi_i \qquad [2]$$

Note that in Eq. [2] the volumetric water content is invariant in time (i.e., steady state), but can vary spatially within the medium. Equation [2] can be further simplified for one-dimensional, steady, saturated water flow as follows:

$$[\partial C/\partial t] = D(\partial^2 C/\partial x^2) - v(\partial C/\partial x) - (\rho/\theta)(\partial S/\partial t) \pm (1/\theta)\ \Sigma\ \phi_i \qquad [3]$$

where $v = (q/\theta)$ is the average pore-water velocity (L/T); and θ is now set equal to the volumetric porosity of the medium.

Assuming that sorption can be represented as an equilibrium process with a linear isotherm ($S = K_pC$), and that first-order kinetics apply to biodegradation ($\phi = -k\theta\ C$), Eq. [3] is restated as follows:

$$R(\partial C/\partial t) = D(\partial^2 C/\partial x^2) - v(\partial C/\partial x) - kC \qquad [4]$$

where $R = [1 + (\rho K_p/\theta)]$ is the retardation factor (dimensionless), K_p is the equilibrium sorption coefficient (L^3/M), and k is the pseudo first-order rate constant (T^{-1}) for biodegradation, which is assumed to occur only in the solution phase.

In terms of completeness, a coupled-processes model must include, at a minimum, the three major processes of interest: transport via advection and hydrodynamic dispersion arising from local-scale solute velocity variations; sorption/desorption; and abiotic/or biotic transformations. We choose here to operationally define coupled-processes models as being "simple" based on the following criteria:

1. *Transport*: Advective-dispersive solute transport (ADT) in a homogeneous medium under steady (usually saturated) flow conditions.
2. *Sorption*: Linear, reversible, isotherm with attainment of instantaneous local equilibrium during transport (LEA).
3. *Transformations*: First-order kinetics with uniformly distributed, nongrowing biomass; that is, "lumped" first-order kinetics (LFK).

A process description at any higher level of sophistication will be considered as being "complex" in our discussion. On this basis, the available models can be conveniently grouped into six categories, depending on whether each of the processes is represented in a simplified or complex fashion. Table 1-2 is a summary of the model categories and their characteristics.

It may be noted that the complexity of process representation is often determined by the modeler's disciplinary bias and also by the intended use of these models. For example, Category 3 models emphasize microbiological growth dynamics, while simplifying the transport and sorption processes. On the other hand, Category 1 models emphasize transport phenomena but simplify the representation of sorption and transformation processes (LEA and LFK). Most of the models in Categories 1 through 5 assume that the porous medium is homogeneous in terms of physical, chemical, and biological processes, a valid assumption for laboratory-scale studies but not likely to be true for almost any "real-world" field-scale application. In recent years, groundwater hydrologists have developed models that permit simulation of multidimensional solute transport in heterogeneous porous media, including layered and randomly varying media. Such models coupled to simplified sorption models (LEA) and complex transformation models (Monod-type kinetics) are included in Category 6.

It is noteworthy that the models grouped in Categories 1 through 6 (Table 1-2) deal with solute transport under steady and primarily saturated water flow conditions. In the crop root zone, water flow in response to episodes of rainfall or irrigation will be under unsaturated, transient flow conditions; whereas in the intermediate vadose zone (below the root zone, but above the water table), water flow is likely to be under unsaturated and steady conditions. Several models developed by soil scientists for pesticides are placed in a special category (Category 7; see Table 1-3) because they include such

Table 1-2. General characteristics of coupled-processes models grouped under six categories.

Process	Description	References
	Category 1	
Transport	1-D flow; ADT† coupled to intraparticle diffusion (mobile-immobile water concept)	Mironenko & Pachepsky, 1984; Lassey, 1988; van Genuchten & Wagenet, 1989; Gamerdinger et al., 1990; Jessup et al., 1989
Sorption	LEA‡	
Degradation	LFK§; degradation in solution and sorbed phases	
	Category 2	
Transport	1-D or 2-D flow; ADT	Eldor & Dagan, 1972; Lindstrom, 1976; Carnahan & Remer, 1984; Mironenko & Pachepsky, 1984; Selim et al., 1976; Cameron & Klute, 1977; Lassey, 1988; van Genuchten & Wagenet, 1989; Angley et al., 1993; Jessup et al., 1989
Sorption	Rate-limited sorption, two-domain sorption, nonlinear isotherms	
Degradation	LFK; degradation in both solution and sorbed phases	
	Category 3	
Transport	1-D or 2-D flow; ADT	Ardakani et al., 1973; Kam, 1974; Parlange et al., 1984; Borden & Bedient, 1986; Borden et al., 1984; Borden et al., 1986; Rifai et al., 1988; Baehr & Corapcioglu et al., 1984; Srinivasan & Mercer, 1988; Corapcioglu et al., 1991; Sykes et al., 1982; Corapcioglu & Haridas, 1985; Bosma et al., 1988; MacQuarrie et al., 1990; Semprini & McCarty, 1991; Kindred & Celia, 1989; Molz et al., 1986; Widdowson et al., 1988a,b
Sorption	LEA	
Degradation	Steady-state biomass or Monod-type growth dynamics	
	Category 4	
Transport	1-D or 2-D flow; ADT	Ying & Weber, 1979; Speitel et al., 1987
Sorption	Mass-transfer constraints	
Degradation	Biomass growth	
	Category 5	
Transport	1-D flow; ADT coupled to mobile-immobile water concepts	Rasmusson, 1982; Brusseau et al., 1989; Brusseau et al., 1992a; van Genuchten, 1974; Gamerdinger et al., 1990
Sorption	Rate-limited sorption	
Degradation	LFK in both sorbed and solution phases	
	Category 6	
Transport	2-D or 3-D flow; randomly variable or layered aquifer, ADT	Widdowson et al., 1987; Molz & Widdowson, 1988; MacQuarrie & Sudicky, 1990; Frind et al., 1989; Miralles-Wilhelm et al., 1990
Sorption	LEA	
Degradation	Biomass distribution (microcolonies); monod-type growth kinetics; consideration of electron acceptors (oxygen)	

† ADT = Advective-dispersive transport.
‡ LEA = Local equilibrium assumption.
§ LEA = Lumped first-order kinetics.

Table 1–3. A comparison of process conceptualization in five models used to describe pesticide behavior in the vadose zone. (Adapted from Pennell et al., 1990.)

	Simulation model				
Process	CMLS Ver. 4.0 (Nofziger & Hornsby, 1987)	GLEAMS Ver. 1.8.54 (Knisel et al., 1989)	LEACHMP Ver. 1.8.54 (Wagenet & Hutson, 1987)	MOUSE Ver. 7x (Steenhuis et al., 1986)	PRZM release 1 (Carsel et al., 1985)
Water flow	Piston displacement of water, instantaneous redistribution between field capacity and wilting point.	Predicts water flow between soil layers based on a storage similar to the "tipping bucket" method.	Solves Richards' equation. Requires moisture release curve data which must be fit to a modified Campbell's function.	Water drainage occurs under unit hydraulic gradient. Unsaturated hydraulic conductivity is calculated using an exponential function.	Water flow based on "tipping bucket" method. Operates between field capacity and wilting point. Instantaneous or time-dependent water redistribution.
Runoff	Runoff not considered.	Runoff based on SCS† curve-number method. Erosion calculated using overland, channel, and impoundent elements, and soil particle characteristics.	Runoff not considered.	Runoff based on SCS curve-number method.	Runoff based on SCS curve-number method. Erosion based on the Universal Soil Loss Equation.
Solute transport	Piston displacement of solute.	Convection transport of solute using water flow between soil layers. Solute can move upward by capillary flow.	Solves the convective-dispersive transport equation.	Piston displacement of solute.	Convective transport of solute based on water flow between soil increments.
Solute dispersion	Tracks a nondispersive solute point.	Numerical dispersion, from convective transport equation, used to simulate actual solute dispersion.	Calculates hydrodynamic dispersion.	Calculates dispersion around midpoint of solute band using an error function.	Numerical dispersion, from convective transport equation, used to simulate actual dispersion.

(continued on next page)

Table 1–3. Continued.

Process	Simulation model				
	CMLS Ver. 4.0 (Nofziger & Hornsby, 1987)	GLEAMS Ver. 1.8.54 (Knisel et al., 1989)	LEACHMP Ver. 1.8.54 (Wagenet & Hutson, 1987)	MOUSE Ver. 7x (Steenhuis et al., 1986)	PRZM release 1 (Carsel et al., 1985)
Sorption	Input solute K_{oc}‡. Input organic C by soil horizon or enter K_d§ by soil horizon.	Input K_{oc} for up to 10 solutes and metabolites. Input organic matter by soil horizon.	Input K_{oc} for solute and two metabolites. Input organic C by soil increment.	Input K_d for solute in the root zone and region below the root zone.	Input solute degradation rate coefficients by soil horizons.
Degradation	Input solute half-life by soil horizon.	Input half-life for each solute or metabolite by horizon. Input one coefficient of transformation for each component.	Input five degradation rate coefficients for three components by soil increments.	Input solute half-life by soil increment.	Input solute degradation rate coefficients by soil horizons.
Evapotranspiration	Input daily PET¶. Water removed from wettest soil horizons in the root zone first.	Potential evaporation calculated from solar radiation and air temperature. Actual ET# is then calculated using leaf area index and soil-water content.	Input weekly PET total. Water removal based on root distribution, root resistance, and soil-water potential.	Input daily PET. Water removal based on soil-water content and ET input parameters.	Input daily ET and crop ET coefficient. Water removal based on root distribution, and soil-water content.
Roots	Input maximum rooting depth. Root biomass constant.	Input maximum rooting depth. Water use is a function of depth based on an exponentially decreasing function.	Root biomass can be constant or increasing. For constant root biomass, input relative root function by soil increment.	Uniform root distribution to a rooting depth that can vary over time.	Root biomass can be constant or increasing. Root distribution decreases linearly to maximum rooting depth.

† SCS = Soil Conservation Service, USDA.
‡ K_{oc} = K_d normalized to soil organic carbon content.
§ K_d = Equilibrium sorption coefficient.
¶ PET = Potential evapotranspiration.
# ET = Evapotranspiration.

transient flow and transport considerations. Even these models, however, deal with one-dimensional flow and use lumped, pseudo first-order kinetics (LFK) for biodegradation. All models listed in Category 7 (Table 1-3) fail to consider multidimensional, transient water flow and solute transport, as might be needed in several agricultural scenarios (e.g., to predict pesticides applied via drip or microjet irrigation systems, i.e., chemigation). None of the models listed in Tables 1-2 and 1-3 link the unsaturated zone and the saturated zone for predicting solute transport and transformations, as might be needed for cases with shallow, fluctuating water tables. Only a limited use of such models has been documented in the published literature.

The development of coupled-processes models of increasing complexity can be anticipated with improved understanding of the linkage of relevant processes at the local- and field-scale; neither mathematical prowess nor computational power are likely to be constraints for future model development. The use or experimental validation of such complex models will most likely be limited by inability to provide the values for the vast array of required model parameters.

Let us consider, as an example, the model proposed by MacQuarrie and Sudicky (1990), which is listed under Category 6 in Table 1-2. This model describes solute transport in a randomly-variable, two-dimensional, steady flow field in a saturated porous medium. This requires the specification of the heterogeneity of the saturated hydraulic conductivity using a spatial autocorrelation function or variogram; this is no easy task and such information is not available in most cases as extensive sampling of the medium is required to characterize variability of hydraulic properties. The biodegradation of the organic substrate in this model is coupled to the availability of an electron acceptor (in this case, dissolved oxygen), and the utilization of the substrate and electron acceptor are described using Monod-type models. Among the inputs required is the *initial* biomass of the *specific* degraders, surely a daunting task given that it may be difficult enough to specify even the *total* biomass. Finally, MacQuarrie and Sudicky (1990) did not model local nonequilibrium conditions for sorption, or the spatial variations in sorptive and microbiological properties. Given such difficulties, process representation in models has been simplified based on the ability to measure the required parameters. The problems in model parameter estimation are further discussed in the following section.

ESTIMATION OF MODEL PARAMETERS

The major parameters needed to run the coupled-processes models can be grouped into three categories: (i) hydrological parameters, including quantitative relationships for soil hydraulic conductivity and water content dependence on soil-water matric potential, and hydrodynamic dispersivity; (ii) chemical parameters, primarily for quantifying equilibrium and nonequilibrium sorption, such as the equilibrium sorption coefficient (K_p; L^3/M) and the first-order mass-transfer rate constant (k_2; T^{-1}); and (iii) microbiological parameters, such as biomass growth dynamics and chemical degradation kinetics.

Extensive amounts of data have been collected and indirect estimation techniques have been proposed for characterizing the hydrological and chemical parameters. For example, hydraulic properties can be estimated using easily measured surrogate parameters such as particle-size distribution (cf., Gupta & Larson, 1979; Woesten & Van Genuchten, 1988; Rawls et al., 1982; Williams et al., 1992). Similarly, equilibrium sorption coefficients (K_p) can be estimated with reasonable certainty, for nonpolar organic compounds, given aqueous solubility, octanol-water partition coefficients, surface area, molecular connectivity, and other parameters (cf., Green & Karickhoff, 1990; Gerstl, 1990). The sorption mass-transfer constants (k_2) can be estimated for a variety of nonpolar organic chemicals from the inverse log-log relationship noted between k_2 and K_p (Brusseau & Rao, 1989a; Brusseau et al., 1991).

Estimation of the model parameters related to biomass growth dynamics of specific degraders and substrate degradation kinetics appears to be difficult. Translation of laboratory-scale studies, with either pure cultures or batch reactors with soils, to predict field-scale behavior has not always been successful. Degradation rates measured in laboratory studies are often substantially different when compared to field-scale results. The problems facing experimental scientists and modelers in this regard have been best summarized (Table 1-4) by Blackburn (1989) in a recent critical review.

Table 1-4. A summary of some relevant issues and factors that influence biodegradation at several spatial scales of interest. (Based on Blackburn's [1989] discussion.)

Spatial scale	Some relevant issues
Subcellular/Genetic level	Presence of appropriate genes (genotype) Gene expression (phenotype) Gene transmission (inheritance) Plasmids
Single-cell/Organism level	Maintenance and growth (cell physiology) Membrane transport (substrate uptake) Attachment to surfaces (location of cell) Cometablic process (no-growth) Critical substrate concentration (induction)
Microbial consortium/Colony level	Size and activity of subpopulations Mixed culture interactions Colony-level physiological activity Community structure
Continuously stirred reactor (CSTR) level	Presence of surfaces Effect of sorption nonequilibrium and reversibility Environmental conditions (pH, concentration, O_2)
Column reactor level	Dynamics of transport Distribution of organisms/colonies Effect of sorption nonequilibrium
Field/Aquifer level	Large-scale spatial heterogeneity in flow Availability of electron acceptors (O_2, NO_3, SO_4) Temporally fluctuating environmental conditions

Blackburn (1989) argues that in predicting microbial processes, deterministic approaches are likely to fail because of the level of complexity present at all spatial scales of interest (this is further discussed next) and due to manifestation of what Gibbs (1976) refers to as the microbiological-version of the Heisenberg Uncertainty Principle. Blackburn (1989) suggests that

> our attempts to infer the state of the mixed-culture biotransformations using measurements from samples taken from the system depend upon our ability to preserve all sample biochemical and ecological characteristics, and the ability to measure those characteristics without bias from the chemical, biochemical, or microbial analytical techniques.

Thus, Blackburn (1989) argues that the very act of sampling and experimentation alters the microbial community structure and resulting rates of degradation, such that the type of model and key parameters may no longer be transferrable to the original system.

Madsen (1991) also discusses the issue of uncertainty in unequivocally establishing that biodegradation of an organic chemical has, in fact, occurred under natural, real-world conditions. The inability to provide chemical mass balance and distinguish between abiotic and biotic transformation creates difficulty in proving in situ biodegradation. Madsen (1991) recommends the following approach to confirm in situ biodegradation: (i) isolate metabolically adapted microorganisms from field sites; (ii) show chemical mass loss exceeds potential losses due to abiotic transformation; and (iii) distinguish abiotic and biotic transformation products (e.g., specific metabolites of each pathway). Following these guidelines, Madsen et al. (1991) successfully demonstrated in situ biodegradation of aromatic hydrocarbons in a shallow, unconfined aquifer contaminated by coal tar components. Other reports of in situ biodegradation that meet the aforementioned criteria are lacking. Madsen (1991) suggests that approaches needed for successful field-scale documentation of in situ biodegradation are likely to depend on unique, site-specific conditions, and will require multidisciplinary efforts.

A hierarchial organization of the spatial scales of interest in biodegradation studies can be formulated by adding one larger spatial scale (field- and aquifer-scale) to those proposed by Blackburn (1989). This scheme is presented in Fig. 1-1 and Table 1-4; note the progression from the smallest scale (molecular level) to the largest (field- or aquifer-scale). Blackburn (1989) states that biotic and abiotic processes responsible for biodegradation operate at all levels and have strong interactions across all spatial scales (see Table 1-4). The time scales for process regulation vary widely, ranging from 10^{-6} to 10^{2} s at the molecular scale to several decades or centuries at the aquifer scale (see Table 1-5).

The foregoing considerations led Blackburn (1989) to conclude that

> we need to study a given (microbial) activity in the context of its operating complexity, but our methods are insufficient to enable system monitoring and analysis in the complex system.

One possible interpretation of such arguments is that biodegradation kinetics can only be characterized by in situ measurements for each spatial

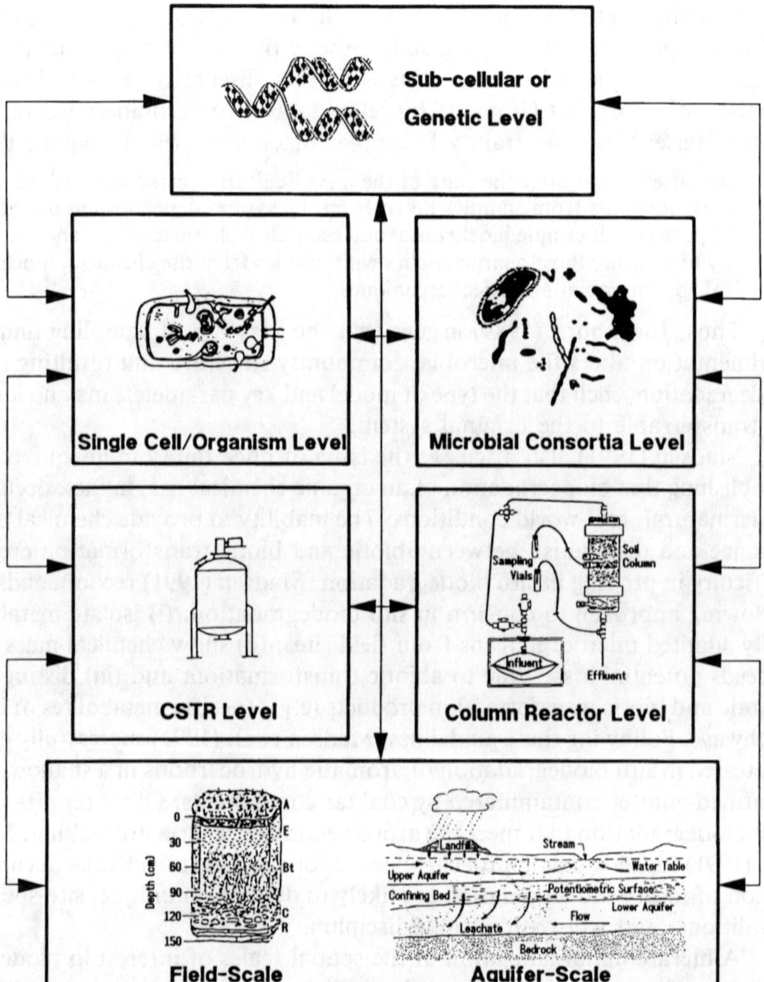

Fig. 1-1. Hierarchial organization of the spatial scales of interest in biodegradation studies. Modified from Blackburn (1989). CSTR = continuously stirred tank reactor.

scale and chemical of interest, without any potential for site-to-site extrapolation of data. This is, of course, an expensive and disheartening proposition for the practitioners, especially since methods for reliable in situ measurement of biodegradation rates are not presently available.

Given such arguments for prevasive uncertainty and complexity in microbial systems, it is a wonder that simple, first-order kinetics representing the pooled effects of all processes leading to biodegradation seems to even approximate the measured data, as has been reported in numerous laboratory and field studies. We will examine this paradoxical situation in the next section.

Table 1–5. Typical range of temporal scales at which process regulation might occur for degradation processes at various spatial scales. (Adapted from Blackburn, 1989.)

Spatial scale	Characteristic response time ($\log_{10}$ seconds)																
	−6	−5	−4	−3	−2	−1	0	1	2	3	4	5	6	7	8	9	10
Subcellular	←	—	—	—	—	—	—	—	→								
Single cell				←	—	—	—	—	—	—	—	—	—	→			
Microbial colony										←	—	—	—	—	—	→	
Continuously stirred tank reactor												←	—	—	—	—	→
Column reactor												←	—	—	—	—	→
Field/Aquifer												←	—	—	—	—	→

COMPARISON OF LABORATORY- AND FIELD-SCALE MEASUREMENTS

In the previous section, we have presented arguments that estimation of microbial degradation rate constants is fraught with numerous problems arising from the biological complexity of the system and scale-related effects. We will now address a related issue and attempt to answer the deceptively simple question: Are degradation rate constants measured in laboratory incubation experiments useful for predicting degradation rates under field conditions?

In answering the above question two specific issues are of significance. First, the temporal variations in several environmental factors (soil temperature, soil-water content, and soil aeration status) at a specific location within the field must be considered. Note that the patterns in such temporal variability might be different at different locations in the field because the soil physical-chemical properties are spatially variable. Second, the spatial variations in microbial properties must be accounted for to ensure that the soil samples used in laboratory studies are indeed representative of the field(s) from which they were collected. The interconnection between these two issues is made clear by another question: Are the spatial variations in degradation rates observed in a field the result of site-to-site variations in the inherent microbial degradative capacity of the soil, or are they an expression of the local variations in soil environmental factors superimposed on a rather uniform microbial degradative capacity? The first explanation assigns field-scale variability in degradation rates to spatial variations in microbial activity (type and population size of specific degraders), while the latter explanation attributes the observed field-scale variability to the effects of variable soil physical-chemical properties on uniform microbial properties. The resolution of these issues, we believe, is of importance in the design of useful laboratory/field experiments and in building appropriate field-scale models for biodegradation. Laskowski et al. (1982) summarized results from several published laboratory studies on pesticide degradation in soils, and concluded that degradation rates for specific pesticides may vary by a factor of 2 to 80 among different soils. Laskowski et al. (1982) do not provide an analysis of the sources of variability in degradation rates.

In an earlier section, we reviewed arguments supporting the view that translation of parameters over spatial scales is likely to fail. Here, we will examine two cases of reported success in using laboratory-scale data to describe field-scale biodegradation of pesticides in soils.

Several investigators have attempted to characterize spatial variability of biodegradation rates by collecting soil samples from multiple locations in a field (size ranging from several square meters to a few hectares), and measuring degradation kinetics under controlled laboratory conditions. In a majority of such studies, the soil samples were air dried, sieved, and rewetted (thus, eliminating likely differences in microbial activity), although a few investigators have used aseptic soil samples as collected from the field (Ou et al., 1986). Coefficients of variation (CV) for pesticide degradation half-lives measured in both types of studies (cf., Rao & Davidson, 1980; Ou et al., 1984; Rao et al., 1986) are surprisingly small (CV $\leq$ 30%), suggesting that the inherent microbial degradative capacity is not likely to vary much, at least in small (about a few hectares) agricultural fields that are managed in a fairly uniform manner (tillage, fertilization, and irrigation).

The foregoing discussion should not be taken to imply that in situ measurements of biodegradation rates will be equally uniform in a field. Soil and environmental factors do vary on a local scale as a result of spatial and temporal variabilities in soil physical-chemical properties and variations in water and solute inputs at the soil surface. As a result, the in situ biodegradation rates may vary widely even though the inherent microbial populations are quite uniform. That this may indeed be the case is supported by a field data set for which dramatic variabilities in in situ denitrification rates in croplands were reported (cf., Ryden & Lund, 1980a,b). We are not aware of similar field studies on field-scale variability of pesticide degradation rates.

Walker et al. (1983) conducted herbicide degradation studies in field plots located at 21 locations in 11 countries. They evaluated the agreement between field-observed simazine [2-chloro-4,6-bis(ethylamino)-*s*-triazine] herbicide loss and that predicted based on a simple, first-order kinetics model. Rate coefficients were obtained from lab-scale incubation studies and were adjusted for site-specific temporal variations in soil temperature and soil-water contents. They state that, "Occasionally there was close agreement between predicted and observed residues, but in general the model underestimated rates of loss in the field." Some examples of their data are shown in Fig. 1-2. Walker et al. (1983) also considered the agreement between field data and model predictions to be sufficient for practical applications, especially given the large variability in measured pesticide residue concentrations and that several other processes (e.g., volatilization or leaching beyond sampling depth) could have contributed to measured loss. These results suggest that a major fraction of field-scale variability in pesticide degradation rates may be explained by a simple accounting of local variations in only two environmental factors: soil temperature and soil-water content.

Pennell et al. (1990) present another example of successful application of lab-measured degradation rate constants to predict field-scale degradation of the nematicide aldicarb [2-methyl-2(methylthio)propionaldehyde O(methyl-carbamoyl) oxime] and its two oxidative metabolites, aldicarb sulfoxide and aldicarb sulfone (Fig. 1-3). Predictions were based on the simulation model LEACHMP (Wagenet & Hutson, 1989), which explicitly couples transient water flow and solute transport in the root zone to sequential oxidation and simultaneous hydrolysis of aldicarb, aldicarb sulfoxide, and aldicarb sulfone. The oxidation and hydrolysis rate constants for aldicarb transformations required in the LEACHMP model were obtained from Ou et al. (1988), who conducted laboratory incubation experiments using soil samples collected from different profile depths at the field site. In surface soils with higher microbial activity, the biological oxidation pathway is dominant, whereas in subsoils with a lower microbial activity abiotic hydrolysis is the primary pathway for aldicarb degradation (Ou et al., 1988).

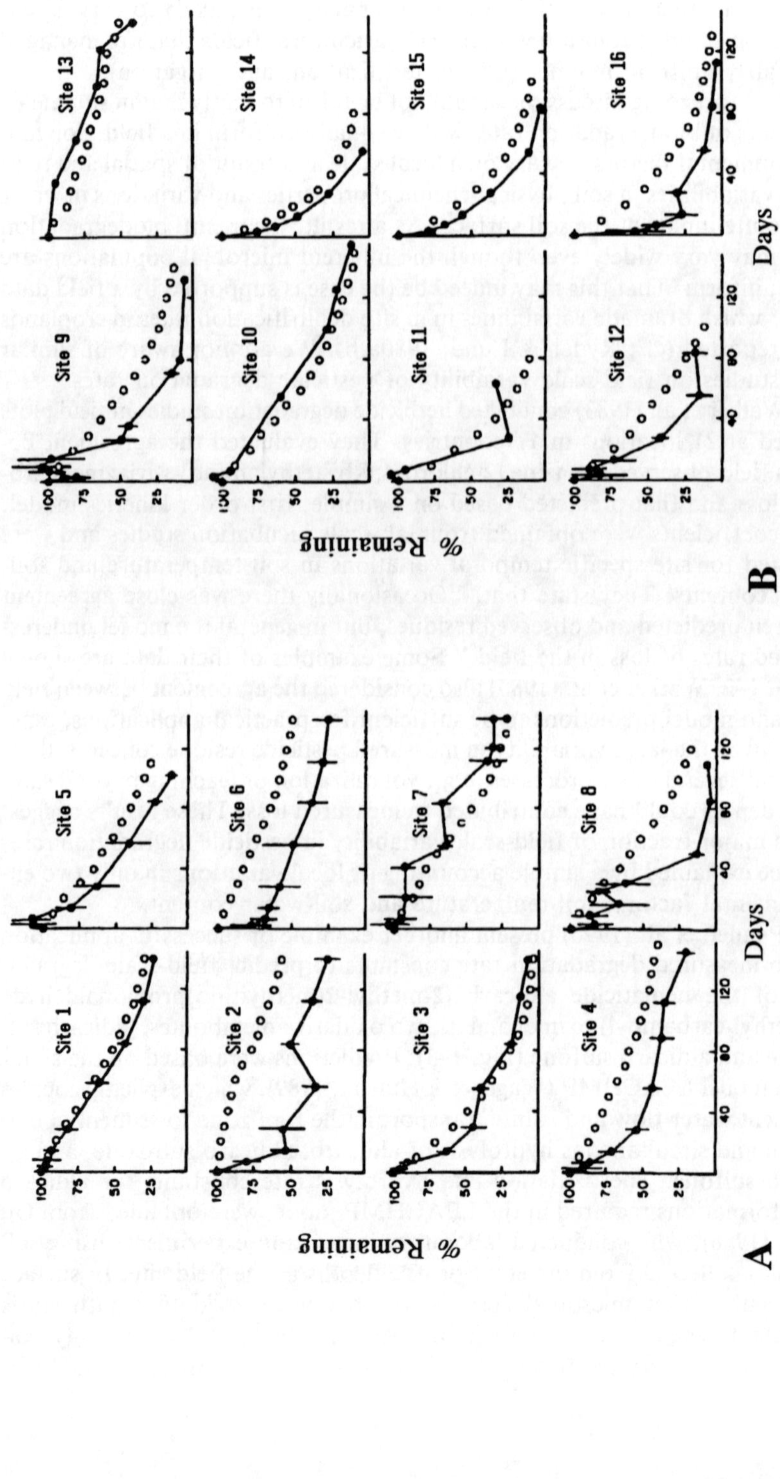
% Remaining
Site 1
Site 2
Site 3
Site 4
Site 5
Site 6
Site 7
Site 8
Site 9
Site 10
Site 11
Site 12
Site 13
Site 14
Site 15
Site 16
Days
A
B

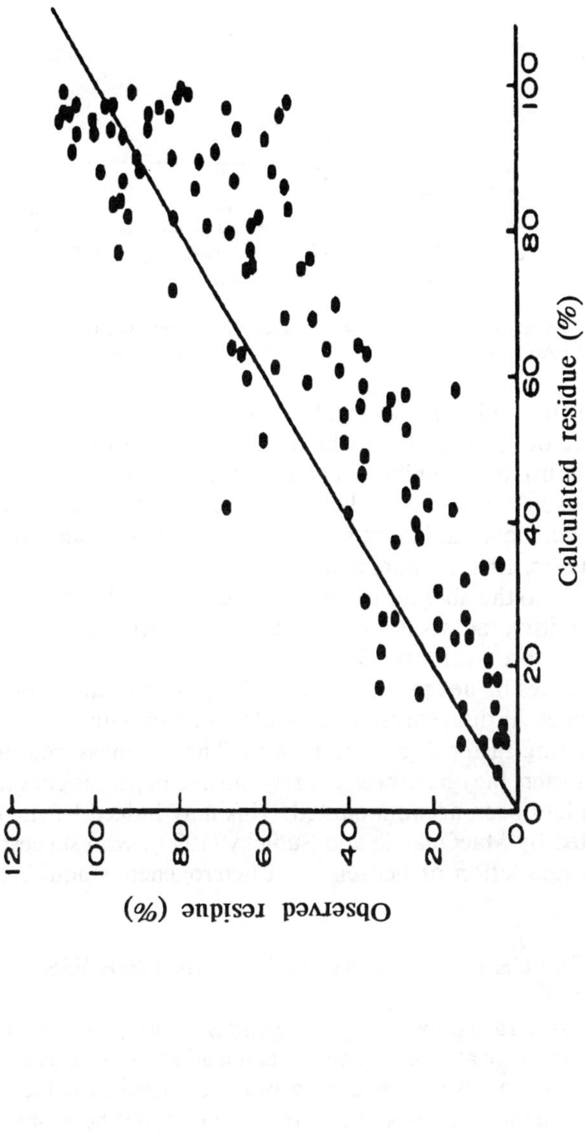

C

Fig. 1-2. (A) and (B) Comparison of field-measured (closed circles connected by lines) and predicted (open circles) losses of simazine herbicide residues. Measured data were collected at 16 field sites in 11 countries. (C) A summary of measured and predicted herbicide residues at all field sites. Compiled from Walker et al. (1983).

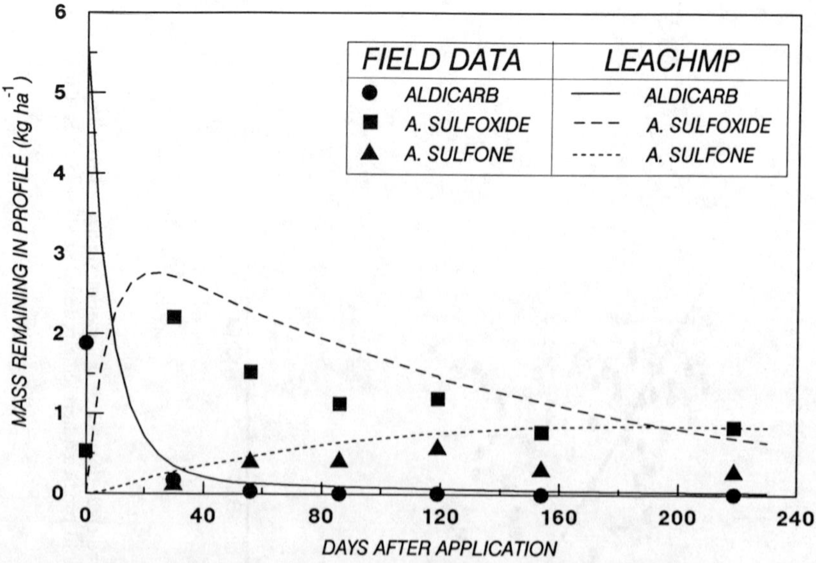

Fig. 1-3. Comparison of field-measured (data points) aldicarb residues and predictions (solid lines) based on the LEACHMP model. Taken from Pennell et al. (1990).

The results of the studies reviewed above are indeed encouraging, but a general optimistic outlook on lab-to-field transferability of biodegradation rate constants must await similar data from several studies under diverse conditions. The reason for caution is based on the reported failures of attempts to bioremediate contaminated soils and aquifers. In many of these reported cases (Madsen, 1991), failure to achieve anticipated levels of cleanup appears to be due not to the absence of appropriate microbial consortia, but due to other rate-limiting factors (such as desorption kinetics or insufficient concentration of electron acceptors). The explanation lies, in part, in recognizing the consequences of not properly decoupling the relevant processes. The impacts of nonequilibrium sorption on biodegradation kinetics were not considered in expecting faster degradation rates. The lab-measured degradation kinetic parameters may have been entirely satisfactory if the complexity of other processes had been accommodated. This may indeed be the case, as was demonstrated by MacQuarrie and Sudicky (1990), who successfully simulated the biodegradation of benzene in a heterogeneous aquifer.

SIGNIFICANCE OF SPATIAL SCALES IN PROCESS

An important issue raised by MacQuarrie and Sudicky (1990) is that indiscriminate lumping of local-scale processes can lead to erroneous simulations and misleading conclusions regarding organic chemical behavior in porous media. For practical reasons, it is often assumed that heterogeneous flow fields can be represented as being macroscopically homogeneous; this approach is usually referred to as the "equivalent porous medium" concept.

The effects of local-scale pore-velocity variations are represented by a *macrodispersion* term, in a manner similar to how local-scale hydrodynamic dispersion is taken to average over the effects of pore-scale velocity variations. MacQuarrie and Sudicky (1990) showed that such an approach can lead to a serious overestimation of substrate degradation rate as a result of the far greater mixing of the substrate and dissolved oxygen plumes predicted to occur at the local scales when a macrodispersion concept is employed. This is a clear demonstration of the importance of appropriately understanding local-scale physical heterogeneity in explaining and predicting macro-scale observations of biodegradation.

For field-scale predictions of solute transport, local-scale equilibrium sorption can, under certain conditions, be assumed as long as the spatial heterogeneities in flow domains are appropriately considered (Brusseau, 1989). However, when a microbial-degradation model is coupled to such a solute-transport model based on the local equilibrium assumption, we may overpredict field-scale degradation rates because the local-scale sorption nonequilibrium might limit degradation rates. Thus, averaging over spatial scales appropriate for one process may not be acceptable for other coupled processes.

SUMMARY

Application of the reductionist approaches, wherein each component of the system is studied in isolation, in investigating environmental behavior of organic chemicals has resulted in accumulation of a considerable database on several key processes. Having made impressive advances in our understanding of the key processes (e.g., transport, transformations, and sorption), it is now important to examine the linkages between these processes. Coupled-process models provide the stimulus for a paradigm shift—from the reductionist approaches to the relational approaches—where an investigation of the interrelations among the processes is considered even more important than the examination of individual processes themselves. These models also provide a conceptual framework for explaining experimental observations and in designing better laboratory experiments and field studies. There is an equal need for conducting innovative field-scale experiments such that databases are available for model testing and improvement. The predictive capabilities of the coupled-process models must await improvements in our ability to estimate the required model parameters at various spatial scales of interest. This is the challenge facing both modelers and experimentalists alike, and only through the collaborative efforts of both groups can this task be addressed.

ACKNOWLEDGMENT

This research was sponsored by the Subsurface Science Program, Office of Health and Environmental Research, U.S. Department of Energy through a contract (No. DE-C06-76RLO 1830) to Battelle PNL. Approved for publication as Florida Agric. Exp. Stn. J. Ser. no. R-02277.

REFERENCES

Alexander, M., R.J. Wagenet, P.C. Baveye, J.T. Gannon, U. Mingelrin, and Y. Tan. 1991. Movement of bacteria through soil and aquifer sand. EPA 600/S2-91-010. Environ. Res. Lab., U.S. Environ. Prot. Agency, Ada, OK.

Angley, J.T., M.L. Brusseau, W.L. Miller, and J.J. Delfino. 1992. Nonequilibrium sorption and aerobic biodegradation of dissolved alkylbenzenes during transport in aquifer material: Column experiments and evaluation of a coupled-process model. Environ. Sci. Technol. 26:1404–1410.

Ardakani, M.S., J.T. Rehbock, and A.D. McLaren. 1973. Oxidation of nitrite to nitrate in a soil column. Soil Sci. Soc. Am. Proc. 37:53–56.

Baehr, A., and M.Y. Corapcioglu. 1984. A predictive model for pollution from gasoline in soils and groundwater. p. 144–156. *In* Petroleum hydrocarbons and organic chemicals in groundwater, Houston, TX. 5–7 Nov. 1984. Natl. Water Well Assoc., Dublin, OH.

Blackburn, J.W. 1989. Is there an "uncertainty principle" in microbial waste treatment? p. 149–161. *In* M. Huntley (ed.) Biotreatment agricultural wastewater. CRC Press, Boca Raton, FL.

Borden, R.C., and P.B. Bedient. 1986. Transport of dissolved hydrocarbons influenced by oxygen-limited biodegradation 1. Theoretical development. Water Resour. Res., 22:1973–1982.

Borden, R.C., M.D. Lee, J.T. Wilson, C.H. Ward, and P.B. Bedient. 1984. Modelling the migration and biodegradation of hydrocarbons derived from a wood-creosoting waste. p. 130–143. *In* Petroleum hydrocarbons and organic chemicals in groundwater, Houston, TX. 5-7 Nov. 1984. Natl. Water Well Assoc., Dublin, OH.

Borden, R.C., P.B. Bedient, M.D. Lee, C.H. Ward, and J.T. Wilson. 1986. Transport of dissolved hydrocarbons influenced by oxygen-limited biodegradation 2. Field application. Water Resour. Res. 22:1983–1990.

Bosma, T.N.P., J.L. Schnoor, G. Schraa, and A.J.B. Zehnder. 1988. Simulation model for biotransformation of xenobiotics and chemotaxis in soil columns. J. Contam. Hydrol. 2:225–236.

Brusseau, M.L. 1989. Nonequilibrium sorption: Investigation at microscopic to megascopic scales. Ph.D. diss. Univ. of Florida, Gainesville (Diss. Abstr. 90-22609).

Brusseau, M.L., and P.S.C. Rao. 1989a. The influence of sorbate-organic matter interactions on sorption nonequilibrium. Chemosphere 18:1691–1706.

Brusseau, M.L., and P.S.C. Rao. 1989b. Sorption nonideality during organic contaminant transport in porous media. Crit. Rev. Environ. Control 19:33–99.

Brusseau, M.L., R.E. Jessup, and P.S.C. Rao. 1989. Modeling the transport of solutes influenced by multi-process nonequilibrium. Water Resour. Res. 25:1971–1988.

Brusseau, M.L., R.E. Jessup, and P.S.C. Rao. 1991. Nonequilibrium sorption of organic chemicals: Elucidation of rate-limiting processes. Environ. Sci. Technol. 25:134–142.

Brusseau, M.L., R.E. Jessup, and P.S.C. Rao. 1992a. Modeling solute transport influenced by multi-process nonequilibrium and transformation reactions. Water Resour. Res. 28:175–182.

Brusseau, M.L., P.S.C. Rao, and C.A. Bellin. 1992b. Modeling coupled processes in porous media: Sorption, transformation, and transport of organic solutes. p. 147–184. *In* R.J. Wagenet et al. (ed.) Interacting processes in soil science. Adv. Soil Sci., Lewis Publishers, Boca Raton, FL.

Cameron, D.R., and A. Klute. Convective-dispersive solute transport with a combined equilibrium and kinetic adsorption model. Water Resour. Res. 13:183–188.

Carnahan, C.L., and J.S. Remer. 1984. Nonequilibrium and equilibrium sorption with a linear sorption isotherm during mass transport through an infinite porous medium: Some analytical solutions. J. Hydrol. 73:227–258.

Carsel, R.F., L.A. Mulkey, M.N. Lorber, and L.B. Baskin. 1985. The pesticide root zone model (PRZM): A procedure for evaluating pesticide leaching threats to groundwater. Ecol. Model 390:46–69.

Corapcioglu, M.Y., and A. Haridas. 1985. Microbial transport in soils and groundwater: A numerical model. Adv. Water Resour. 8:188–200.

Corapcioglu, M.Y., M.A. Hossain, and M.A. Hossain. 1991. Methanogenic biotransformation of chlorinated hydrocarbons in groundwater. J. Environ. Eng. 117:47–65.

Dashman, T., and G. Stotzky. 1986. Microbial utilization of amino acids and a peptide bound on homoionic montmorillonite and kaolinite. Soil Biol. Biochem. 18:5–14.

Eldor, M., and G. Dagan. 1972. Solutions of hydrodynamic dispersion in poruos media. Water Resour. Res. 8:1316–1331.

Frind, E.O., E.A. Sudicky, and J.W. Molson. 1989. Three-dimensional simulation of organic transport with aerobic biodegradation. p. 89–96. *In* L.M. Abriola (ed.) Groundwater contamination. Int. Assoc. for Hydrologic Sci. Publ. 185.

Gamerdinger, A.P., R.J. Wagenet, and M.Th. van Genuchten. 1990. Application of two-site/two-region models for studying simultaneous nonequilibrium transport and degradation of pesticides. Soil Sci. Soc. Am. J. 54:957–963.

Gerstl, Z. 1990. Estimation of organic chemical sorption by soils. J. Contam. Hydrol. 6:357–375.

Gibbs, C.F. 1976. Methods and interpretation in measurement of degradation rate. Proc. of the Int. Biodegradation Symp., 3rd. p. 127–140. *In* J.M. Sharpley and A.M. Kaplan (ed.) Applied Sci., London.

Goldstein, R.M., L.M. Mallory, and M. Alexander. 1985. Reasons for possible failure of inoculation to enhance biodegradation. Appl. Environ. Microbiol. 50:977–983.

Gordon, A.S., and F.J. Millero. 1985. Adsorption mediated decrease in the biodegradation rate of organic compounds. Microb. Ecol. 11:289–298.

Green, R.E., and S.W. Karickhoff. 1990. Sorption estimates for modeling. p. 79–102. *In* H.H. Cheng (ed.) Pesticides in the soil environment. SSSA Book Ser. 2. SSSA, Madison, WI.

Gupta, S.K., and W.E. Larson. 1979. Estimating soil water retention characteristics from particle size distribution, organic matter percent, and bulk density. Water Resour. Res. 15:1633–1635.

Huang, S., and M. Chou. 1990. Kinetic model for microbial uptake of insoluble solid-state substrate. Biotechnol. Bioeng. 35:547–558.

Jessup, R.E., M.L. Brusseau, and P.S.C. Rao. 1989. UFBTC: Interactive programs for modeling one-dimensional solute transport. Florida Agric. Exp. Stn. Rep.

Kam, T.K. 1974. Analytical and numerical solutions of convective diffusive equations of solutes subject to linear and Michaelis-Menton types of kinetics in soils. Ph.D. diss. Univ. of California, Berkeley.

Kindred, J.S., and M.A. Celia. 1989. Contaminant transport and biodegradation, 2, Conceptual model and test simulations. Water Resour. Res. 25:1149–1159.

Laskowski, D.A., C.A.I. Goring, P.J. McCall, and R.L. Swann. 1982. Terrestrial environment, p. 198–240. *In* R.A. Conway (ed.) Environmental risk analysis for chemicals. Van Nostrand Reinhold Co., New York.

Lassey, K.R. 1988. Unidimensional solute transport incorporating equilibrium and rate-limited isotherms with first-order loss: 1. Model conceptualizations and analytic solutions. Water Resour. Res. 24:343–350.

Lehman, R.G., J.R. Miller, D.D. Fontaine, D.A. Laskowski, J.H. Hunter, and R.C. Cordes. 1992. Degradation of a sulfonamide herbicide as a function of soil sorption. Weed Res. 32:197–205.

Lindstrom, F.T. 1976. Pulsed dispersion of trace chemical concentrations in a saturated sorbing porous medium. Water Resour. Res. 12:229–238.

MacQuarrie, K.T.B., and E.A. Sudicky. 1990. Simulation of biodegradable organic contaminants in groundwater. 2. Plume behavior in uniform and random flow fields. Water Resour. Res. 26:223–240.

MacQuarrie, K.T.B., E.A. Sudicky, and E.O. Frind. 1990. Simulation of biodegradable organic contaminants in groundwater. 1. Numerical formulation in principle directions. Water Resour. Res. 26:207–223.

Madsen, E.L. 1985. Effects of chemical speciation on microbial mineralization of organic compounds. Ph.D. diss. Cornell Univ., Ithaca, NY (Diss. Abstr. 85-04534).

Madsen, E.L. 1991. Determining in situ biodegradation: Facts and challenges. Environ. Sci. Technol. 25:1663–1673.

Madsen, E.L., J.L. Sinclair, and W.C. Ghiorse. 1991. In situ biodegradation: Microbiological patterns in a contaminated aquifer. Science 252:830–833.

Mihelcic, J.R. 1988. Microbial degradation of polycyclic aromatic hydrocarbons under denitrification conditions in soil-water suspensions. Ph.D. diss. Carnegie Mellon Univ., Pittsburgh, PA (Diss. Abstr. 88-17721).

Mihelcic, J.R., and R.G. Luthy. 1988a. Degradation of polycyclic aromatic hydrocarbon compounds under various redox conditions in soil-water systems. Appl. Environ. Microbiol. 54:1182–1187.

Mihelcic, J.R., and R.G. Luthy. 1988b. Microbial degradation of acenaphthalene and napthalene under denitrification conditions in soil-water systems. Appl. Environ. Microbiol. 54:1188–1198.

Mihelcic, J.R., and R.G. Luthy. 1988c. The potential effects of sorption processes on the microbial degradation of hydrophobic organic compounds in soil-water suspensions. p. 1–14. *In* Physiochemical and biological detoxification of hazardous wastes. Proc. Int. Conf. 3–5 May 1988. Hazardous Substance Manage. Res. Ctr., Atlantic City, New Jersey.

Miller, M.E., and M. Alexander. 1991. Kinetics of bacterial degradation of benzylamine in a montmorillonite suspension. Environ. Sci. Technol. 25:240–245.

Miller, R.M., and R. Bartha. 1989. Evidence from liposome encapsulation for transport-limited microbial metabolism of solid alkanes. Appl. Environ. Microbiol. 55:269–274.

Miralles-Wilhelm, F., L.W. Gelhar, and V. Kapoor. 1990. Effects of heterogeneities on field-scale biodegradation in groundwater. Trans. Am. Geophys. Union. 71:1304.

Mironenko, E.V., and Y.A. Pachepsky. 1984. Molz, F.J., and M.A. Widdowson. 1988. Internal inconsistencies in dispersion-dominated models that incorporate chemicals and microbial kinetics. Water Resour. Res. 24:615–619.

Molz, F.J., M.A. Widdowson, and L.D. Benefield. 1986. Simulation of microbial growth dynamics coupled to nutrient and oxygen transport in porous media. Water Resour. Res. 22:1207–1216.

Myrold, D.D., and J.M. Tiedje. 1985. Diffusional constraints on denitrification in soil. Soil Sci. Soc. Am. J. 49:651–657.

Nofziger, D.L., and A.G. Hornsby. 1987. Chemical movement in layered soils: User's manual. Cir. 780. Florida Coop. Ext. Ser. Inst. of Food and Agric. Sci., Univ. of Florida, Gainesville.

Ogram, A.V., R.E. Jessup, and P.S.C. Rao. 1985. Effects of sorption on biological degradation rates of 2,4-dichlorophenoxyacetic acid in soils. Appl. Environ. Microbiol. 49:582–587.

Ou, L.T., P.S.C. Rao, K.S.V. Edvardsson, R.E. Jessup, A.G. Hornsby, and R.L. Jones. 1988. Aldicarb degradation in sandy soils from different depths. Pestic. Sci. 23:1–12.

Ou, L.T., P.S.C. Rao, and W.B. Wheeler. 1984. Estimation of parameters for modeling the behavior of selected pesticides and orthophosphate. p. 48–88. *In* P.S.C. Rao et al. (ed.) EPA-600/3-84-019. U.S. Environ. Protection Agency, Athens, GA.

Ou, L.T., J.E. Thomas, K.S.V. Edvardsson, P.S.C. Rao, and W.B. Wheeler. 1986. Aerobic and anaerobic degradation of aldicarb in aseptically collected soils. J. Environ. Qual. 15:356–363.

Parlange, J.Y., J.L. Starr, D.A. Barry, and R.D. Braddock. 1984. Some approximate solutions of the transport equation with irreversible reactions. Soil Sci. 137:434–442.

Pennell, K.D., A.G. Hornsby, R.E. Jessup, and P.S.C. Rao. 1990. Evaluation of five simulation models for predicting aldicarb and bromide behavior under field conditions. Water Resour. Res. 26:2679–2693.

Rao, P.S.C., K.S.V. Edvardsson, L.T. Ou, R.E. Jessup, P. Nkedi-Kizza, and A.G. Hornsby. 1986. Spatial variability of pesticide sorption and degradation parameters. p. 100–115. *In* W.Y. Garner et al. (ed.) Evaluation of pesticides in groundwater. ACS Symp. Ser. 28. Am. Chem. Soc., Washington, DC.

Rao, P.S.C., and J.M. Davidson. 1980. Estimation of pesticide retention and transformation parameters required in nonpoint source pollution models. p. 23–67. *In* M.R. Overcash and J.M. Davidson (ed.) Environmental impact of nonpoint source pollution. Ann Arbor Sci. Publ., Ann Arbor, MI.

Rasmusson, A. 1982. Transport processes and conversion in an isothermal fixed-bed catalytic reactor. Chem. Eng. Sci. 37:411–415.

Rawls, W.J., D.L. Brakensiek, and K.E. Saxton. 1982. Estimation of soil water properties. Trans. Am. Soc. Agric. Eng. 25:1316–1320, 1328.

Rifai, H.S., P.B. Bedient, J.T. Wilson, K.M. Miller, and J.M. Armstrong. 1988. Biodegradation modeling at aviation fuel spill site. J. Environ. Eng. 114:1007–1029.

Rijnaarts, H.H.M., A. Bachmann, J.C. Jumelet, and A.J.B. Zehnder. 1990. Effect of desorption and intraparticle mass transfer on the aerobic biomineralization of α-hexachloroclohexane in contaminated calcareous soil. Environ. Sci. Technol. 24:1349–1354.

Robinson, K.G., W.S. Farmer, and J.T. Novak. 1990. Availability of sorbed toluene in soils for biodegradation by acclimated bacteria. Water Res. 24:345–350.

Ryden, J.C., and L.J. Lund. 1980a. Nitrous oxide evolution from irrigated land. J. Environ. Qual. 9:387–393.

Ryden, J.C., and L.J. Lund. 1980b. Nature and extent of directly measured denitrification losses from some irrigated vegetable crop production units. Soil Sci. Soc. Am. J. 44:505–511.

Scow, K.M., and M. Alexander. 1992. Effects of diffusion on the kinetics of biodegradation: Experimental results with synthetic aggregates. Soil Sci. Soc. Am. J. 56:128–134.

Selim, H.M., J.M. Davidson, and R.S. Mansell. Evaluation of a two-site adsorption-desorption model for describing solute transport in soils. p. 444–448. *In* Proc. Summer Computer Simulation Conf., Washington, DC.

Semprini, L., and P.L. McCarty. 1991. Comparison between model simulations and field results for in-situ biorestoration of chlorinated aliphatics: 1. Biostimulation of methanotrophic bacteria. Groundwater 29:365–374.

Smith, S.C., C.C. Ainsworth, S. Traina, and R.J. Hicks. 1992. The effect of sorption on the biodegradation of quinoline. Soil Sci. Soc. Am. J. 56:737–746.

Speitel, G.E., K. Dovantzis, and F.A. DiGiano. 1987. Mathematical modeling of biogeneration in GAC columns. J. Environ. Eng. 113:32–48.

Srinivasan, P., and J.W. Mercer. 1988. Simulation of biodegradation and sorption processes in groundwater. Groundwater 26:475–487.

Steen, W.C., D.F. Paris, and G.L. Baughman. 1980. Effects of sediment sorption on microbial degradation of toxic substances. p. 477–482. *In* R.A. Baker (ed.) Contaminants and sediments. Vol. 1. Fate and transport, case studies, modeling, toxicity. Ann Arbor Sci. Publ., Ann Arbor, MI.

Steenhuis, T.S., S. Pacenka, and K.S. Porter. 1986. MOUSE: A management model for evaluating groundwater contamination from diffuse surface sources aided by computer graphics. Appl. Agric. Res. 2:227–289.

Steinberg, S.M., J.J. Pignatello, and B.L. Sawhney. 1987. Persistence of 1,2-dibromomethane in soils: entrapment in intraparticle micropores. Environ. Sci. Technol. 21:1201–1208.

Stotzky, G., and L.T. Rem. 1966. Influence of clay minerals on microorganisms. I. Montmorillonite and kaolinite on bacteria. Can. J. Microbiol. 12:547–563.

Stucki, G., and M. Alexander. 1987. Role of dissolution rate and solubility in biodegradation of aromatic compounds. Appl. Environ. Microbiol. 53:292–297.

Subba-Rao, R.V., and M. Alexander. 1982. Effect of sorption on mineralization of low concentrations of aromatic compounds in lake water sediments. Appl. Environ. Microbiol. 44:659–668.

Sykes, J.F., S. Soyupak, and G.J. Farquhar. 1982. Modeling of leachate organic migration and attenuation in groundwaters below sanitary landfills. Water Resour. Res. 18:135–145.

van Genuchten, M.Th. 1974. Mass transfer studies in sorbing porous media. Ph.D. diss. New Mexico State Univ., Las Cruces (Diss. Abstr. 75-11005).

van Genuchten, M.Th., and R.J. Wagenet. 1989. Two-site/two-region models for pesticide transport and degradation: Theoretical development and analytical solutions. Soil Sci. Soc. Am. J. 53:1303–1310.

van Loosdrecht, M.C.M., J. Lyklema, W. Norde, and A.J.B. Zehnder. 1990. Influences of interfaces on microbial activity. Microbiol. Rev. 54:75–87.

Wagenet, R.J., and J.L. Hutson. 1987. LEACHM: A finite-difference model for simulating water, salt, and pesticide movement in the plant root zone. Continuum 2. 1987. New York State Resources Inst., Cornell Univ., Ithaca, NY.

Walker, A., R.J. Hance, J.G. Allen, G.G. Briggs, Y-L. Chen, J.D. Gaynor, E.J. Hogue, A. Malquor, K. Moody, J.R. Moyer, W. Pestemer, A. Rahman, A.E. Smith, J.C. Streibig, N.T.L. Torstensson, L.S. Widyanato, and Z. Zandvoort. 1983. EWRS herbicide-soil working group: Collaborative experiment on simazine persistence in soils. Weed Res. 23:373–383.

Weber, E.J. 1991. Studies of benzidine-based dyes in sediment-water systems. Environ. Toxicol. Chem. 10:609–618.

Widdowson, M.A., F.J. Molz, and L.D. Benefield. 1987. Development and application of a model for simulating microbial growth dynamics coupled to nutrient and oxygen transport in porous media. p. 28–51. *In* Proc. of the NWWA Conf. on Solving Ground Water Problems with Models, Denver, CO. 10–12 Feb. 1987. Natl. Water Well Assoc., Dublin, OH.

Widdowson, M.A., F.J. Molz, and L.D. Benefield. 1988a. A numerical transport model for oxygen- and nitrate-based respiration linked to substrate and nutrient availability in porous media. Water Resour. Res. 24:1553–1565.

Widdowson, M.A., F.J. Molz, and L.D. Benefield. 1988b. Modeling multiple organic contaminant transport and biotransformations under aerobic and anaerobic (denitrifying) conditions in the subsurface. p. 397–416. *In* Proc. of Conf. on Petroleum Hydrocarbons and Organic Chemicals in Groundwater, Houston, TX. 9–11 Nov. 1988. Natl. Water Well Assoc., Dublin, OH.

Williams, R.D., L.R. Ahuja, and J.W. Naney. 1992. Evaluation of methods to estimate soil water characteristics from soil texture, bulk density, and limited data. Soil Sci. 153:172–184.

Woesten, J.H.M., and M.Th. Van Genuchten. 1988. Using texture and other soil properties to predict the unsaturated soil hydraulic functions. Soil Sci. Soc. Am. J. 52:1762–1770.

Wu, S., and P.M. Gschwend. 1986. Sorption kinetics of hydrophobic organic compounds to natural sediments and soil. Environ. Sci. Technol. 20:717–725.

Wszolek, P.C., and M. Alexander. 1979. Effect of desorption rate on the biodegradation of *n*-alkylamines bound to clay. J. Agric. Food Chem. 27:410–414.

Ying, W., and W.J. Weber. 1979. Bio-physiochemical adsorption model systems for wastewater treatment. J. Water Pollut. Control Fed. 51:2661–2667.

2 Degradation of Atrazine and Metolachlor in Subsoils from an Atlantic Coastal Plain Watershed[1]

Richard M. Johnson and Jeffry J. Fuhrmann

University of Delaware
Newark, Delaware

A thorough understanding of the factors affecting herbicide degradation is necessary to maximize agricultural benefits and minimize adverse environmental impacts. Although herbicide persistence has been studied primarily in surface soils, the need to evaluate persistence in subsoils has become increasingly apparent. Harris et al. (1969) observed decreased atrazine (2-chloro-4-ethylamino-6-isopropylamino-*s*-triazine) degradation with increasing soil depth. When averaged over 11 locations, 61% more atrazine was recovered at a 38-cm depth as compared to 8 cm. The authors attributed these trends to higher organic matter, temperature, and better aeration at the 8-cm depth. Adams and Thurman (1991) reported similar trends in atrazine transformation with soil depth. Locke and Harper (1991) noted a significantly higher evolution of both $^{14}CO_2$ and CO_2 in surface soils as compared to subsoils amended with metribuzin [4-amino-6-(1,1-dimethylethyl)-3-(methylthio)-1,2,4-triazin-5(4H)one]. Moorman and Harper (1989) reported significantly less metribuzin degradation in subsoils as compared to surface soils (5% vs. 15–20%, respectively) over 91 d.

The objective of this experiment was to determine the degradation potential of atrazine and metolachlor [2-chloro-*N*-(2-ethyl-6-methylphenyl)-*N*-(2-methoxy-1-methylethyl) acetamide] in three contrasting subsoils common to the Delaware Coastal Plain. Both herbicides are used extensively in grain-production systems in the region and have been detected in Delaware groundwater (Ritter et al., 1987).

MATERIALS AND METHODS

Soils

The three subsoils used in this study were sampled from a C3 horizon (0.84–1.07 m) of a Klej loamy sand (mesic, coated Aquic Quartzipsamment),

---

[1] Published as Miscellaneous Paper No. 1405 of the Delaware Agricultural Experiment Station.

Table 2-1. Selected chemical and physical properties† of three Delaware Atlantic Coastal Plain subsoils.

Soil series	Organic matter	Effective exchange capacity	Exchangeable acidity	Soil separates		
				Sand	Silt	Clay
	$g\ kg^{-1}$	$cmol_c\ kg^{-1}$		%		
Klej	0.4	0.61	ND‡	92	3	5
Runclint	1.2	4.43	2.39	45	27	28
Matawan	1.9	6.98	6.48	17	35	48

† Organic matter (Nelson & Sommers, 1982); effective exchange capacity, and exchangeable acidity (Thomas, 1982); soil separates (Day, 1965).
‡ ND = not detectable.

a Btg2 horizon (0.97–1.27 m) of a Runclint loamy sand (mesic, coated Typic Quartzipsamment) and a Btg2 horizon (0.79–1.12 m) of a Matawan sandy loam (fine-loamy, siliceous, mesic Aquic Hapludult) (Table 2–1). The field capacity moisture contents were estimated by pressure plate techniques (0.03 MPa) to be 140, 200, and 280 g kg^{-1} for the Klej, Runclint, and Matawan soils, respectively. Soils were dried only enough to pass a 2-mm sieve prior to use.

Herbicides and Metabolites

Technical and ring-labeled herbicides, as well as metabolites, were obtained from Ciba Geigy, Greensboro, NC. The specific activity of the ring-labeled atrazine was 53.5 μCi mg^{-1}, the radiochemical purity was 94.1%, and the chemical purity was 96.5%. The specific activity of metolachlor was 71.5 μCi mg^{-1}, the radiochemical purity was 97.0%, and the chemical purity was 99.9%. The atrazine metabolites investigated were: deethylatrazine, deisopropylatrazine, and hydroxyatrazine. Metolachlor degradation was monitored by assaying for 4-(2-ethyl-6-methylphenyl)-5-methyl-3-morpholinone, a dechlorinated metabolite.

Degradation Studies

Soil (50 g oven-dry basis) was weighed into 250-mL biometer flasks (Bartha & Pramer, 1965). Stock solutions of technical and ring-labeled atrazine and metolachlor were prepared in methanol (MeOH) and added to the soil to obtain a herbicide concentration of 2 mg kg^{-1} and a specific activity of 1 μCi $flask^{-1}$. Once the MeOH had evaporated, the soils were brought to field capacity with distilled water, stoppered, and incubated for 73 d at 25 ± 1 °C in the dark. Soil blanks (no added herbicide) and biometer blanks (no added soil) were also included.

Evolved CO_2 and $^{14}CO_2$ were trapped in 10 mL of 0.1 *M* NaOH. When the trapping solution was sampled, an additional 10 mL of 0.1 *M* NaOH was used to rinse the sidearm flask. This solution was then combined with the original trapping solution for analysis. To determine CO_2 concentrations, 2 mL of 1 *M* $BaCl_2$ and four drops of phenolphthalein indicator were added to the combined solutions. The solutions were then titrated with 0.1 *M* HCl. To determine $^{14}CO_2$ concentrations, a 0.2-mL aliquot was taken from the solution and placed in 3 mL of scintillation fluid (Scintiverse TM E, Fisher Scientific, Pittsburgh, PA)[2] and counted for three replications of 10 min on a Searle Isocap/300 liquid scintillation spectrophotometer.

Extraction Procedures

When the experiment ended, the soil in the biometers was transferred to 250-mL Erlenmeyer flasks and extracted with 150-mL MeOH on a rotary shaker (150 rpm) for 2 h. The extract was then filtered through Whatman no. 42 filter paper. The soil was washed with an additional 50 mL of MeOH, and the filtrates were combined. The extracted soil samples were dried overnight. A 0.2-mL aliquot of the MeOH extract was placed in 3 mL of scintillation fluid and counted as previously described. A 1.0-g sample of the extracted soil was placed in 3 mL of scintillation fluid and similarly counted to obtain an approximation of the unextracted ^{14}C.

Metabolite Analysis

The MeOH soil extracts were analyzed for the parent herbicides and associated metabolites by reverse-phase high-performance thin layer chromatography (RHPTLC). The extracts were evaporated to dryness under vacuum at 60 °C, redissolved in MeOH, transferred to 20-mL vials, and redried under N_2 gas. The redried samples were dissolved in 200 μL of MeOH, spotted onto RHPTLC plates, and developed using $MeOH:H_2O$ (70:30 for atrazine and 85:15 for metolachlor). The extracted samples were chromatographed along with standards to enable identification of parent herbicide and metabolites. Spots on developed plates were visualized under UV radiation. Identified spots were removed by scraping, placed in 3 mL of scintillation fluid, and counted as previously described.

RESULTS AND DISCUSSION

$^{14}CO_2$ Evolution

Only low levels of $^{14}CO_2$ evolution were observed for both atrazine and metolachlor in this study, indicating that the herbicides were not appreciably mineralized in the soils investigated (Table 2-2). The herbicides used in this study were uniformly ring-labeled, thus ring cleavage would be necessary for evolution of $^{14}CO_2$ to occur.

[2] Mention of trademark or proprietary products in this chapter does not constitute a guarantee or warranty of the product by the Univ. of Delaware and does not imply its approval to the exclusion of other products that may also be suitable.

Table 2-2. Recovery of ^{14}C-atrazine and ^{14}C-metolachlor and total CO_2 evolution for three Delaware Atlantic Coastal Plain subsoils.

Soil series	MeOH extract	Retained by soil	$^{14}CO_2$	Total	Total CO_2 evolution
	Recovery of ^{14}C				
	% of added label				mg $flask^{-1}$
		Atrazine			
Klej	85	3.4	0.31	88	13.3
Runclint	90	4.0	0.23	94	17.7
Matawan	97	6.3	0.28	103	25.0
		Metolachlor			
Klej	65	1.9	0.42	67	17.3
Runclint	66	2.3	1.59	70	16.1
Matawan	67	2.7	0.21	70	29.6
		Herbicide blanks			
Klej	NA†	NA	NA	NA	14.5
Runclint	NA	NA	NA	NA	15.8
Matawan	NA	NA	NA	NA	25.2
Soil blank	NA	NA	NA	NA	13.0

† NA = not applicable.

CO_2 Evolution

Although differences in total CO_2 evolution were noted among the soils examined, there was no consistent effect due to herbicide addition (Table 2–2). This may reflect the low rate of herbicide application and the relatively high levels of background soil respiration.

Extraction

The extraction data revealed that the majority of the applied herbicides was recovered in the MeOH extracts, but that the efficiency of recovery was greater for atrazine than for metolachlor (Table 2–2). Only a small amount of ^{14}C-labeled herbicide remained in the soil as estimated by directly counting the extracted soil.

Metabolites

Analysis of MeOH extracts from the soils receiving atrazine showed that between 89.5 and 91.6% of the extractable ^{14}C was present as the parent compound at the termination of the experiment (Table 2–3). Low levels of atrazine metabolites were detected, indicating that some transformation of the applied atrazine had occurred. The metabolites identified were hydroxyatrazine (4.6–5.6% of total counts) > deethylatrazine (2.1–3.8%) > deisopropylatrazine (1.1–1.6%).

Analysis of the MeOH extracts of the metolachlor-treated soils indicated that a small percentage (3.8–4.6%) of the parent compound was transformed to 4-(2-ethyl-6-methylphenyl)-5-methyl-3-morpholinone (Table 2–3).

Table 2–3. Relative abundance of atrazine and metolachlor and their degradation products in methanol extracts of three Delaware Atlantic Coastal Plain subsoils following a 73-d laboratory incubation.

Herbicide applied	Compound measured	Soil series: Klej	Runclint	Mattawan
		—— % of extracted ^{14}C ——		
Atrazine	Atrazine	89.5	91.6	90.9
	Hydroxyatrazine	5.6	4.6	5.4
	Deethylatrazine	3.8	2.6	2.1
	Deisopropylatrazine	1.1	1.2	1.6
Metolachlor	Metolachlor	96.2	95.4	95.7
	4-(2-ethyl-6-methylphenyl)-5-methyl-3-morpholinone	3.8	4.6	4.3

CONCLUSION

The results indicate that the mineralization potential of atrazine and metolachlor is low in the subsoils investigated from this coastal plain watershed. Thus, herbicide that is leached to these subsoils may be sufficiently persistent to represent a long-term contamination threat to groundwater. Transformation of both herbicides to degradation products was detected, but only at low levels.

REFERENCES

Adams, C.D., and E.M. Thurman. 1991. Formation and transport of deethylatrazine in the soil and vadose zone. J. Environ. Qual. 20:540–547.

Bartha, R., and D. Pramer. 1965. Features of a flask and method for measuring the persistence and biological effects of pesticides in soil. Soil Sci. 100:68–70.

Day, P.R. 1965. Particle fractionation and particle size analysis. p. 545–567. *In* C.A. Black et al. (ed.) Methods of soil analysis. Part 1. Agron. Monogr. 9. ASA, Madison, WI.

Harris, C.I., E.A. Woolson, and B.E. Hummer. 1969. Dissipation of herbicides at three soil depths. Weed Sci. 17:27–31.

Locke, M.A., and S.S. Harper. 1991. Metribuzin degradation in soil: I. Effects of soybean residue amendment, metribuzin level, and soil depth. Pestic. Sci. 31:221–237.

Moorman, T.B., and S.S. Harper. 1989. Transformation and mineralization of metribuzin in surface and subsurface horizons of a Mississippi delta soil. J. Environ. Qual. 18:302–306.

Nelson, D.W., and L.E. Sommers. 1982. Total carbon, organic carbon and organic matter. p. 539–579. *In* A.L. Page et al. (ed.) Methods of soil analysis. Part 2. 2nd ed. Agron. Monogr. 9. ASA and SSSA, Madison, WI.

Ritter, W.F., A.E.M. Chirnside, and R.W. Scarborough. 1987. Pesticide leaching in a coastal plain soil. Am. Soc. Agric. Eng., St. Joseph, MI.

Thomas, G.W. 1982. Exchangeable cations. p. 159–166. *In* A.L. Page et al. (ed.) Methods of soil analysis. Part 2. 2nd ed. Agron. Monogr. 9. ASA and SSSA, Madison, WI.

3 Biodegradation of Atrazine and Alachlor in Subsurface Sediments

Mark Radosevich, Jennifer J. Crawford, Samuel J. Traina, Kye-Heon Oh, and Olli H. Tuovinen

The Ohio State University
Columbus, Ohio

Atrazine (2-chloro-4-ethylamino-6-isopropylamino-*s*-triazine) and alachlor (2-chloro-2′,6′-diethyl-*N*-[methoxymethyl]acetanilide) are commonly used for broadleaf weed control in the USA. Despite their wide use, little is known about their environmental fate in vadose zone and aquifer environments. An interdisciplinary, multi-agency project (Management Systems Evaluation Area, MSEA) was initiated in January 1991 to evaluate the effects of various agricultural management systems on water quality in a buried-valley aquifer system in southern Ohio. The Ohio MSEA site (Fig. 3–1) is located on a 650-acre farm located approximately 2 mi south of Piketon, OH. The site overlies the Scioto Buried Valley (or alluvial) Aquifer. This aquifer has been characterized as shallow, unconfined, and permeable with high recharge rates. Alluvial aquifers are generally vulnerable to groundwater contamination because the short flow path to the water table decreases the potential for attenuation processes such as sorption and biodegradation. Given the vulnerability of alluvial aquifers, the research in this chapter was initiated to evaluate the biodegradation potential for atrazine and alachlor in the vadose zone and aquifer sediments underlying this MSEA site. Biodegradation of 2,4-D (2,4-dichlorophenoxyacetic acid) was also evaluated in a limited number of MSEA samples. The chemical structures of the three herbicides are shown in Fig. 3–2.

MATERIALS AND METHODS

Sample Collection

Sediment samples from above and below the water table were collected during the installation of 22 multiport monitoring wells between January and March 1991 at the Ohio MSEA site. Both site and regional wells were included in sediment sampling. Site wells were installed to monitor effects of specific agricultural practices on groundwater quality; regional wells were installed for regional groundwater analysis. Samples were retrieved with a

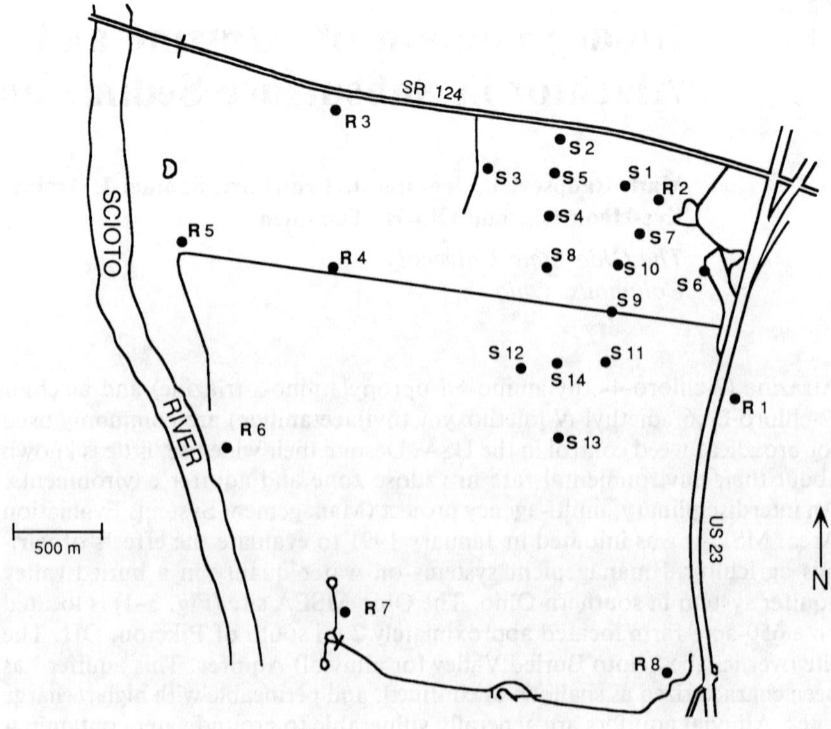

Fig. 3–1. Site (S) and regional (R) wells at the Ohio Management Systems Evaluation Area (MSEA) site.

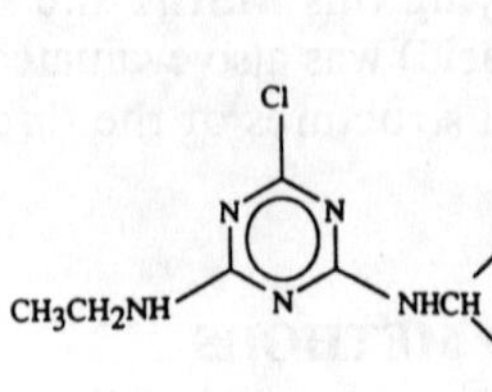

Atrazine

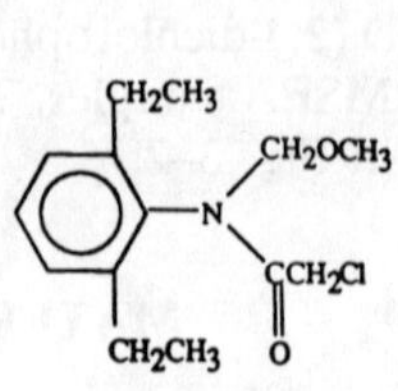

Alachlor

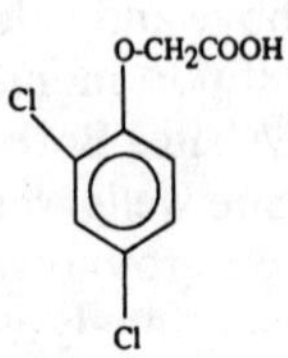

2,4-D

Fig. 3–2. Chemical structures of atrazine, alachlor, and 2,4-D.

flame sterilized split-spoon sampler (11.4 cm in diameter by 91.2 cm in length). The sample core was cut in half lengthwise with sterile instruments. Samples for microbiological analyses were taken from the center of the core and stored in sterile mason jars on ice for the first 24 h and refrigerated at 4.0 ± 0.5 °C thereafter until analyzed.

Processing Management Systems Evaluation Area Samples for Enumeration and Isolation

Each aquifer sample was aseptically diluted 10-fold in sterile 0.1% (w/v) sodium pyrophosphate solution and placed on a shaker for 30 min. Serial dilutions were prepared in phosphate-buffered saline (1.6 g of KH_2PO_4, 0.84 g of K_2HPO_4, 9.0 g of NaCl L^{-1}). Diluted samples were spread-plated in duplicate on peptone-tryptone-yeast-glucose (PTYG) agar which contained (per liter) peptone, 0.05 g; tryptone, 0.05 g; yeast extract, 0.1 g; D-glucose, 0.1 g; $MgSO_4 \cdot 7H_2O$, 0.6 g; $CaCl_2 \cdot 2H_2O$, 0.07 g; and Bacto Agar, 15.0 g. Plates were incubated at 25 °C for up to 14 d. For isolation, single colonies were streaked on PTYG and stored on PTYG slants at 4 °C.

Sample Enrichment for Herbicide Degradation

Shake flask enrichments were prepared to screen the subsurface sediment samples for the microbiological degradation of atrazine and alachlor. The enrichment medium had the following composition (per liter): K_2HPO_4, 0.5 g; $(NH_4)_2SO_4$, 0.5 g; $MgSO_4 \cdot 7H_2O$, 0.5 g; $FeCl_3 \cdot H_2O$, 10 mg; $CaCl_2 \cdot H_2O$, 10 mg; $MnCl_2$, 0.1 mg; $ZnSO_4$, 0.01 mg; pH 6.8. Atrazine or alachlor was added to a final concentration of 0.1 m*M*. The media (100 mL in 250-mL shake flasks) were autoclaved for 20 min at 121 °C. Initial experiments indicated that both herbicides were stable under these conditions.

Six enrichments, each inoculated with 2 to 3 g of sediment material, were prepared for each sample received. Sterile controls were inoculated prior to autoclaving. Three cultures contained atrazine and three alachlor as the sole C and energy source. For both herbicides, one enrichment flask was shaken, one was incubated statically, and one was autoclaved to serve as a sterile control. The shake flask cultures were incubated under static conditions after the first 21 d of shaking. All static incubations were incubated in the dark. The incubation temperature for all enrichments was 25 ± 2 °C.

The weight of each flask was initially recorded. Periodically, the flasks were weighed and adjusted to their initial weights with sterile deionized water to replace losses due to evaporation. Volatile losses of the test herbicides, estimated with sterile controls, were negligible.

Selected samples were also used to enrich for microorganisms capable of degrading 2,4-D. These samples were incubated in shake flasks containing 0.9 m*M* 2,4-D as the sole organic C and energy source at 20 ± 1 °C.

Analytical Methods

The degradation was evaluated on the basis of the loss of parent compound in test cultures. Samples (0.5 mL) for atrazine and alachlor analysis were extracted with methanol (1 mL) in microcentrifuge tubes and centrifuged at 15 400∘*g* for 15 min. Atrazine and alachlor were analyzed by high-pressure liquid chromatography (HPLC) using an ABI 400 Solvent Delivery System[1] equipped with an ABI 783A UV detector set at 230 nm for atrazine analysis and 220 nm for alachlor analysis. Fifty microliter samples were injected onto a C-18 Econosphere column (5 m, 20-cm long) with a Spectra Physics SP8875 autosampler and eluted with an isocratic mobile phase consisting of (v/v) 55% acetonitrile:45% water for atrazine analysis and 70% acetonitrile:30% water for alachlor analysis at a flow rate of 1.0 mL min^{-1}. The UV methodology for 2,4-D analysis (A_{max} 283 nm) has been previously described (Oh & Tuovinen, 1991).

Alternate Enrichment Strategies

Three additional atrazine-enrichment strategies were initiated with composite sediment samples. The first composite (heretofore denoted as type one) received glucose (1.0 g L^{-1}) as an additional C and energy source. The second type of enrichment was based on using a medium that contained atrazine in the minerals salts solution with ammonium sulfate as the N source. The third type was supplemented with glucose in a N-free mineral salts medium to select for organisms capable of using atrazine as the sole N source. The fourth type was also N-free and contained atrazine as the sole source of C, energy, and N. These enrichments were incubated aerobically in the dark at 25 °C. The composite sediment samples were allowed to equilibrate with the enrichment media for 24 h before samples (0.5 mL) were removed and extracted with 1.0 mL of methanol and analyzed by HPLC to provide an initial solution-phase concentration of atrazine in each flask. The enrichments were incubated 21 to 28 d between sampling events. Enrichments that exhibited a 25% loss of parent compound were subcultured into identical media and the sampling process was repeated.

RESULTS AND DISCUSSION

Bacterial Enumeration

The heterotrophic plate counts of the subsurface samples at the Ohio MSEA site are summarized in Table 3–1. The counts ranged from <100 colony-forming units (CFU) g^{-1} (site R2, 15.2–15.8 m) to 3×10^7 CFU g^{-1} (site S1, 1.5–2.1 m). The heterotrophic population density did not appear to be depth dependent. There was no apparent relationship between the plate counts and solution pH, Eh, and temperature, or dissolved oxygen, inorganic N (ammonium, nitrite, and nitrate), phosphate, and dissolved C con-

[1]Mention of trademark or proprietary products in this chapter does not constitute a guarantee or warranty of the product by The Ohio State University and does not imply its approval to the exclusion of other products that may also be suitable.

Table 3–1. Heterotrophic plate counts of sediment samples.

Core sample	Depth interval, m	Avg. log CFU† g^{-1}	Core sample	Depth interval, m	Avg. log CFU g^{-1}	Core sample	Depth interval, m	Avg. log CFU g^{-1}
R1	1.5–2.2	5.65	R7	1.5–1.7	6.57	S7	1.5–2.0	5.75
R1	3.6–4.4	5.47	R7	1.7–2.3	5.86	S7	4.3–4.9	5.43
R1	7.9–8.8	6.70	R7	4.3–4.6	5.56	S7	6.8–7.1	5.15
R1	11.5–12.5	7.08	R7	7.9–8.4	5.42			
R1	21.9–22.5	6.10	R7	13.4–14	5.97	S8	1.8–2.3	5.21
R1	24.3	5.65	R7	17.9–18.5	6.02	S8	4.3–5.2	5.23
			R7	20	6.55	S8	7.9–8.8	4.48
R2	2.6–3.0	5.10						
R2	3.6–4.6	5.11	R8	1.2–1.8	5.36	S9	1.5–2.3	5.39
R2	12.2–12.8	5.57	R8	4.0–4.3	4.14	S9	3.8–4.0	3.90
R2	15.2–15.8	<2.00	R8	4.3–4.6	4.13	S9	3.6–4.1	5.15
R2	17.6–18.2	4.15	R8	8.5–8.8	5.39	S9	7.6–7.9	5.92
R3	2.1–2.7	7.39	S1	1.5–2.1	7.48	S10	Surface	6.74
R3	4.0–4.6	4.01	S1	2.1–2.4	6.24	S10	1.5–2.3	6.26
R3	8.5–9.0	4.68	S1	3.3–4.3	5.00	S10	5.2–5.6	5.23
			S1	7.6–8.2	5.50	S10	13.1–13.7	4.91
R4	3.3–4.3	5.80				S10	17.6–18.1	5.41
R4	4.0–4.3	4.42	S2	1.5–2.1	6.78			
R4	8.8–9.7	5.57	S2	3.6–4.4	5.23	S11	1.5–2.4	5.15
R4	13.8–14.3	3.00	S2	7.3–7.75	5.99	S11	4.3–4.7	5.71
R5	1.5–2.4	6.19	S3	1.5–1.8	6.01	S12	1.8–2.7	5.24
R5	2.0–2.4	4.43	S3	1.5–2.4	6.11	S12	3.6–4.6	4.69
R5	11.6–12.2	5.72	S3	5.3–5.8	4.48	S12	8.2–8.7	5.44
R6	1.5–2.1	6.20	S4	1.5–1.8	5.70	S13	1.5–2.3	5.38
R6	4.6–4.9	5.69	S4	4.0–4.3	5.89	S13	4.3–4.7	5.38
R6	11.9–12.3	5.03	S4	4.0–4.3	4.82	S13	7.3–7.75	5.24
R6	18.2–18.8	4.88	S4	7.6–7.9	6.27			
						S14	2.0–2.4	5.72
			S5	2.6–3.0	6.35	S14	3.3–4.3	5.56
			S5	3.6–4.6	5.54	S14	11.6–12.5	6.15
			S5	13.7–14.6	6.44	S14	18.4–18.8	5.99

† CFU = colony-forming units.

Table 3–2. Summary of physical and chemical parameters for water samples retrieved from monitoring wells in April 1991.

Statistical measure	Parameter							
	Temperature	pH	$Eh_{Ag/AgCl}$	Dissolved O_2	Dissolved C	NH_4^+	$NO_2^- + NO_3^-$	Phosphate
	°C		mV	mg L^{-1}	mg L^{-1}	mg N L^{-1}	mg L^{-1}	mg L^{-1}
Mean	12.3	7.2	135	3.5	6.04	0.06	1.67	0.060
Median	12.3	7.2	152	3.1	1.18	0.07	1.38	0.004
Maximum	14.6	9.6	297	9.6	181	0.16	5.38	0.653
Minimum	9.1	6.8	−97	0.1	0.53	0.01	0.18	0.001

centration. Table 3-2 summarizes the range and central tendencies of these physical and chemical parameters. These parameters were measured at near sample depths in monitoring wells during April 1991.

Several core samples (R2, R4, R5, R7, S11, and S10) gave rise to pigmented colonies. Pigmented organisms were recovered from samples taken at >11-m depths at some well sites (R5 and S10) and 3 to 4 m at the others. Based on colonial characteristics, 10 or more types of bacteria were recovered with PTYG from most samples. Only one colony type was recovered from core sample S4 and two from S14.

These data indicate that there was no discernible trend of heterotrophic bacterial distribution in sediments collected from a relatively large area of site and regional wells at the onset of the MSEA study.

Biodegradation of Atrazine and Alachlor

The biological degradation of atrazine and alachlor was assessed in 81 sediment samples and two drilling water samples. Degradation was measured by following the loss of parent compound. Atrazine biodegradation detected in seven, and alachlor degradation in four, of the 83 samples evaluated (Table 3-3). These samples did not show any specific pattern of depth and spatial position at the MSEA site. The biodegradation of these compounds was slow,

Table 3-3. Summary of atrazine and alachlor biodegradation in subsurface sediment samples.

Core sample	Depth interval, m	Percent degradation	Incubation time, days
	Atrazine		
R7	1.7–2.3	28	124
R7	13.4–14.0	16	120
R7	20.1	27	120
R1	24.3	23	124
S10	Surface	49	139
S13	7.3–7.8	96	122
S5	13.7–14.6	16	136
	Alachlor		
Drilling water		39	167
R1	21.9–22.5	57	169
S4	4.0–4.3	17	138
S10	5.2–5.6	19	160

Core sample	Enrichment type	Percent degradation	Incubation time, days
	Atrazine composites		
R1	Glucose as C source	44	45
R7	Glucose as C source	50	90
R7	Glucose as C source; atrazine as N source	23	41
S13	Glucose as C source; atrazine as N source	96	40
S13	Atrazine as sole C and N source	100	33
S14	Glucose as C source; atrazine as N source	19	33

with approximate half-lives of >150 d. The slow degradation in initial enrichment cultures is consistent with previous reports that 12 to 28% mineralization of ring-labelled atrazine occurred in surface soil microcosms after 180 d of incubation (Winkelmann & Klaine, 1991).

Biodegradation of 2,4-D

The biodegradation of 2,4-D was evaluated with samples only from R1, R2, R4, R5, S5, and S14. Of the 22 samples examined, five showed complete biodegradation of 2,4-D within 30 d of incubation, based on the loss of the parent compound. These samples were from R2 (12.2–12.8 m); R5 (11.6–12.2 m); S5 (13.7–14.6 m); and S14 (3.3–4.3, 11.6–12.5, and 18.4–18.9 m). An additional three samples yielded 2,4-D degrading cultures after approximately 4 mo of incubation (R4 8.8–9.7 m; S5 2.6–3.0 and 3.7–4.6 m).

Alternative Enrichment Strategies for Atrazine Biodegradation

The biochemical pathway of atrazine degradation has been summarized by Erickson and Lee (1989). Only the alkyl chains in the atrazine molecule contain C that can be oxidized for energy. Thus, the dealkylated atrazine ring (*s*-triazine) does not serve as an energy source but it is possible that some organisms degrade the ring structure through cometabolism or as a source of N. To account for cometabolic effects and atrazine-N utilization, additional enrichment cultures were set up, based on adding glucose as an additional source of C and using only atrazine as the sole source of N. These enrichment conditions were tested with composite core samples.

Biodegradation of atrazine was observed in composite samples from three cores (Table 3–3). The extent of degradation ranged from 19% (S14) to nearly 100% (S13) in N-free, glucose-amended enrichment cultures.

CONCLUSION

The heterotrophic plate counts suggested a fairly even distribution of bacteria at all depths of the core sampling sites examined at the onset of the Ohio MSEA field study. This conclusion is contrasted with the patchy, scanty presence of microorganisms capable of degrading atrazine or alachlor. The data, although limited, also suggest that the biodegradation of 2,4-D occurs more widely in these sediments compared with atrazine and alachlor. None of the enrichment cultures yielded metabolites detectable by the HPLC and UV-spectroscopy employed in the study.

It appears that atrazine and alachlor biodegradation in subsurface zones is slow due to the lack of microorganisms capable of degrading these compounds. The incubation temperature used in this study, 25 ± 2 °C, was greater than the in situ aquifer temperatures of 9.7 to 14.3 °C measured at this MSEA site. Thus, it is likely that in situ biodegradation of atrazine and alachlor will be even less extensive than was measured in the laboratory incubations.

ACKNOWLEDGMENTS

We thank M. Jagucki, U.S. Geological Survey, and A. Springer and D. Beak, The Ohio State University, for sample collection at the MSEA site and for placing the data summarized in Table 3-2 at our disposal. We are grateful to M. Prichard and D. Stoeckel for technical assistance. M.R. acknowledges with thanks a National Needs Graduate Fellowship from the USDA. Partial funding for the work was provided by USDA/Cooperative State Research Service under the President's Initiative on Water Quality.

REFERENCES

Erickson, L.E., and K.H. Lee. 1989. Degradation of atrazine and related *s*-triazines. Crit. Rev. Environ. Contr. 19:1–14.

Oh, K.-H., and O.H. Tuovinen. 1991. Detection and identification of substituted phenols as intermediates of concurrent bacterial degradation of the phenoxy herbicides MCPP and 2,4-D. FEMS Microbiol. Lett. 79:141–146.

Winkelmann, D.A., and S.J. Klaine. 1991. Degradation and bound residue formation of atrazine in a western Tennessee soil. Environ. Toxicol. Chem. 10:335–345.

4 Diazinon Degradation and Dissipation in the Root Zone

R. A. McLaughlin, R. H. Ross, and K. Balu
CIBA-GEIGY Corp.
Greensboro, North Carolina

Diazinon [*O,O*-diethyl-*O*-(2-isopropyl-[soil] 6-methyl-4-pyrimidinyl)phosphorothioate] is an organophosphate insecticide registered for numerous uses on crops, ornamentals, and turf. As part of the U.S. Environmental Protection Agency reregistration process for diazinon, CIBA-GEIGY initiated studies in 1988 at six locations around the USA to study the behavior of diazinon and its four major environmental degradation products in soil under field conditions. Specifically, diazinon and the metabolites were monitored for degradation rate, occurrence, and movement through the soil profile.

MATERIALS AND METHODS

Table 4–1 provides the information on the sites and treatment regimes used for this study. The locations were selected to reflect a wide range of climates. To maximize leaching potential, all plots were located on light-textured soils and rainfall was supplemented with irrigation to maintain above-average amounts of precipitation (see Fig. 4–5 through 10 for water inputs). The soil profile became more sandy with depths to 120 cm at all sites except the Madera site, which had a slight decrease (to 69%) in sand content.

The applications were performed at the maximum rate allowed under the 1988 label for each formulation and crop. All applications were made using calibrated sprayers with the nozzles directed to the ground, except the granular applications, which were made using hand-held applicators.

Soil sampling occurred before and several hours after each application. Sampling occurred approximately 1, 3, 7, 14, 28, 60, 90, 120, 180, 270, 365, 454, and 540 d after the last application. Sampling was halted if samples from two successive intervals had no detectable residues of diazinon or metabolites. Soil samples were taken to a depth of 120 cm using a hydraulic probe mounted on a truck or tractor. Each soil sample was contained within an acetate liner in the probe. Five samples were taken from each of three

 Sorption and Degradation of Pesticides and Organic Chemicals in Soil. SSSA Special Publication no. 32.

Table 4–1. Information on the sites and treatments used to study the dissipation of diazinon and its metabolites in soil.

Location	City	Formu-lation	Crop	Soil	pH	Percent organic matter	% sand/ % silt	Applications (no. × kg a.i. ha^{-1})
New York	Phelps	AG500	Apple, newly planted	Sandy loam	6.4	2.4	81/13	7 × 3.7
California	Reedley	AG500	Citrus, 12-yr old	Loamy sand	7.7	0.1	85/15	4 + 4 + 6.8 + 4 + 9.3
Illinois	Gladstone	14G	Corn	Sand	6.6	1.6	93/3	4 × 2.5
California	Madera	14G	Corn	Sandy loam	5.6	0.2	71/21	4 × 2.5
Florida	Windermere	50W	Citrus, 30-yr old	Sand	5.4	1.4	94/2	2 × 3.7 + 3 × 6.2
California	Polar	50W	Apple, newly planted	Sandy loam	7.4	0.92	70/18	7 × 3.7

plots and subsequently composited by plot and 15-cm increments up to 60 cm and by 30-cm increments from 60 to 120 cm, resulting in three samples per depth for each interval. A single control plot, located away from the treated plots and maintained identically to the treated plots minus the diazinon treatments, was sampled in a similar manner. Samples were maintained cold during sampling and then frozen until analysis.

To more accurately determine the degradation kinetics of diazinon and its' metabolites, a bare soil plot received a single application of diazinon and was sampled similarly to the crop plots. These plots were kept free of vegetation throughout the course of the study. The analytical results from the topsoil sample (0–15 cm) are presented in this chapter to describe the degradation and dissipation of diazinon residues.

Soil samples were shipped frozen to one of three analytical laboratories and stored frozen for subsequent analyses. Three laboratories were required due to the volume of samples (approximately 1000 per site) for analysis. The composited samples were homogenized mechanically and a 50-g subsample was taken for analysis. The subsample was shaken with 100 mL of 9:1 acetonitrile/water for 30 min, then filtered. The filtrate was partitioned by adding 50 mL of saturated NaCl solution and 100 mL dichloromethane. The lower aqueous phase was partitioned again with 100 mL 1:1 acetonitrile/dichloromethane, then with 50 mL of dichloromethane. All organic phases were passed through anhydrous Na_2SO_4 and evaporated to dryness. The residue was reconstituted in either acetone for gas chromatography or mobile phase for liquid chromatography.

All samples were analyzed for diazinon and the four major soil diazinon metabolites, which were determined from earlier studies (Laanio et al., 1972)[1]. The analyses for diazinon and diazoxon [*O,O*-diethyl-*O*-(2-isopropyl-6-methyl-4-pyrimidinyl)phosphorate] were performed using gas chromatography on a 15 m × 0.53 mm column coated with a 5 μm DB-1 film (J & W Scientific, Folsom, CA)[2] and N-P (flame thermionic) detection. The exact instrument conditions varied between laboratories, but generally an oven ramp from 150 to 250 °C was used followed by a "burnout" at a higher temperature for several minutes. One laboratory also determined G-27550 (2-isopropyl-4-methyl-6-hydroxypyrimidine), demethylated G-27550 (2-ethyl-4-methyl-6-hydroxypyrimidine), and GS-31144 [2-(2-propanol)-4-methyl-6-hydroxypyrimidine] by this method; the remaining two laboratories used liquid chromatography. The exact liquid chromatography conditions also varied, but both labs used a Supelcosil LC-8-DB column, a mobile phase of 90% phosphate buffer (0.5 g L^{-1} H_2PO_4, pH 7.5):10% methanol, and detection at 225 to 230 nm.

Quantitation was performed using external standards for each compound. The limit of quantitation was set at the minimum concentration that could be consistently recovered (70–120%) from freshly fortified samples

[1]The chemical structures of the analytes are shown in Fig. 4-1.

[2]Mention of trademark or proprietary products in this manuscript does not constitute a guarantee or warranty of the product by the U.S. Department of Agriculture and does not imply its approval to the exclusion of other products that may also be suitable.

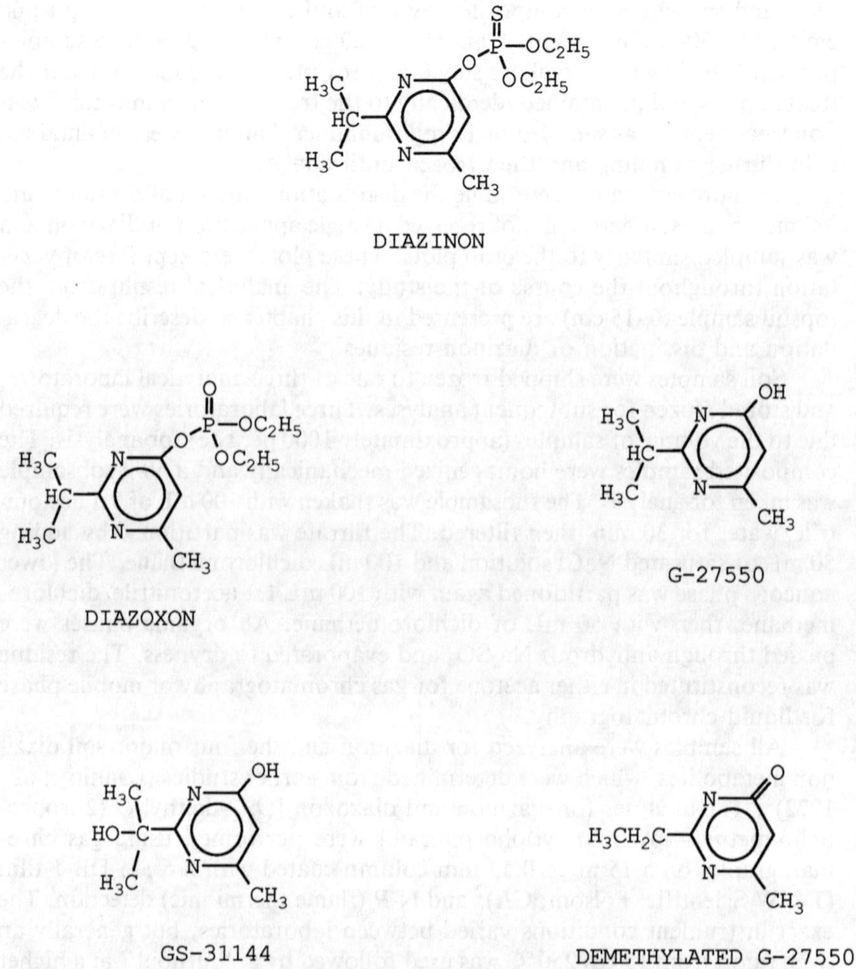

Fig. 4–1. Structures of diazinon and its four major soil metabolites.

from the control plots. This varied by compound and laboratory, but was either 0.01 or 0.02 mg kg^{-1}. All results were corrected for procedural recoveries of <100% and for soil moisture.

RESULTS

The half-lives of diazinon ranged from 3 to 13 d, and for total residues (diazinon + metabolites) the half-lives ranged from 6 to 71 d (Table 4–2). The usual pattern of degradation for diazinon followed normal first-order kinetics, with a rapid decline over the first week followed by much slower dissipation for 1 to 2 mo (Fig. 4–2 through 4).

Table 4–2. Half-lives for diazinon and diazinon total residues for all sites and both cropped and bare soil plots.

Formulation	Location	Crop application (total kg a.i. ha^{-1})	Crop half-life (days)		Bare ground application (kg a.i. ha^{-1})	Bare ground half-life (days)	
			Diazinon	Total†		Diazinon	Total‡
AG500	New York	25.9	80	73	11.2	45	46
AG500	California	28.1	7	6	12.3	13	17
14G	Illinois	9.9	5	96	9	6	11
14G	California	9.9	6	10	9	7	9
50W	Florida	19.7	5	6	11.2	8	8
50W	California	25.9	18	52	11.2	6	61

† Total = Diazinon + all metabolites at all soil depths.
‡ Total = Diazinon + all metabolites in the 0 to 15 cm soil samples.

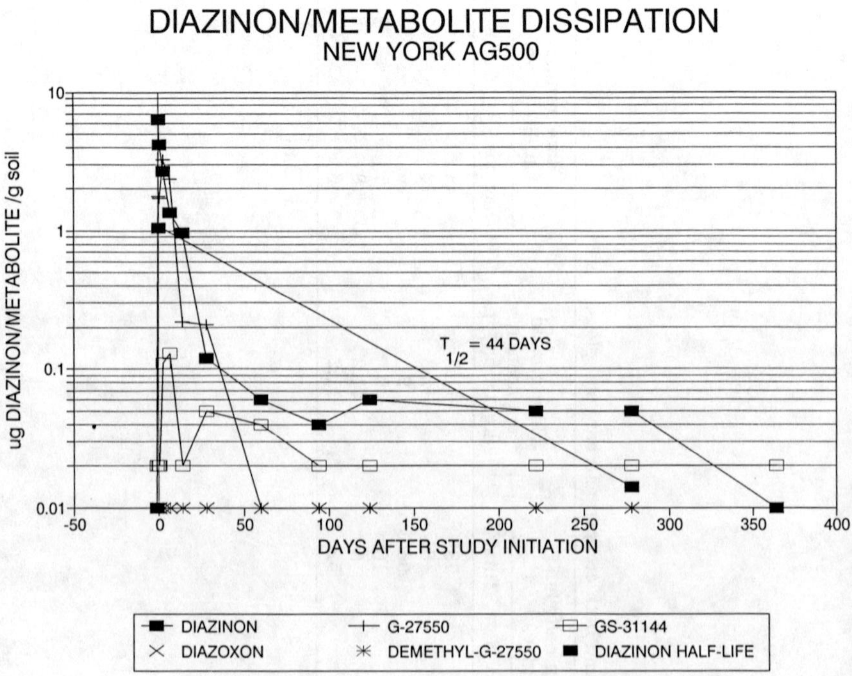

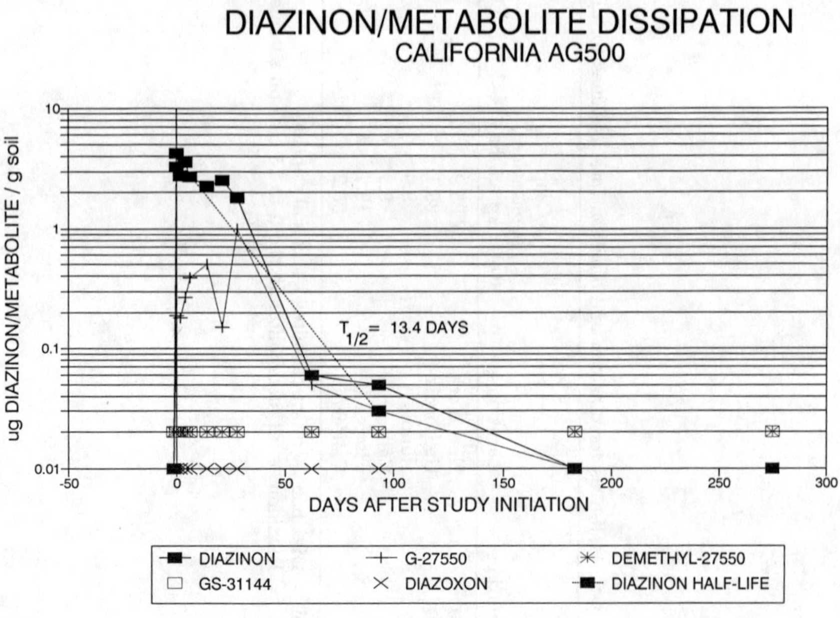

Fig. 4–2. Decline curves for diazinon and its four major soil metabolites in soil from New York and California bare soil plots treated with the 0.48 kg diazinon L^{-1} formulation of diazinon.

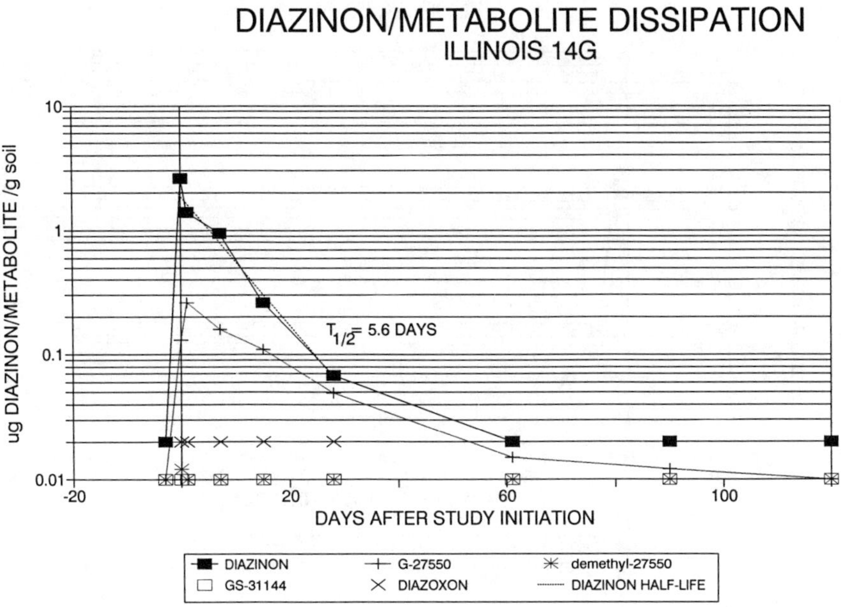

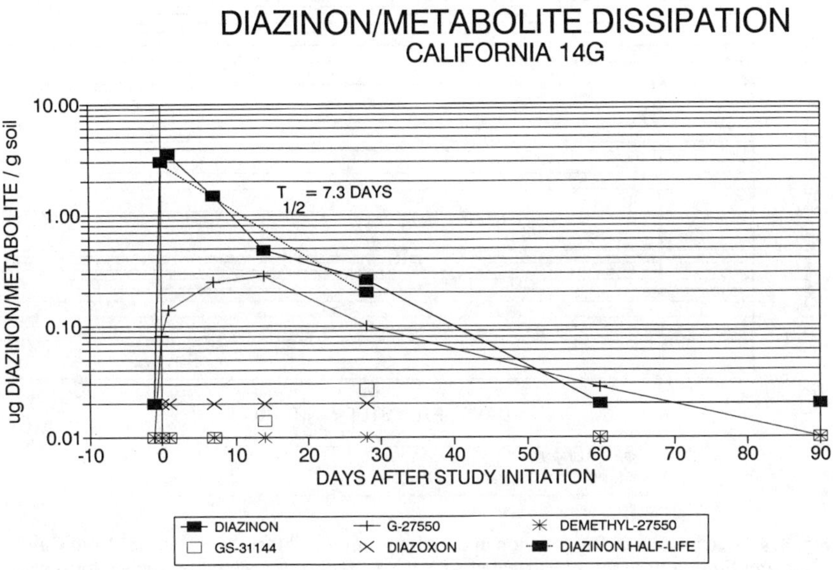

Fig. 4-3. Decline curves for diazinon and its four major soil metabolites in soil from Illinois and California bare soil plots treated with the 14% diazinon granular formulation of diazion.

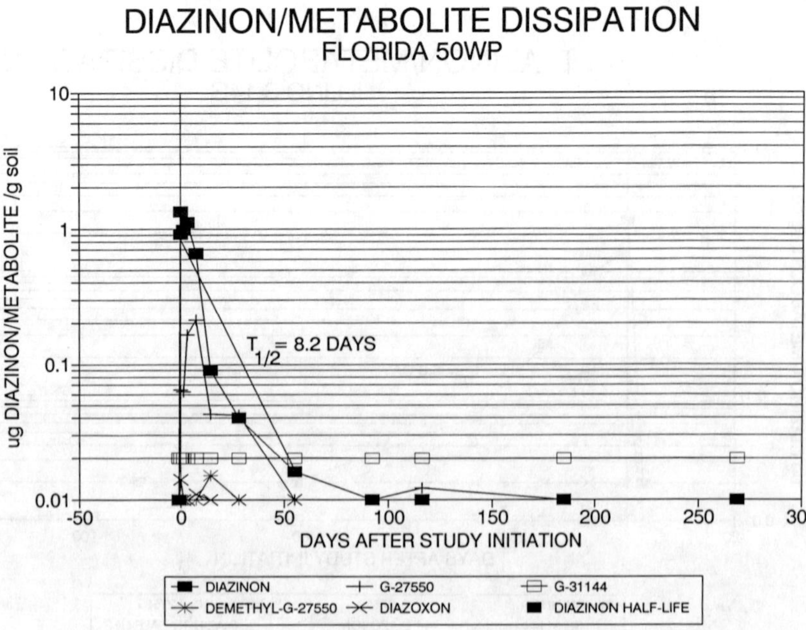

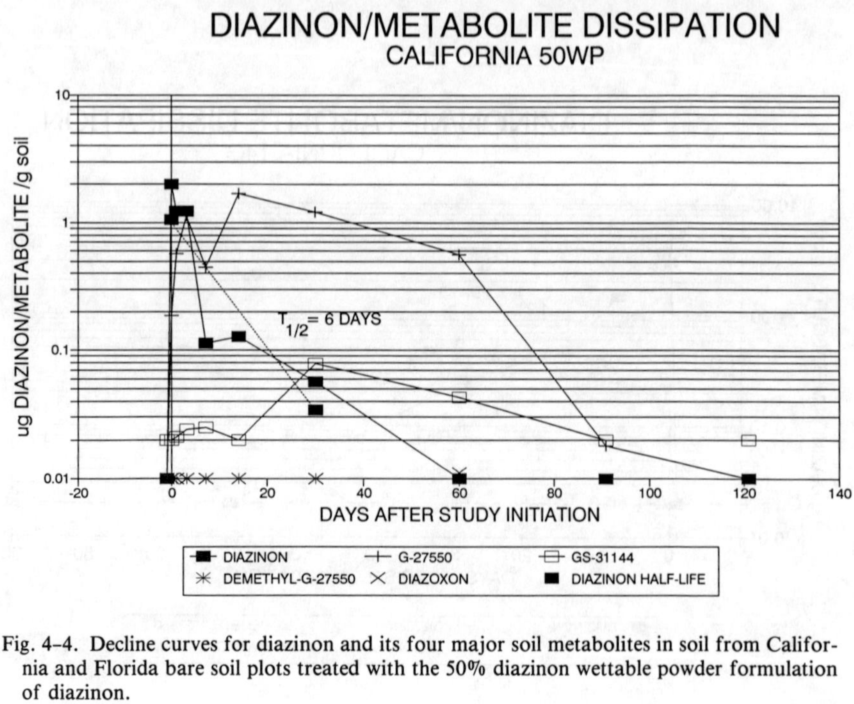

Fig. 4–4. Decline curves for diazinon and its four major soil metabolites in soil from California and Florida bare soil plots treated with the 50% diazinon wettable powder formulation of diazinon.

The major metabolite observed was G-27550, the pyrimidine moiety formed after the hydrolysis of the phosphate ester linkage, with occasional detections of GS-31144, and few detections of diazoxon or demethylated G-27550. G-27550 appeared soon after application, occurring often even in 0-d samples. It exhibited a pattern of decline similar to that of diazinon, with somewhat greater stability and residence time. The half-life of G-27550 cannot be determined directly in this study because of the constant formation of new G-27550 from degrading diazinon. However, the rate of disappearance was similar to that of diazinon in most studies (Fig. 4–2 through 4).

In spite of the sandy nature of the soils in the study and the high application rates, diazinon showed no propensity for leaching (Fig. 4–5 through 10). In four of six studies it was not detected below 0.30 m, and in the fifth (California apples) only trace amounts were detected in the 0.30 to 0.45 m depth and none below that. The Florida citrus study had detections at low concentrations down to the 0.45 to 0.60 m depth, but none detected from 0.60 to 1.20 m. This Florida study also had the coarsest soil texture.

The major metabolite, G-27550, was somewhat more mobile in soil than diazinon (Fig. 4–5 through 10). The greatest depth of detection ranged from 0.30 m in the California corn (*Zea mays* L.) and New York apple (*Malus* spp.) studies to 1.20 m in the Iowa corn, California citrus, and Florida citrus locations. The observed concentrations were quite low at 0.90 to 1.20 m for these three studies, with maximum concentrations of 0.01 to 0.04 mg kg^{-1}. All studies were continued until there were no residues of diazinon or any metabolite at any sample depth.

DISCUSSION

Diazinon had a relatively short half-life in the soil across all climates, rates, and formulations. The half-lives found in this study were similar to the published values of 7 to 28 d (Bartsch, 1974; Kuhr & Tashiro, 1978; Bro-Rasmussen et al., 1970; Sattar, 1990). It also exhibited little potential for leaching, which is partly due to its rapid dissipation. A bimodal degradation appeared to occur in several studies, with a relatively rapid degradation within the first several weeks followed by slow degradation at low concentrations. This pattern of degradation has been noted elsewhere (Kono & Tamashita, 1974; Bro-Rasmussen et al., 1970; Read, 1969).

G-27550 was the major metabolite observed in this study, which is expected as it is the hydrolysis product of the cleavage of the thiophosphate group from the pyrimidinyl ring (Laanio et al., 1972; Bartsch, 1974). While G-27550 appeared to have a half-life similar to diazinon, it was more mobile in several of the studies. However, only low concentrations of G-27550 were detected at the lower depths.

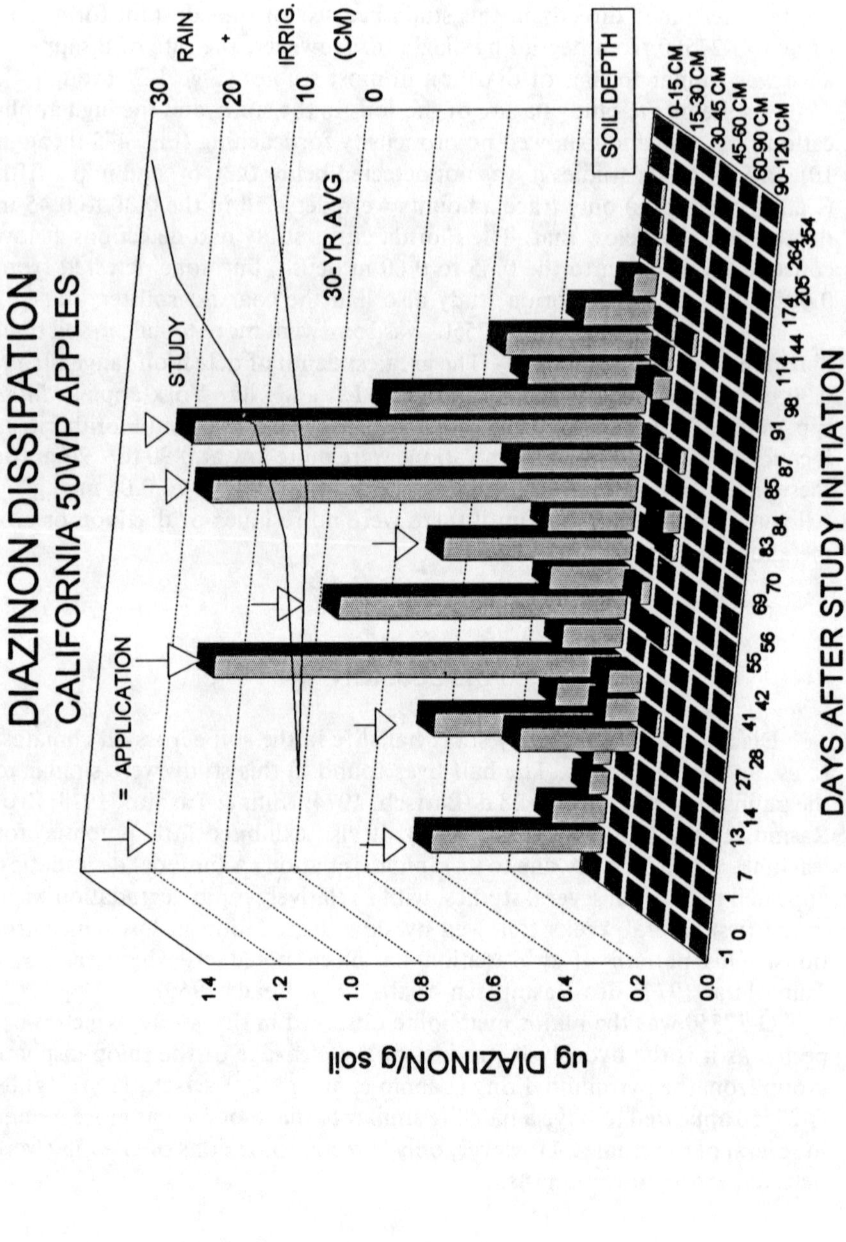

DIAZINON DISSIPATION
CALIFORNIA 50WP APPLES
= APPLICATION
STUDY
30-YR AVG
RAIN + IRRIG. (CM)
30
20
10
0
ug DIAZINON/g soil
1.4
1.2
1.0
0.8
0.6
0.4
0.2
0.0
SOIL DEPTH
0-15 CM
15-30 CM
30-45 CM
45-60 CM
60-90 CM
90-120 CM
-1 0 1 7 13 14 27 28 41 42 55 56 69 70 83 84 85 87 91 98 114 144 174 205 264 354
DAYS AFTER STUDY INITIATION

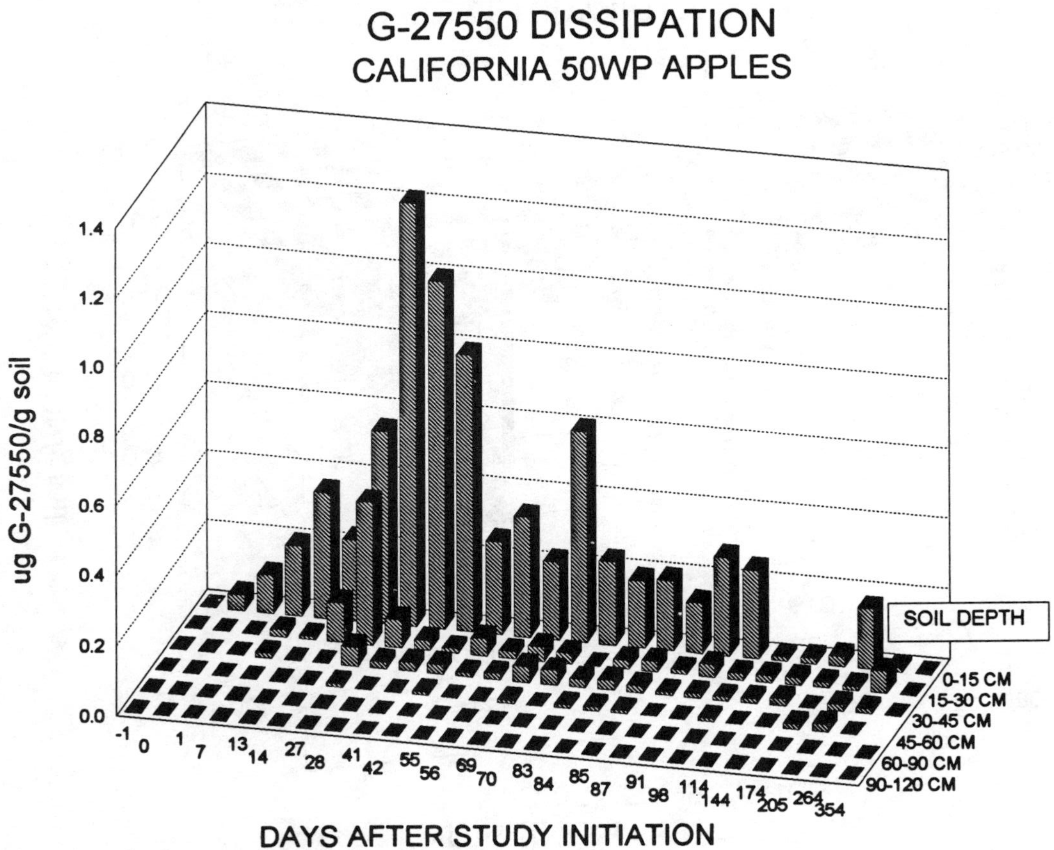

Fig. 4–5. Dissipation of diazinon and G-27550 in soil in a California apple orchard treated with a total of 25.9 kg a.i. ha^{-1} diazinon in the 50% wettable powder formulation.

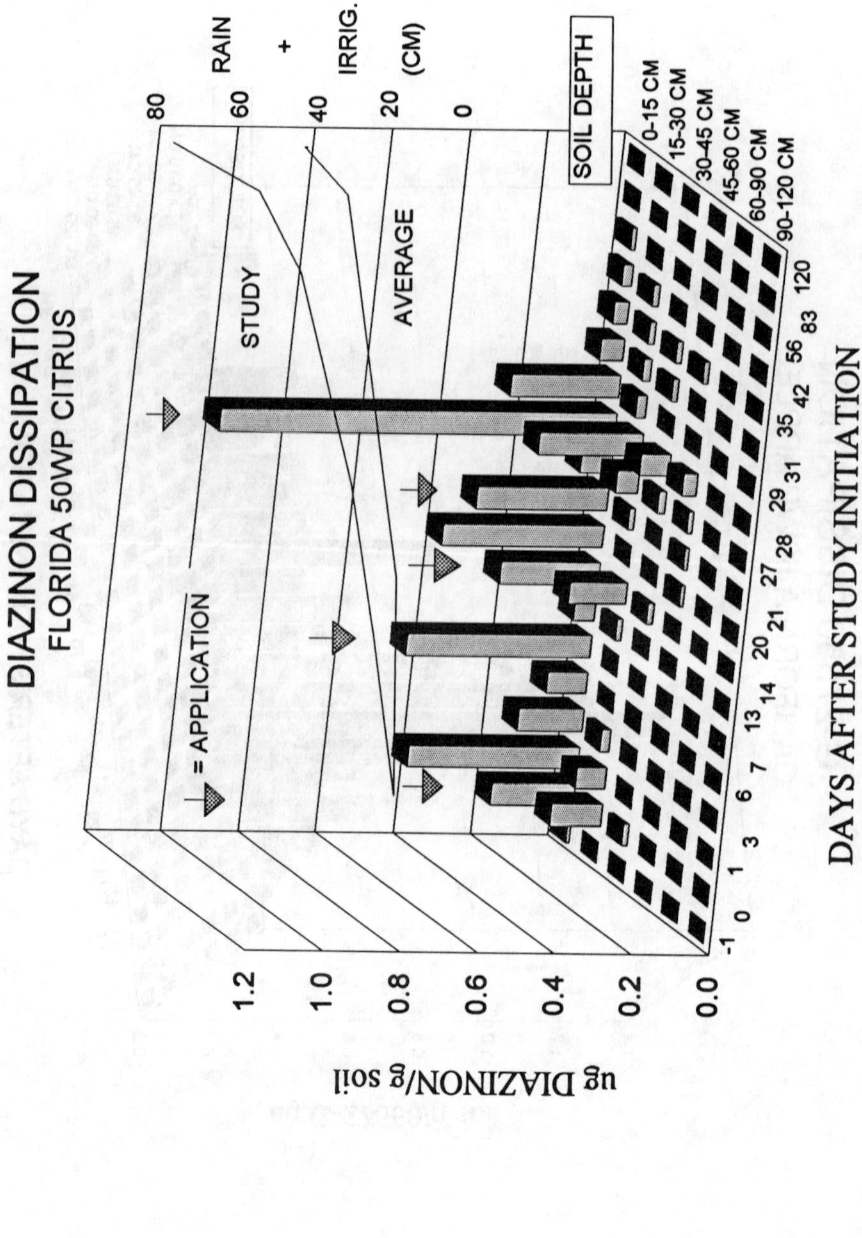
DIAZINON DISSIPATION
FLORIDA 50WP CITRUS
= APPLICATION
STUDY
AVERAGE
RAIN
+
IRRIG.
(CM)
80
60
40
20
0
ug DIAZINON/g soil
1.2
1.0
0.8
0.6
0.4
0.2
0.0
SOIL DEPTH
0-15 CM
15-30 CM
30-45 CM
45-60 CM
60-90 CM
90-120 CM
-1
0
1
3
6
7
13
14
20
21
27
28
29
31
35
42
56
83
120
DAYS AFTER STUDY INITIATION

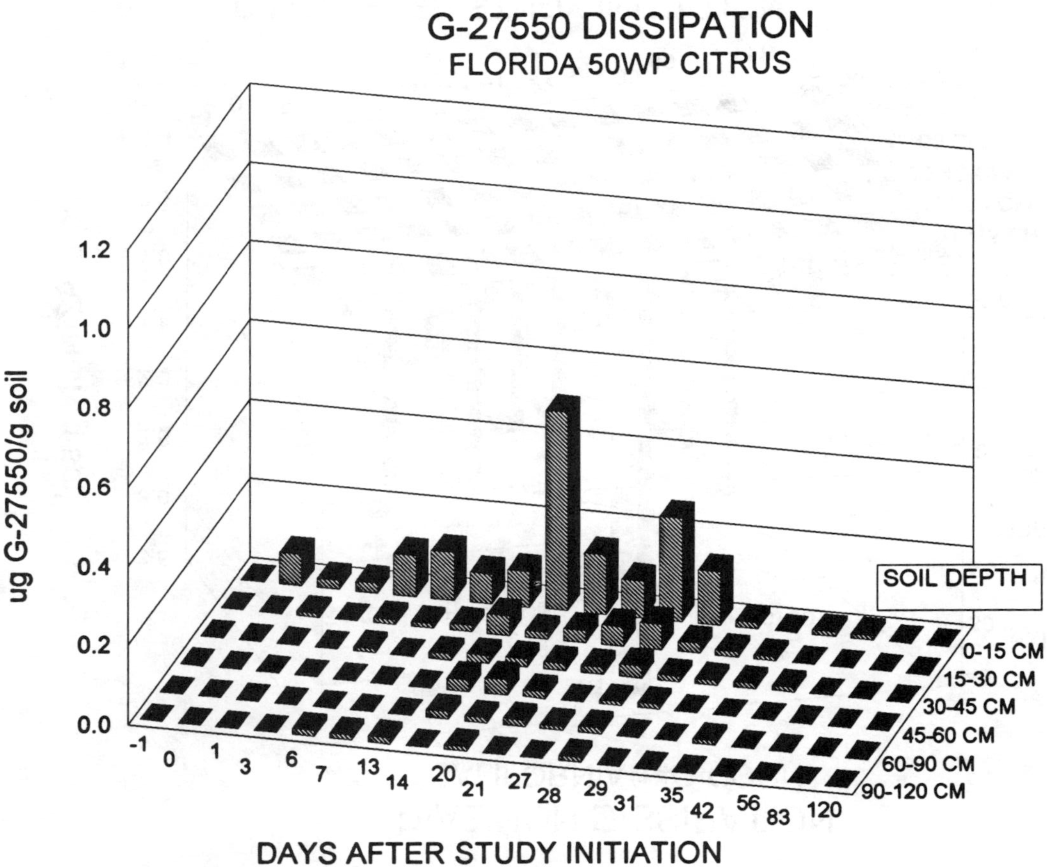

Fig. 4–6. Dissipation of diazinon and G-27500 in soil in a Florida citrus orchard treated with a total of 19.7 kg ai.i ha $^{-1}$ diazinon in the 50% wettable powder formulation.

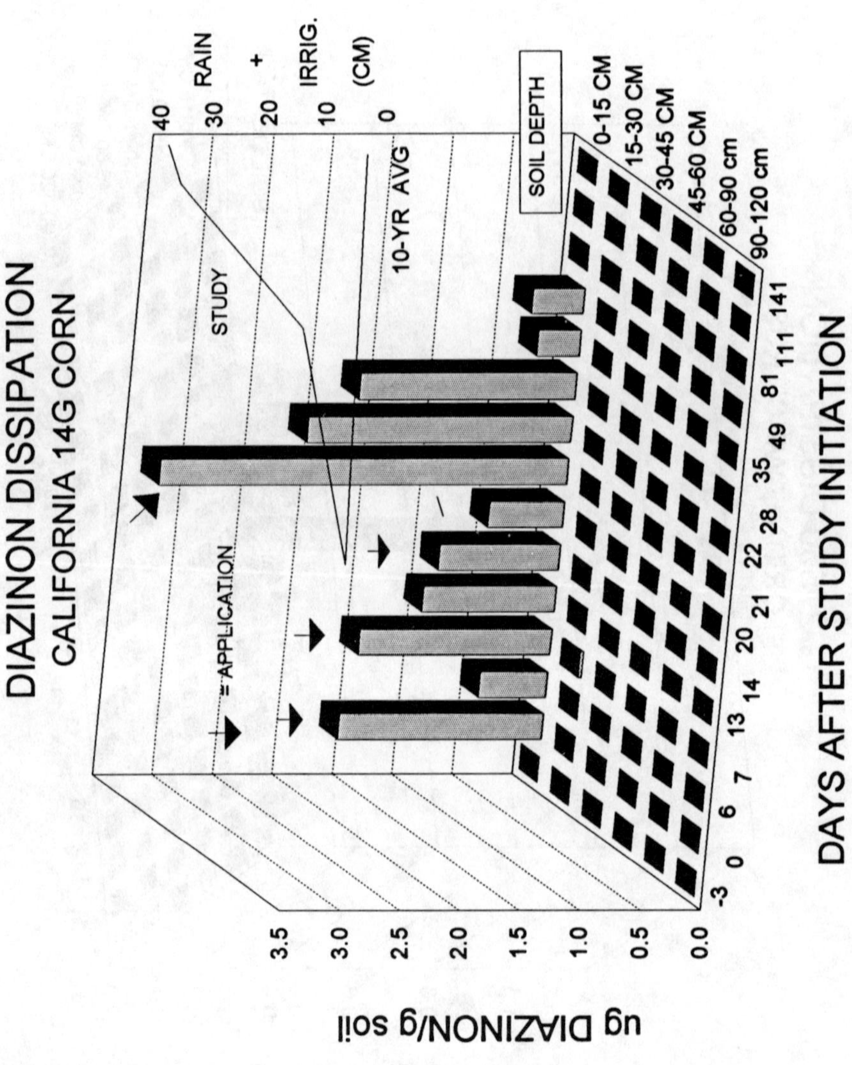
DIAZINON DISSIPATION
CALIFORNIA 14G CORN
= APPLICATION
STUDY
10-YR AVG
RAIN
+
IRRIG.
(CM)
40
30
20
10
0
SOIL DEPTH
0-15 CM
15-30 CM
30-45 CM
45-60 CM
60-90 cm
90-120 cm
3.5
3.0
2.5
2.0
1.5
1.0
0.5
0.0
ug DIAZINON/g soil
-3
0
6
7
13
14
20
21
22
28
35
49
81
111
141
DAYS AFTER STUDY INITIATION

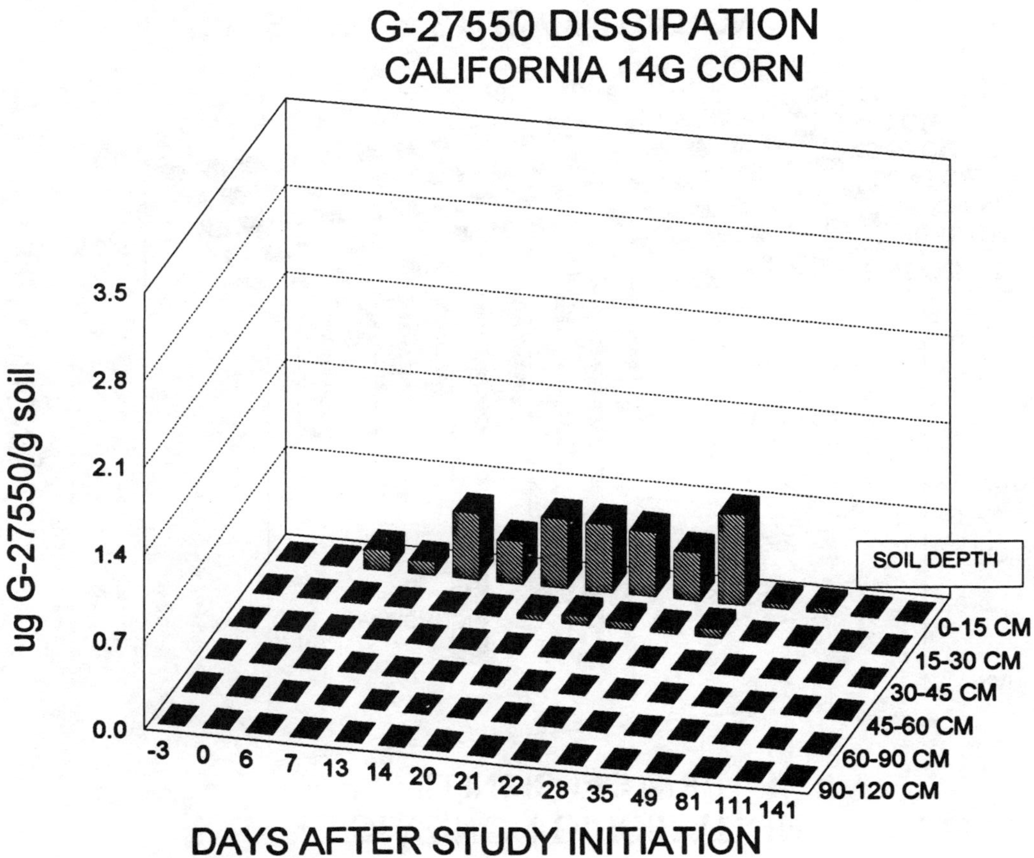

Fig. 4–7. Dissipation of diazinon and G-27550 in soil in a California corn crop treated with a total of 9.9 kg a.i. ha^{-1} diazinon in 14% granular formulation.

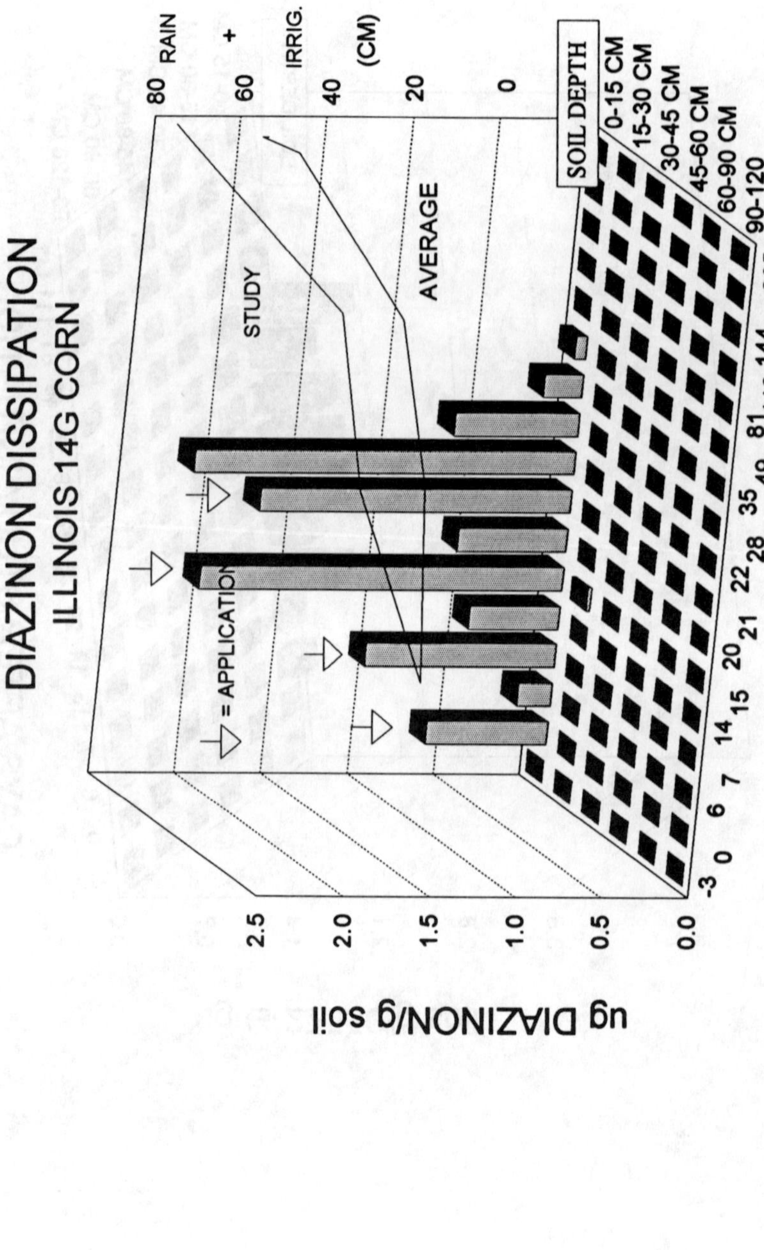
DIAZINON DISSIPATION
ILLINOIS 14G CORN
ug DIAZINON/g soil
0.0
0.5
1.0
1.5
2.0
2.5
RAIN + IRRIG. (CM)
80
60
40
20
0
STUDY
AVERAGE
= APPLICATION
SOIL DEPTH
0-15 CM
15-30 CM
30-45 CM
45-60 CM
60-90 CM
90-120
-3 0 6 7 14 15 20 21 22 28 35 49 81 112 144 306 365
DAYS AFTER STUDY INITIATION

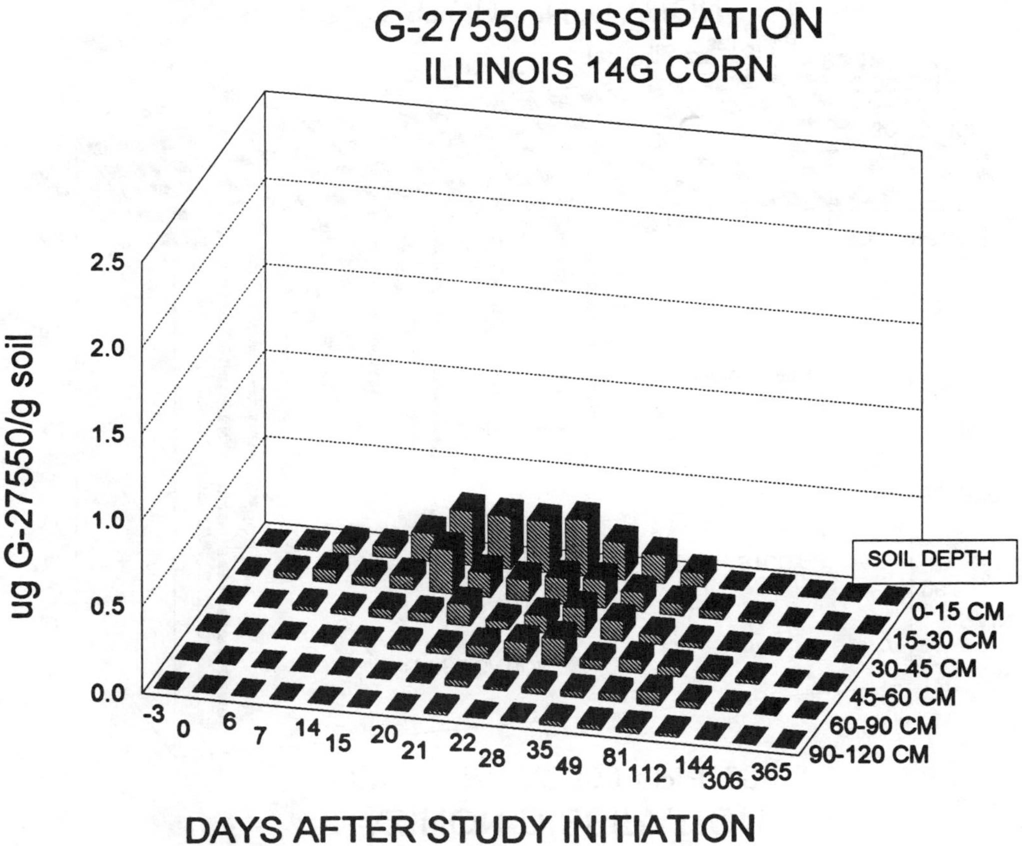

Fig. 4–8. Dissipation of diazinon and G-27550 in soil in an Illinois corn crop treated with a total of 9.9 kg a.i. ha^{-1} diazinon in 14% granular formulation.

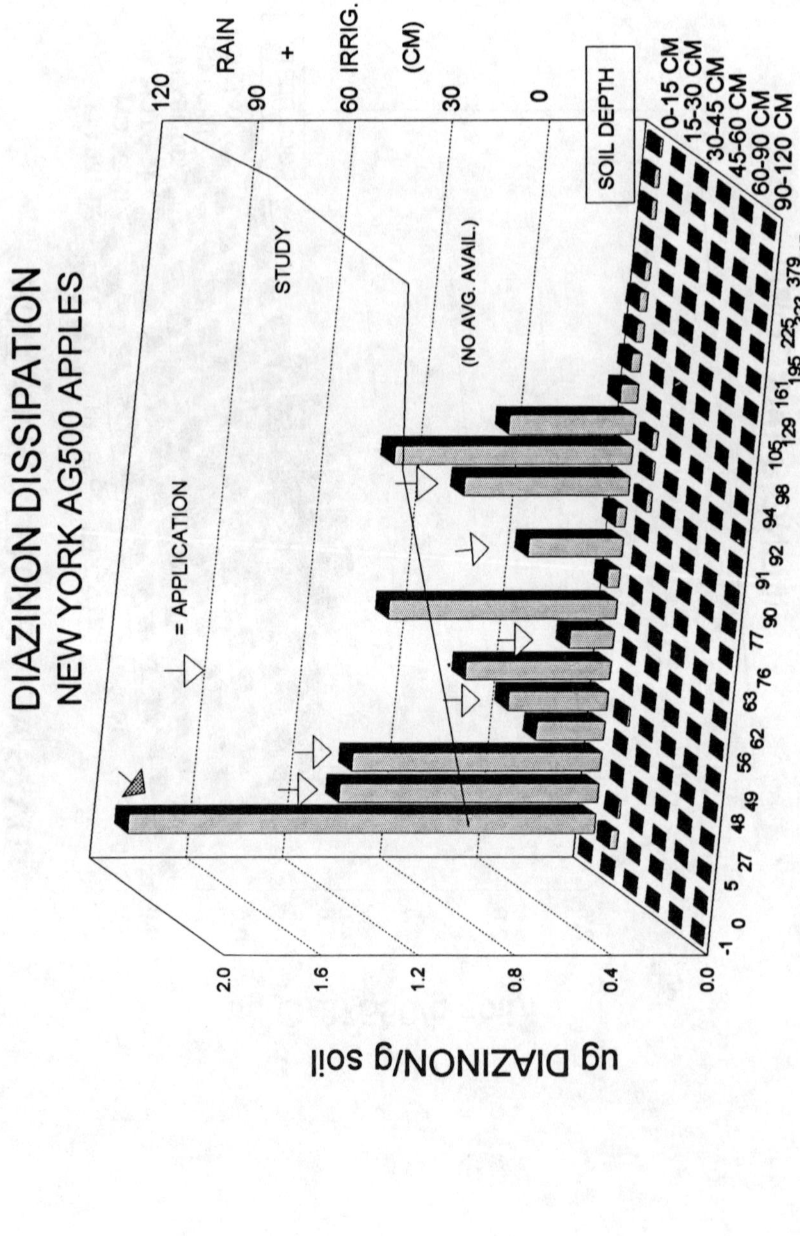
DIAZINON DISSIPATION
NEW YORK AG500 APPLES
= APPLICATION
STUDY
(NO AVG. AVAIL.)
RAIN
+
IRRIG.
(CM)
120
90
60
30
0
SOIL DEPTH
0-15 CM
15-30 CM
30-45 CM
45-60 CM
60-90 CM
90-120 CM
ug DIAZINON/g soil
2.0
1.6
1.2
0.8
0.4
0.0
-1
0
5
27
48
49
56
62
63
76
77
90
91
92
94
98
105
129
161
195
225
323
379
465
DAYS AFTER STUDY INITIATION

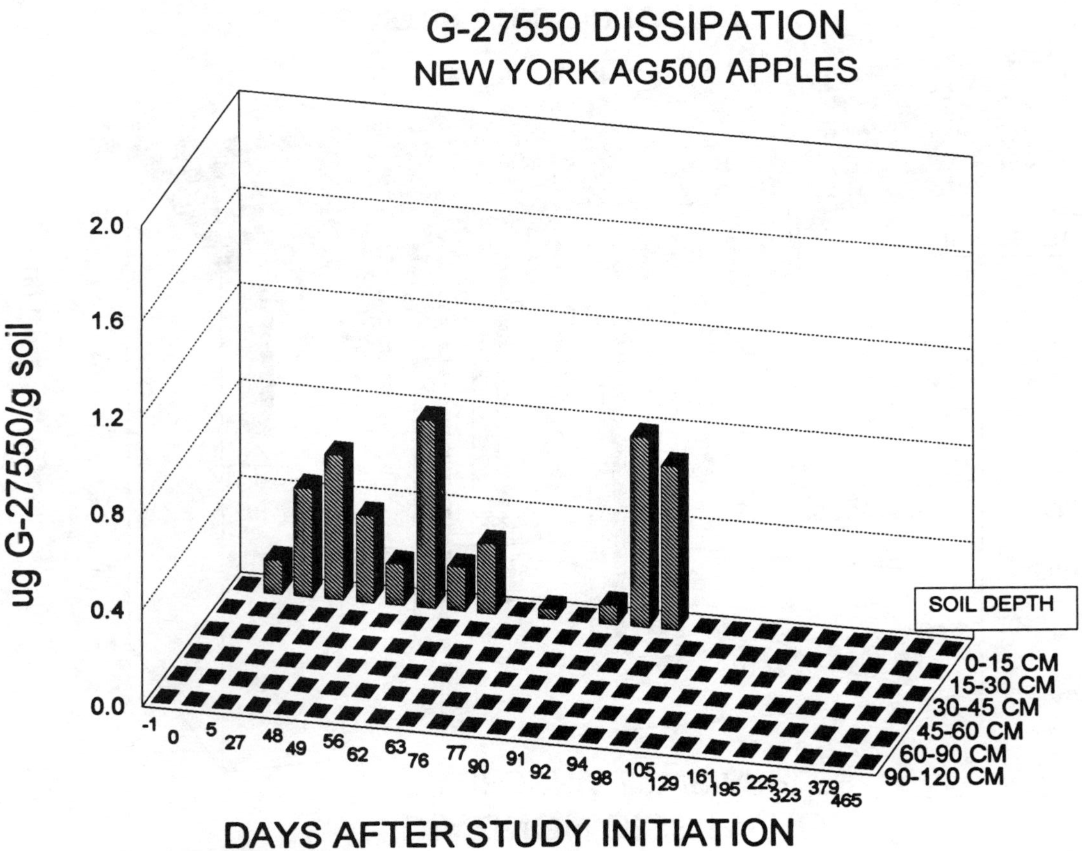

Fig. 4–9. Dissipation of diazinon and G-27550 in soil in a New York apple orchard treated with a total of 25.9 kg a.i. ha^{-1} diazinon in 0.48 kg L^{-1} formulation.

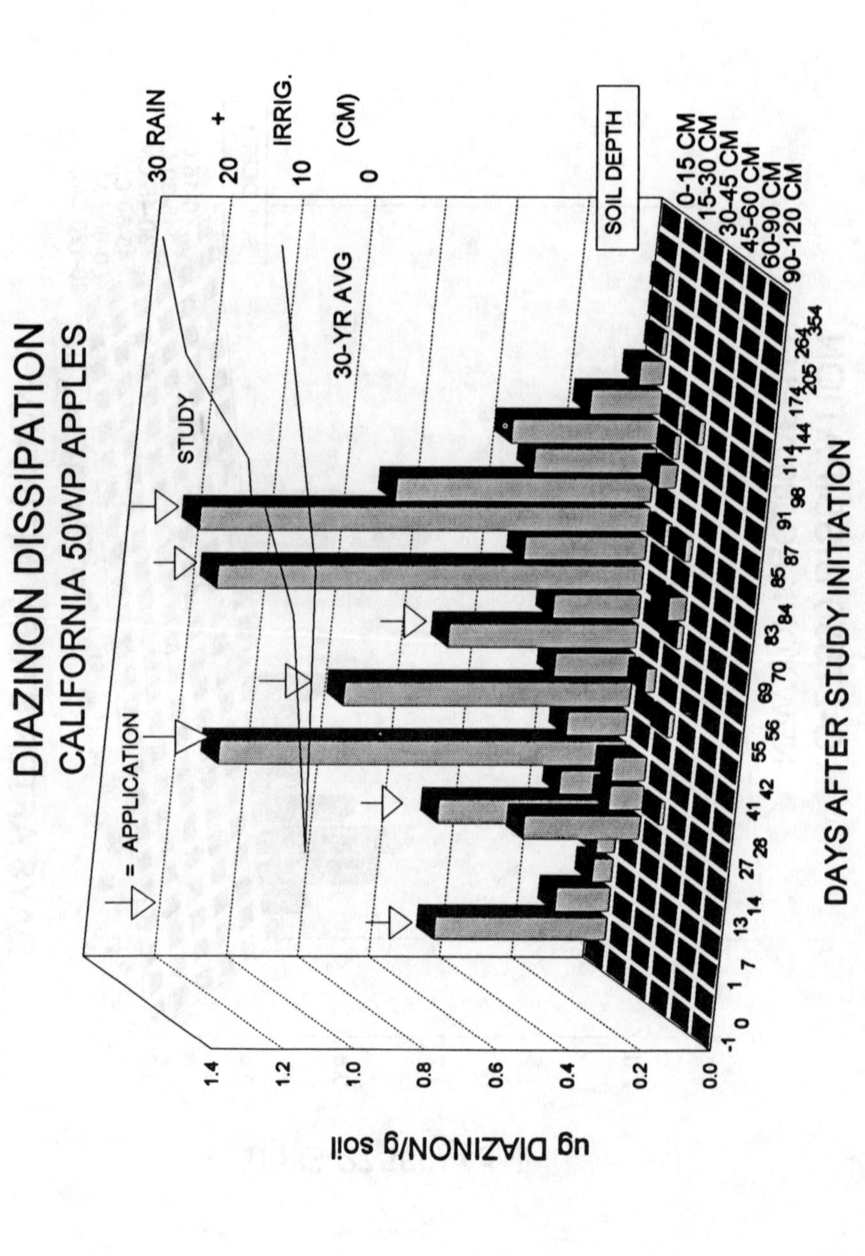
DIAZINON DISSIPATION
CALIFORNIA 50WP APPLES
= APPLICATION
STUDY
30-YR AVG
30 RAIN
20 +
IRRIG.
10
(CM)
0
ug DIAZINON/g soil
0.0
0.2
0.4
0.6
0.8
1.0
1.2
1.4
SOIL DEPTH
0-15 CM
15-30 CM
30-45 CM
45-60 CM
60-90 CM
90-120 CM
-1 0 1 7 13 14 27 28 41 42 55 56 69 70 83 84 85 87 91 98 114 144 174 205 264 354
DAYS AFTER STUDY INITIATION

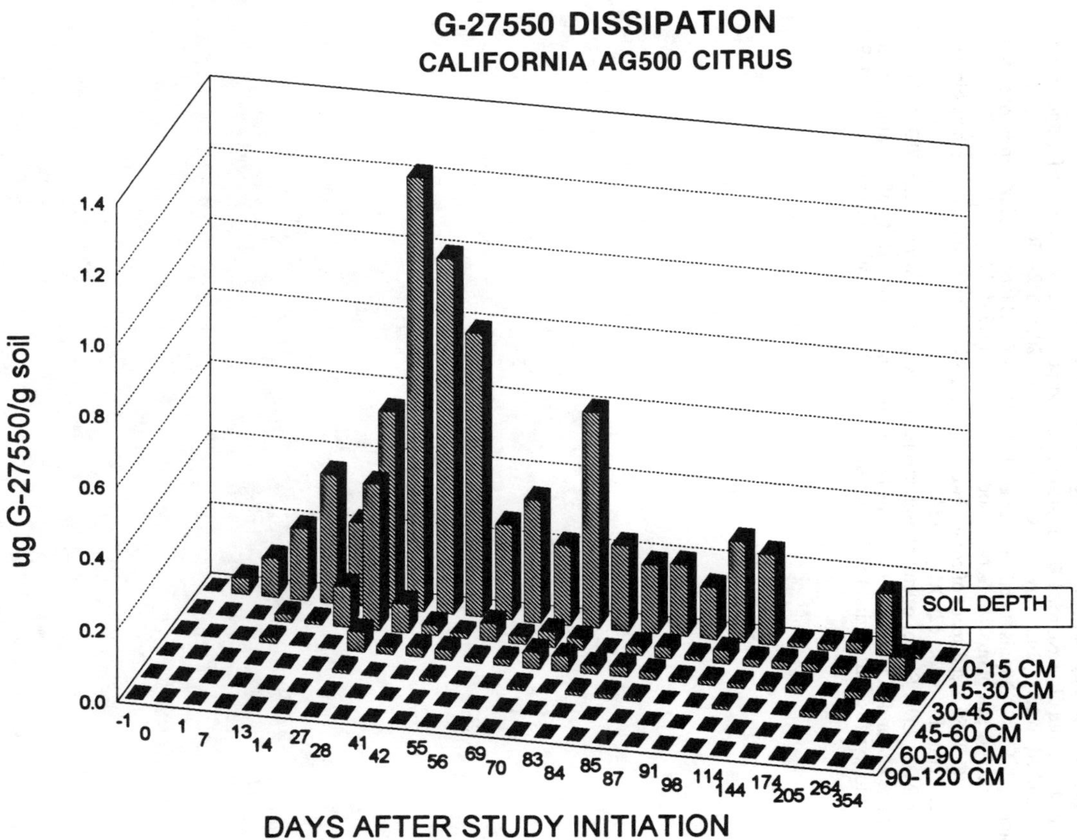

Fig. 4–10. Dissipation of diazinon and G-27550 in soil in a California citrus orchard treated with a total of 28.1 kg a.i. ha^{-1} diazinon in 0.48 kg L^{-1} formulation.

REFERENCES

Bartsch, E. 1974. Diazinon II: Residues in plants, soils, and water. Residue Rev. 51:37–68.

Bro-Rasmussen, F., F. Noddegaard, and K. Voldum-Clausen. 1970. Comparison of the disappearance of organophosphorus insecticides from soil in laboratory and in outdoor experiments. Pestic. Sci. 1:179–182.

Getzin, L.W., and I. Rosefield. 1966. Persistence of diazinon and zinophos in soils. J. Econ. Entomol. 59(3):512–516.

Kono, S., and M. Yamashita. 1974. Decrease in residual amounts of diazinon in cabbages and soil after its application. Sci. Pestic. Contr. 39(4):119–125.

Kuhr, R.J., and H. Tashiro. 1978. Distribution and persistence of chlorpyrifos and diazinon applied to turf. Bull. Environ. Contam. Toxicol. 20(5):652–656.

Laanio, T.L., G. Dupuis, and H.O. Esser. 1972. Fate of ^{14}C-labelled diazinon in rice, paddy soil, and pea plants. J. Agric. Food Chem. 20(6):1213–1219.

Read, D.C. 1969. Persistence of some newer insecticides in mineral soils measured by bioassay. J. Econ. Entomol. 62(6):1338–1342.

Sattar, M.A. 1990. Fate of organophosphorus pesticides in soils. Chemosphere 20(3–4):387–396.

Tashiro, H., and R.J. Kuhr. 1978. Some factors influencing the toxicity of soil application of chlorpyrifos and diazinon to European chafer grubs. J. Econ. Entomol. 71(6):904–907.

5 Effect of Soil Depth on Carbofuran and Aldicarb Degradation

G. A. Buyanovsky, A. M. Gajda, R. J. Kremer, and G. J. Pieczonka

University of Missouri and USDA-ARS
Columbia, Missouri

Preferred pathways, especially in heavy clay soils with high shrink-swell potential, are suspected to be a primary route for the movement and accumulation of pollutants deep within the profile and close to the groundwater table (White et al., 1986).

Conservation tillage systems rely heavily or exclusively on chemical weed and insect control. Although these systems have proven effective in reducing soil erosion and thus the amount of sediments in runoff, soils under some conservation tillage systems, such as no-tillage, are more susceptible to the development of deep cracks that potentially can enhance downward movement of chemicals.

The fate of pesticides transported to the lower soil profile is unclear. Most of the experiments to establish "apparent half-life" of pesticides have been done with the biologically active surface layer. It is well known, however, that microbial counts for deeper parts of the soil profile are usually 10^2 to 10^3 times smaller than for the upper 0 to 10 cm layer. Locke and Harper (1991) attributed depth differences in metribuzin degradation to reductions in microbial activity with increasing soil depth.

Due to reduced biological activity, the persistence of pesticides in the lower soil profile could be much longer than is usually taken into account, and the probability of groundwater contamination greatly increased. A project is in progress with an overall goal to test the hypothesis that measurable contaminant transport via preferred pathways in some Missouri soils is possible. As part of the project, a laboratory experiment has been conducted to estimate the mineralization rates of two widely used pesticides within a soil profile. Carbofuran (2,3-dihydro-2,2-dimethyl-7-benzofuranyl methylcarbamate) and aldicarb [2-methyl-2-(methylthio) propionaldehyde *O*-methylcarbamoyloxime], are broad-spectrum, soil-applied systemic pesticides for control of mites, nematodes, and insects. The two chemicals are distinct

 Sorption and Degradation of Pesticides and Organic Chemicals in Soil. SSSA Special Publication no. 32.

in their leaching characteristics and water solubility (carbofuran, 0.7 g L^{-1}; aldicarb, 6 g L^{-1}).

MATERIALS AND METHODS

The ^{14}C-labeled carbofuran and aldicarb were obtained from the FMC Corp. (Princeton, NJ) and Rhone-Poulenc Ag Co. (Research Triangle Park, NC), respectively. Samples of Menfro silt loam (fine-silty, mixed, mesic Typic Hapludalf) were taken from depths of 0 to 10 cm, 20 to 30 cm, 50 to 60 cm, and 90 to 100 cm. To preserve the microbial community, soil was refrigerated immediately after sampling.

Solutions of ^{14}C-labeled carbofuran and aldicarb in ethyl acetate were amended with nonlabeled chemicals (technical grade) to provide a concentration in the soil of 20 mg kg^{-1} and were injected into the soil already placed in an Erlenmeyer flask. Water content was adjusted to 0.25 to 0.30 cm^3 cm^{-3}. Before the experiment was finished, a glucose solution was added to the soil in incubation flasks at the rate of 0.5 g of glucose 100 g^{-1} of soil, to check the possibility of cometabolism.

Incubation was carried out in a system with a constant flow of CO_2-free, humidified air (Buyanovsky et al., 1988). Each flask had a separate scrubbing tower with glass beads filled with 1.0 *N* NaOH. The solution was changed every 3 to 5 d during the first month of incubation, and every 7 to 10 d afterward. An aliquot of the absorbant was taken for radioactivity measurements, and the rest was titrated after addition of $BaCl_2$.

Residues of pesticides remaining in soil after incubation were extracted with methylene chloride. The extract was evaporated, dissolved in methanol, and analyzed by HPLC with an ultraviolet (UV) detector at 210/280 nm, on C-18 column with acetonitrile/water gradient mobile phase (Sparacino & Hines, 1976).

Populations of microorganisms capable of degrading pesticides were determined after incubation by dilution plating on a medium containing mineral salts and 200 mg L^{-1} of carbofuran or 150 mg L^{-1} of aldicarb as the only C source. TTC (2,3,4-triphenyl-tetrazolium chloride) was incorporated into the medium at a rate of 25 mg L^{-1} as an indicator of metabolically active microorganisms (Mueller et al., 1989).

RESULTS AND DISCUSSION

Microbial respiration for incubated soil increased rapidly during the first 3 to 5 d, and reached a maximum after 10 to 15 d, then rapidly decreased and stabilized for a long period (Fig. 5–1). Such dynamics are normal for disturbed soil incubated in laboratory conditions (Buyanovsky et al., 1988). Total amount of C-CO_2 evolved from the soil of the 0- to 10-cm depth was three times greater than from the soil of 50- to 60-cm depth. This may be easily explained by the apparent difference in microbial biomass and organ-

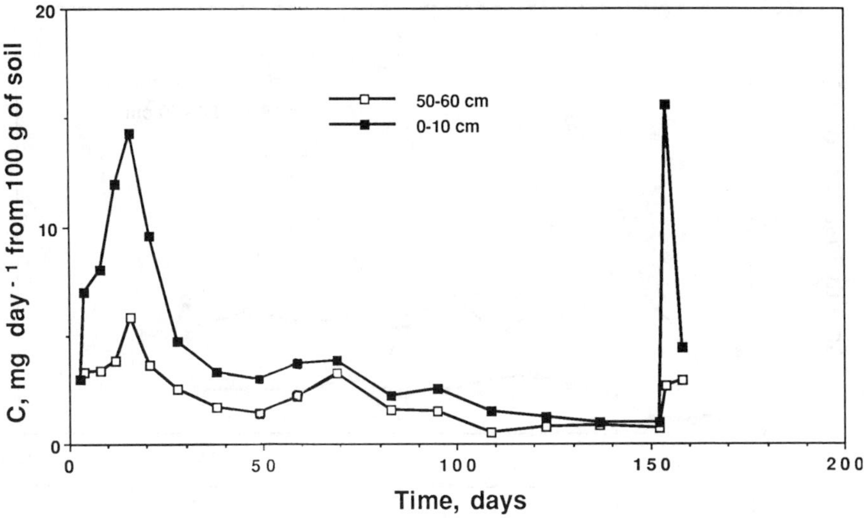

Fig. 5–1. Dynamics of CO_2-C evolution during incubation of Menfro soil. Glucose was added at Day 150.

ic C content, which was 1.8% in the upper layer and <0.5% below 50 cm. No difference in CO_2 evolution was noted between the flasks amended with pesticides and control.

Figure 5–2A and 2B, represents the relative daily activity losses (becquerel, Bq) from soil incubated with radiolabeled carbofuran and aldicarb, respectively. Only results for the 0- to 10-cm and 50- to 60-cm depths are shown because the dynamics of mineralization in soil from 20 to 30 cm was similar to that from 0 to 10 cm, and the activities for a depth of 50 to 60 cm were similar to the 90- to 100-cm depth.

Evolution of $^{14}CO_2$ from carbofuran in soil at both depths was low during the first week and started to rise after the general respiration maximum passed. A significant difference in degradation rates and dynamics between the upper and lower parts of the profile was noted. Daily losses of ^{14}C from 0 to 10 cm of soil increased rather rapidly after 7 d of incubation and reached a maximum by Day 50. After this, the degradation rate dropped and then stabilized. The $^{14}CO_2$ evolution from the lower depth was extremely low during the first month of incubation, and only after this period started to increase. Carbofuran mineralization, as indicated by $^{14}CO_2$ evolution, did not exceed 43% of the initial amount in the soil from the surface layer (Fig. 5–3A). Approximately the same amount of labeled C was found in unextractable form (Table 5–1). Total degradation in the lower part of the profile was in the range of 5 to 8%.

In a similar experiment with aldicarb, we found a much more rapid adjustment of microbial population to a new source of C than in the case with carbofuran. A high degradation rate of aldicarb in the soil from the surface layer was observed from the beginning of incubation (Fig. 5–2B). Soil from the lower layers showed much smaller aldicarb degradation rates (Fig. 5–3B).

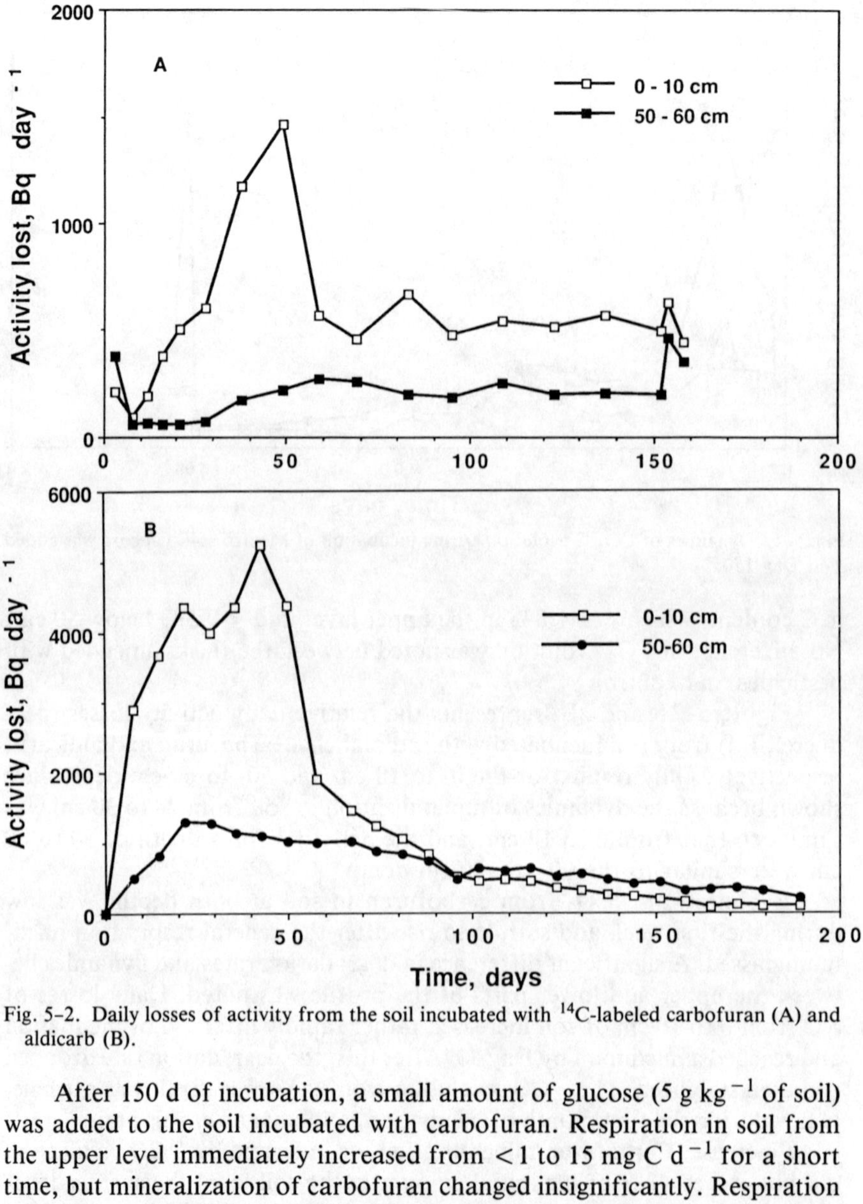

Fig. 5–2. Daily losses of activity from the soil incubated with ^{14}C-labeled carbofuran (A) and aldicarb (B).

After 150 d of incubation, a small amount of glucose (5 g kg^{-1} of soil) was added to the soil incubated with carbofuran. Respiration in soil from the upper level immediately increased from <1 to 15 mg C d^{-1} for a short time, but mineralization of carbofuran changed insignificantly. Respiration in soil from the lower part of the profile increased from 0.5 to 2.5–3.0 mg C d^{-1}, but $^{14}CO_2$ evolution approximately doubled relative to the time when the degradation rate reached its maximum (see Fig. 5–2A). This indicated that the role of cometabolism in biodegradation was insignificant in the upper layers. In soil below 50 cm, the availability of C facilitated biodegradation. Similar results were obtained with aldicarb.

The number of presumptive carbofuran-degrading microorganisms varied from 7 to 9 × 10^5 g^{-1} of soil in the upper part of the soil profile

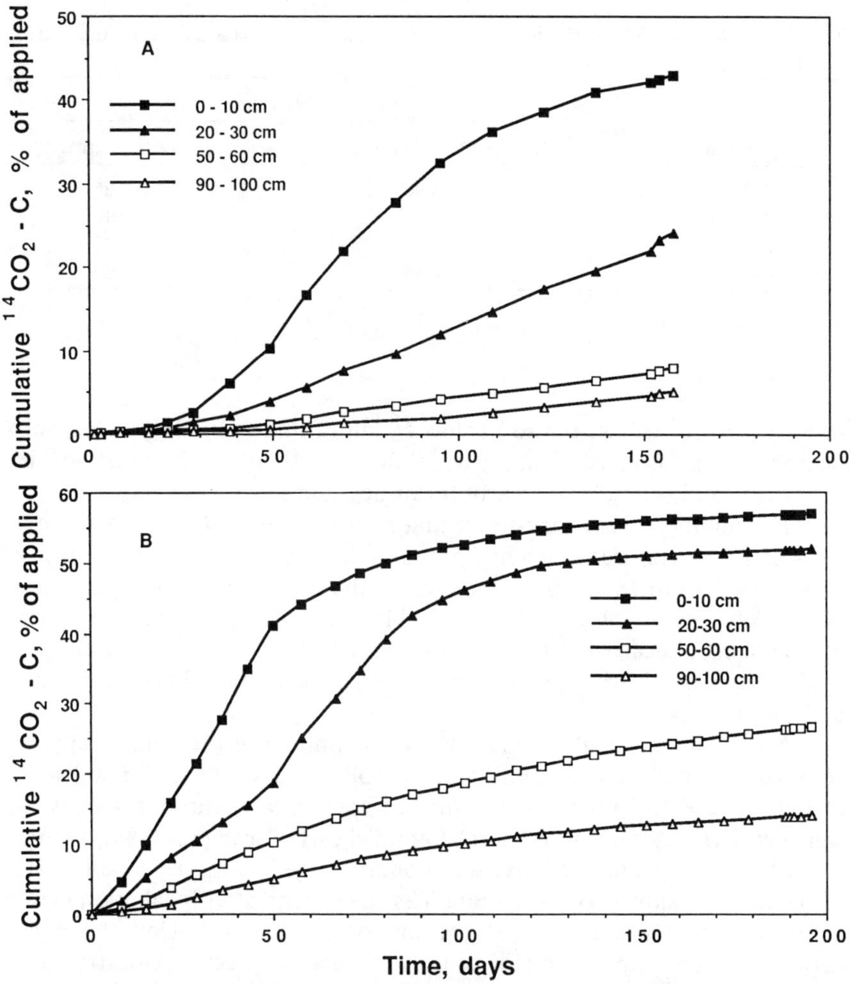

Fig. 5-3. Cumulative curves of $^{14}CO_2$-C evolution from the soil incubated with carbofuran (A) and aldicarb (B).

Table 5-1. Concentration of carbofuran and hydroxycarbofuran in Menfro soil after 5 mo of incubation.

Soil depth	Degraded and released as $^{14}CO_2$	Extracted as carbofuran	Extracted as hydroxy-carbofuran	Nonextractable ^{14}C residues
cm		%		
0-10	43.1	0.2	<0.1	42.5
20-30	23.3	4.6	<0.1	22.8
50-60	8.5	30.7	<0.1	32.0
90-100	4.9	40.3	<0.1	34.2

Table 5-2. Number of pesticide-metabolizing organisms in the soil after incubation (log g^{-1}).

	Medium	
Depth, cm	With carbofuran	With aldicarb
0-10	5.96	7.25
20-30	5.83	7.00
50-60	4.80	6.32
90-100	4.86	5.60

to 6 to 7 × 10^4 g^{-1} in the soil below 50 cm (Table 5-2). Only colonies of microorganisms isolated from the upper horizons showed color reaction with TTC, thus indicating active carbofuran degradation.

The number of aldicarb-degrading microorganisms (mostly actinomycetes and bacteria) was much higher than that of carbofuran degraders and varied from 10 to 18 × 10^6 g^{-1} of soil in the upper layers to 0.4 to 2.1 × 10^6 g^{-1} in the soil below 50 cm (Table 5-3). The population of fungi metabolizing aldicarb was found to be about 5 × 10^4 g^{-1} of soil in the surface horizon. Aldicarb is probably more readily used by soil organisms than carbofuran.

After 5 mo of incubation, only a small amount of the initially applied carbofuran could be extracted from the soil of upper layers. In soil from 50 to 60 and 90 to 100 cm, 30 and 40% of applied carbofuran, respectively, were extracted. A substantial part of applied carbofuran (22-42%), as indicated by residual radioactivity, was bound to the soil matrix (Table 5-1). No residues of aldicarb or its metabolites were found after incubation in soil from the 0- to 10- and 20- to 30-cm horizons, but in soil below the 50-cm depth, aldicarb sulfone and aldicarb sulfoxide were detected. Unextractable residues of aldicarb were found to be very high in the lower profile (Table 5-3).

Table 5-3. Concentration of aldicarb and its metabolites in Menfro soil after 5 mo of incubation.

		Concentration			
Depth	Degraded and released as $^{14}CO_2$	Aldicarb	Aldicarb sulfone	Aldicarb sulfoxide	Non-extractable ^{14}C residues
cm	%		ng g^{-1}		%
0-10	56.8	<50	<100	<50	13.2
20-30	51.9	<50	<100	<50	8.7
50-60	26.8	<50	127	172	46.6
90-100	14.0	<50	463	535	64.3

CONCLUSION

1. Microbial communities in the surface (0–10 cm) layer and at the depth below 50 cm differ not only in their general activity, but also in their ability to degrade pesticides.
2. The microbial community of the upper layer had a much shorter adjustment time for added chemicals than populations from deeper parts of the profile.
3. The total amount of insecticides degraded in the soil from the upper 30 cm was three to five times greater than in the soil below 50 cm, thus persistence of a pesticide could increase dramatically when it reaches the lower part of the soil profile (below 50 cm).

ACKNOWLEDGMENT

The study was funded by the USDA Water Quality Program under grant no. 89-COOP-1-4723.

REFERENCES

Buyanovsky, G.A., G.J. Pieczonka, G.H. Wagner, and M.L. Fairchild. 1988. Degradation of captan under laboratory conditions. Bull. Environ. Contam. Toxicol. 40:689–695.

Locke, M.A., and S.S. Harper. 1991. Metribuzin degradation in soil: I—Effects of soybean residue amendment, metribuzin level, and soil depth. Pestic. Sci. 31:221–237.

Mueller, J.G., H.D. Skipper, E.G. Lawrence, and E.L. Kline. 1989. Bacterial stimulation by carbamothioate herbicides. Weed Sci. 37:424–427.

Sparacino, C.M., and J.W. Hines. 1976. High-performance liquid chromatography of carbamate pesticides. J. Chromatogr. Sci. 14:549–556.

White, R.E., J.S. Dyson, Z. Gerstl, and B. Yaron. 1986. Leaching of herbicides through undisturbed cores of a structured clay soil. Soil Sci. Soc. Am. J. 50:277–283.

6 Effect of Sorption-Desorption and Diffusion Processes on the Kinetics of Biodegradation of Organic Chemicals in Soil

Kate M. Scow
University of California
Davis, California

Heterogeneity is the rule in most natural environments. It is rare, except perhaps among parasites, for an organism to be in continuous contact with its food source and not have to expend energy to maintain itself while searching for a suitable C and energy source. Microorganisms are no exception. Except in laboratory cultures, microbial populations almost never have unlimited access to the substrates and nutrients they require for growth and metabolism. Soil, as a habitat for microorganisms, is extremely heterogeneous, and it has microscale variation in both physical structure and chemical composition. This heterogeneity creates a multitude of microenvironments, and each microenvironment has a unique set of factors governing microbial growth and activity in that location. Conventional environmental measurements of microbial activity are made at a scale too large to be representative of these processes. As a result, different microenvironmental conditions are lumped together with consequent masking of the factors that may be important in controlling rates of microbial processes. The interactions between abiotic and biotic components in soil are particularly complex and usually neglected in studies of microbial metabolism in soil. The assumption that microbial populations in soil behave the same as in solution culture is part of the reason why quantitative descriptions of microbial reactions frequently do not predict processes in soil.

Sorption from the soil solution or atmosphere onto or into mineral and organic matrices contributes to nonuniform distribution of chemicals in soil. Numerous agricultural and industrial chemicals are subject to some degree of sorption to soil. Strongly sorbed chemicals are often quite persistent in the environment and, collectively, present one of the most difficult challenges in the remediation of contaminated sites. Although sorbed pollutants are temporarily retained on organic matter or clay and thus may be considered to be "removed" from the system, the potential exists for transport through runoff to surface waters, desorption and leaching into groundwater, or facilitated transport on soil colloids through the soil solution. Evidence for the eventual mobility of sorbed chemicals includes the finding that many groundwater aquifers are contaminated with pesticides and organic chemicals that have relatively high-sorption partition coefficients (Sawhney, 1989; Zoeteman et al., 1980; Elder et al., 1991).

Sorption has a profound impact on a chemical's spatial distribution and its availability to microorganisms in soils. Even compounds with low-sorption partition coefficients may exist primarily in the sorbed phase in soil at field moisture levels (Pignatello, 1989). Because an important variable in biodegradation is the available concentration of the chemical substrate undergoing degradation (Simkins & Alexander, 1984; Hamaker, 1972a), a process like sorption that significantly lowers the solution concentration should influence the biodegradation rate. It is not surprising, however, that a sorbent as complex and undefined as organic matter would have multiple, and possibly contradictory, effects on microbial processes. These other effects will be discussed in greater detail in the section entitled Other Effects of Sorbent-Chemical Interactions on Microbial Populations.

The purpose of this chapter is to summarize the current knowledge on how the sorption and desorption of organic chemicals in soil influences their biodegradation, review several approaches to describing the kinetics of biodegradation of sorbed chemicals, and then consider other potential impacts of sorbate-sorbent interactions on microbial populations. The volume of information on sorption and biodegradation is far too extensive to cover in a single review. Therefore, this chapter will emphasize issues primarily, but not exclusively, relevant to surface soils and will focus on the interaction between organic matter and nonionic hydrophobic chemicals.

INTRODUCTION TO THE PROBLEM

Research to date on the biodegradation of sorbed chemicals in soil has largely been empirical in nature and not required specific knowledge of the properties and mechanisms underlying the interactions of sorbate, sorbent, and microbial populations. Even without an understanding of specific mechanisms, it is obvious that the rate at which a sorbed chemical can be metabolized must be strongly influenced by the locations of sorbed chemicals and microorganisms within the soil matrix, the distances between the chemical and microbial populations, and the accessibility of the chemical to microorganisms. Information on the properties of soil organic matter, interactions between organic chemicals and organic matter, and the distribution of microbial populations in soil will be briefly reviewed to provide background on the problem of biodegradation of sorbed chemicals.

Properties of Soil Organic Matter

Soil organic matter is a complex and heterogeneous mixture of primarily humic substances (humic and fulvic acids and humin), polysaccharides, lignin, simple carbohydrates, lipids, proteins, and organic acids (Hayes, 1985; Stevenson, 1982). This organic matter is associated with clays, iron, and aluminum hydroxides, as well as other inorganic constituents. Divalent and polyvalent cation bridges, ligand exchange reactions, and coulombic attractions link humic substances to negatively charged inorganic colloids (Hayes, 1985). Microbial biomass is intimately associated with organic matter. Many analytical methods for organic chemicals cannot distinguish microbial biomass from the nonliving fraction of organic matter.

To date, there is no consensus on the exact structure of organic matter, although several hypotheses have been proposed (Alison, 1968; Stevenson, 1982; Schnitzer & Khan, 1978). What most of these hypotheses share in common is that organic matter is an amorphous, three-dimensional polymer composed of substituted aromatic and aliphatic hydrocarbon core materials with hydrophobic side chains and a variety of polar substituents (Hayes, 1985; Hayes et al., 1989). During the formation of organic matter from plant residues, differences in organic matter structure and characteristics may arise as a function of the local conditions. In soils with high clay content, there is a greater tendency to preserve potentially labile compounds such as amino acids (Sörensen, 1972, 1975; Balesdent et al., 1988) and in the presence of smectite clays there may be greater condensation of phenolic rings into larger aromatic compounds (Mortland & Halloran, 1976). Rearrangements and reactions within the humus structure continue after its formation in all soils (Stevenson, 1982; Haider, 1992).

Clearly, soil organic matter is far too complicated, and its true identity too uncertain, to allow realistic description in kinetic models of sorption and biodegradation. Thus, most models that consider organic matter as a sorbent assume that its structure is simple. In equilibrium approaches, organic matter has been treated as a dimensionless "phase" that behaves like an organic solvent (Chiou, 1989). Organic matter has been conceptualized as an organic polymer through which the solute diffuses to sorption sites (Brusseau & Rao, 1989b). For aggregated soils or sediment, sorption kinetics of hydrophobic chemicals can be described by assuming spherical aggregates having an outer and inner surface area, with access to the inner sites controlled by diffusion of the sorbate (Wu & Gschwend, 1986; Mihelcic & Luthy, 1991). The model chosen is sometimes a function of the type of sorption process being considered and the environment being considered (e.g., surface soil vs. vadose zone). Pignatello (1989) remarked that the distinction between phase and surface models blurs when one considers that organic matter is a nonordered, tangle of polymer segments with extended inner surfaces along the voids, cavities that are enclosed in the polymers, and with boundaries between the different polymeric material comprising it. The distinction between the phase and surface models may be considerably important, however, when trying to integrate microbiology into the picture.

Sorption of Organic Chemicals: An Overview

Extensive research has been conducted on the sorption of organic chemicals and no attempt will be made here to summarize this massive body of knowledge. Several reviews on the topic have been published, including those by Chiou (1989), Karickhoff (1980), Brusseau and Rao (1989a), Harmon et al. (1989), Pignatello (1989), and Calvet (1989). The purpose of this section is to briefly introduce some of the important concepts of sorption and desorption in soil systems.

There are a variety of associations between organic chemicals and solid phases. The term *sorption* is general and encompasses all types of mechanisms involved in reversible associations of chemicals with the solid phase. *Adsorption* is the specific term for the condensation of vapors or solutes on surfaces or pores of a solid by physical or chemical bonding forces (Chiou, 1989). Different types of sorption processes include "hydrophobic sorption" (or partitioning), van der Waals-London interactions, charge-transfer (e.g., H bonding), ligand exchange, and ion-exchange reactions. The latter three processes are more site directed and associated with specific functional groups than are the first two phenomena (Pignatello, 1989). Many researchers have recognized that the organic fraction of soils dominates the sorption of nonionic organic compounds at organic C levels of >0.1%, whereas below this value interactions with mineral surfaces may be more important (Pignatello, 1989). Multiple mechanisms of sorption may be involved in the association of a particular chemical with one soil.

Hassett and Banwart (1989) distinguish between sorption driven by enthalpy-related forces related to bonding of the sorbate to surfaces and the lack of solvation of the sorbate in the soil solution (e.g., hydrophobic sorption) which is entropy related. Chiou (1989) considered hydrophobic sorption to be distinct from adsorption because it results in the homogeneous distribution of the sorbate throughout the entire volume of the sorbent.

Traditionally, hydrophobic sorption has been conceptualized as an equilibrium phenomenon that is rapid and reversible (Chiou, 1989). It is now recognized that the rate-limiting step in both hydrophobic sorption and exchange reactions is often a mass transfer process and sorption/desorption is a kinetic not an equilibrium phenomenon. A lack of consideration of the kinetics of adsorption and desorption processes has contributed to the shortcomings of models predicting the transport of organic chemicals in soil (van Genuchten et al., 1974; Davidson & Chang, 1972). Sorption is better represented as a two-phase process with an initial fast stage followed by a longer slow phase controlled by mass transfer of the solute to internal adsorption sites (Cameron & Klute, 1977; Karickhoff, 1980; McCall & Agin, 1985). The kinetics appear to be reflected in the sorption partition coefficient (e.g., the higher the partition coefficient, the slower the kinetics of sorption) (Karickhoff & Morris, 1985; Wu & Gschwend, 1986). Brusseau & Rao (1989b) found an inverse log-log linear relationship between a chemical's equilibrium sorption partition coefficient and its sorption rate constant for nonpolar organics that lack reactive functional groups and another relationship for chemicals with reactive functional groups. In many cases, there appears to be an irreversible and resistant-to-biodegradation fraction that increases in size over time, and even weakly sorbed nonpolar organic compounds are subject to this phenomenon (Pignatello, 1989; McCall & Agin, 1985).

Brusseau and Rao (1989a) distinguish between transport- and sorption-related nonequilibrium behavior of organic chemicals. Transport-related nonequilibrium occurs when the process of sorption itself is instantaneous but transport of the solute to the sorptive domain is limited by diffusion through porous particles. Sorption-related nonequilibrium occurs with solute diffusion within a polymeric matrix of sorbent organic matter. The former may be more appropriate to mineral-dominated porous media of the subsurface; the latter to soils with higher organic matter contents. Ball et al. (1991) and Brusseau et al. (1991) disagree about the distinction between the two processes, but both concur that microscale diffusive mass transport processes control rates of sorption and desorption. The low effective diffusion coefficients measured for sorbed organic chemicals in soil reflect the close relationship between sorption and diffusion. Effective diffusion coefficients in sediment of between 1×10^{-9} and 1×10^{-12} $cm^2\ s^{-1}$ were measured for sorbing compounds in sediment (Wu & Gschwend, 1986) and 1×10^{-14} $cm^2\ s^{-1}$ for aged residues of ethylene dibromide in soil (Pignatello et al., 1987). These values are considerably lower than "typical" diffusion coefficients usually associated with organic chemicals in soil (Hamaker, 1972b).

The development of quantitative kinetic descriptions of the biodegradation of sorbed chemicals has lagged behind the rapid advances in the area of sorption. This lag is inevitable in part because of the strong dependency of the biodegradation rate on the concentration of chemical in solution. The kinetics of sorption and desorption govern the solution concentration of sorbed chemicals and is important information required for a rigorous analysis of biodegradation kinetics. Even with an understanding of sorption-desorption dynamics, further advances in the kinetics of biodegradation may be slow because so little is known about the spatial distribution of soil microbial populations, especially that of particular species or groups able to degrade pollutants (see section entitled Distribution of Microbial Populations in Soil).

Distribution of Microbial Populations in Soil

Given the complexity of organic matter, it is a challenge simply to imagine where microorganisms are located throughout the organic fraction of soil. Our knowledge of the spatial distribution of microbial populations in soil is remarkably limited. Only a small fraction <0.2%, of the surface area of soil organic matter or mineral surfaces is covered by microorganisms (Hissett & Gray, 1976; Foster, 1988). In sediments, 3 to 15% of the particle surface was estimated to be covered by cells (Tsernoglou & Anthony, 1971). In soil, higher populations of organisms are associated with plant roots or the rhizosphere than with the bulk soil (Nedwell & Gray, 1987). Presence of microbial populations depends on the chemical properties of the surfaces. Gray et al. (1968) found that although organic particles constituted <15% of the total surface area in a dune soil, 60% of the total population of bacteria counted were associated with organic material. Siala et al. (1974) found a strong association between bacteria and surfaces of humic material and iron coatings.

Microorganisms in nature are frequently associated in groups, although individual cells are also found in soil. Microcolonies of bacteria in soil are small relative to those formed on the rich nutrient conditions of agar plates and range from a few to 150 cells per microcolony (Kilbertus, 1980; Gray, 1967). Bacteria, either singly or in colonies, are usually associated with soil pores of diameters ≥ 2 μm (Kilbertus, 1980). This distribution may significantly influence the availability of nutrients to microorganisms in soils. For example, in silt loam soils approximately 38% of the total pore space is contained in pores with effective diameters <2 mm (from Fig. 3 in Papendick & Campbell, 1981). Therefore, at field moisture levels a significant fraction of the soil solution is theoretically inaccessible to many microorganisms.

At the soil aggregate scale, the distribution of microbial populations is nonuniform and different groups of organisms are favored in different locations within the aggregate. Because the interiors of many soil aggregates are often anaerobic, even in well-aerated soils (Myrold & Tiedje, 1985), it is improbable that aerobic microorganisms are active deep within large aggregates. Kilbertus (1980) found that only 6% of the inner volume of soil aggregates was colonized by microorganisms. Under aerobic conditions, CO_2 production and microbial biomass C was found to be greatest in microaggregates (<0.25 mm) and biomass declined with increasing aggregate diameters up to 20 mm (Seech & Beauchamp, 1988). In an artificial system in which a *Pseudomonas* sp. was immobilized in alginate or polyacrylamide-hydrazide beads of 2.5 to 3.0 mm diam., the bacteria grew only within the outer layer of the beads (Bettmann & Rehm, 1984). This confinement to the outer layer was attributed to either diffusional limitation of oxygen or their C source into the bead or rapid consumption of these resources by the outer layer of microorganisms before diffusion of the substrate into the beads. Hattori (1966) proposed a technique to separate organisms living inside from those outside of soil aggregates. This technique assumed that soil which was hand-shaken in buffer solution released organisms living on the outside of aggregates, whereas soil disrupted by mild sonic oscillation presumably released organisms within the aggregates. Use of this technique led to the observation that a higher proportion of gram-negative bacteria is found inside soil aggregates, whereas sporeformers, actinomycetes, and fungi are more commonly found outside of aggregates (Hattori, 1973; Hattori & Hattori, 1976).

Our current knowledge of the quantitative distribution of soil microorganisms is severely constrained by existing methodologies. Microscopy, including transmission and scanning electron microscopy (SEM) (Waid, 1973; Waid & Suan, 1974) and fluorescent microscopy (Dommergues & Mangenot, 1970; Casida, 1971), is the most commonly used technique for determining microbial distribution in soil. All microscopic techniques are hampered by disruptive sample preparations or by obstruction of the view of microorganisms by soil particles. Physiological techniques, such as measurement of respiration, are usually not sensitive enough to detect activity at microscopic scales. Hopefully, with the rapid development of immunological and molecular techniques, we will soon have significantly more sensitive tools with which to look at questions of microscale distribution of microbial populations in porous media.

EFFECTS OF SORPTION ON THE BIODEGRADATION OF ORGANIC CHEMICALS IN SOIL

There has been a low, yet consistent, level of interest in the biodegradation of sorbed chemicals, particularly agricultural pesticides, over the past several decades. In recent years, interest has dramatically increased and the focus has shifted to industrial pollutants as well. This shift results, in part, from difficulties encountered in implementing bioremediation technologies in contaminated soils.

Biodegradation rates of chemicals in soil are frequently far lower than rates measured in solution cultures of microorganisms (Alexander & Scow, 1989; Hamaker, 1972a). Seemingly contradictory studies can be found in the literature that indicate rates and final extents of metabolism may be either lower or higher for sorbed chemicals (van Loosdrecht et al., 1990). These contradictions probably result from differences among studies in the impact of sorbents on microbial populations and in how the organic chemical is sorbed. Many of the contradictory studies were conducted in aquatic systems where alternate impacts of sorbents may be more important than in systems dominated by porous media (see section entitled Other Effects of Sorbent-Chemical Interactions on Microbial Populations).

Studies in Soil and Soil Slurries

It is generally believed that only that fraction of a chemical dissolved in the solution can be metabolized by microorganisms in soil. Chemicals for which solution-phase utilization has been demonstrated include 2,4-D (2,4-dichlorophenoxy acetic acid) (Ogram et al., 1985), α-hexachlorocyclohexane (Rijnaarts et al., 1991), and naphthalene (Mihelcic & Luthy, 1991). Thus, a decrease in the biodegradation rate of chemicals is a commonly observed effect of increased organic matter content in soil.

If desorption reactions are limited by diffusive mass transport (see section entitled Sorption of Organic Chemicals: An Overview) and biodegradation rates are limited by desorption, then the biodegradation of sorbed chemicals should be limited by diffusion. There is considerable evidence that microbial processes in soil are limited by diffusion of substrates to sites where organisms are located. Numerous reports have been made of elevated values of K_m, the Michaelis-Menten constant, for microbial transformations in soil and the high values appear to result from slower diffusion of the substrate to the degraders in soil than solution, rather than from a lower affinity of soil microorganisms for the substrate (Firestone, 1982). Focht and Shelton (1987) found that the K_m value for *Pseudomonas alcaligenes* growing in soil on 3-chlorobenzoate was 33 times higher than when the bacterium was growing in solution, whereas the μ_{max} values and yield were the same. The apparent K_m values reported for nitrate reduction in soil ranged from 13 to 1300 times those measured in solution (Myrold & Tiedje, 1985). The K_m for urease activity in shaken cultures was one-half that of the K_m value measured in static cultures, apparently because more of the urease that was associated with the soil matrix came in contact with its substrate in the shake cultures (McLaren, 1978). A model simulation of the effect on the apparent K_m value for biodegradation with increased diffusional constraints and sorption of a chemical is demonstrated in the section entitled Models Linking Biodegradation and Mass Transfer Processes.

Because of the difficulty in separating diffusion from sorption in soil, few studies on the biodegradation of sorbed chemicals explicitly address the two processes independently. Many older studies include surveys of the biodegradation rate of a particular pesticide in several soils with different organic matter contents. Many of the studies found an inverse relationship between degradation rate and organic matter content (Hamaker,1972a). Sorption of pyrazon [5-amino-4-chloro-2-phenyl-3-(2H)-pyridazinone] (Smith & Meggitt, 1970), *N'*(4-chlorophenoxy)phenyl-*N,N*-dimethylurea (Geissbühler et al., 1963), and atrazine (2-chloro-4-ethylamino-6-isopropylamino-*s*-triazine) and diuron [*N'*-(3,4-dichlorophenyl)-*N,N*-dimethylurea] (McCormick & Hiltbold, 1966) has been implicated as protecting at least some fraction of these chemicals from microbial decomposition. The mineralization rate of glyphosate in several soils declines as the amount of the herbicide bound by the colloids increases (Nomura & Hilton, 1977), suggesting preferential or sole use of the chemical in solution. The relationship between the fraction sorbed and the degradation rate is not always proportional. For example, phenol had a lower degradation rate in a soil than what would be expected based on the concentration of the chemical in solution (Scott et al., 1983).

The soil fumigant, 1,2-dibromoethane (EDB), may persist in soil at nanogram per gram levels for 20 yr, even though it is volatile and relatively water soluble. The persistence of EDB has been attributed to its entrapment in micropores because residual EDB can be liberated by pulverization of the soil (Steinberg et al., 1987; Pignatello et al., 1987). Addition of fresh EDB resulted in rapid degradation whereas chemical that had been in the soil for years was degraded at a very slow rate, if at all (Fig. 6–1).

Some of the investigations of the biodegradation of sorbed chemicals have been conducted using soil slurries, in part because the questions asked necessitated their use (e.g., to compare the activity of organisms in the absence and presence of soil). Ogram et al. (1985) measured the biodegradation of 2,4-D in soil and clay suspensions inoculated with a pure culture of bacterium able to degrade the pesticide. By analyzing patterns in the first-order rate constants obtained from the disappearance curves under different conditions, the authors found the trends to be best explained by a model that assumed only the solution phase of the chemical could be metabolized and that both sorbed and free bacteria were able to degrade the pesticide equally. The biodegradation of aged residues of α-hexachlorocyclohexane

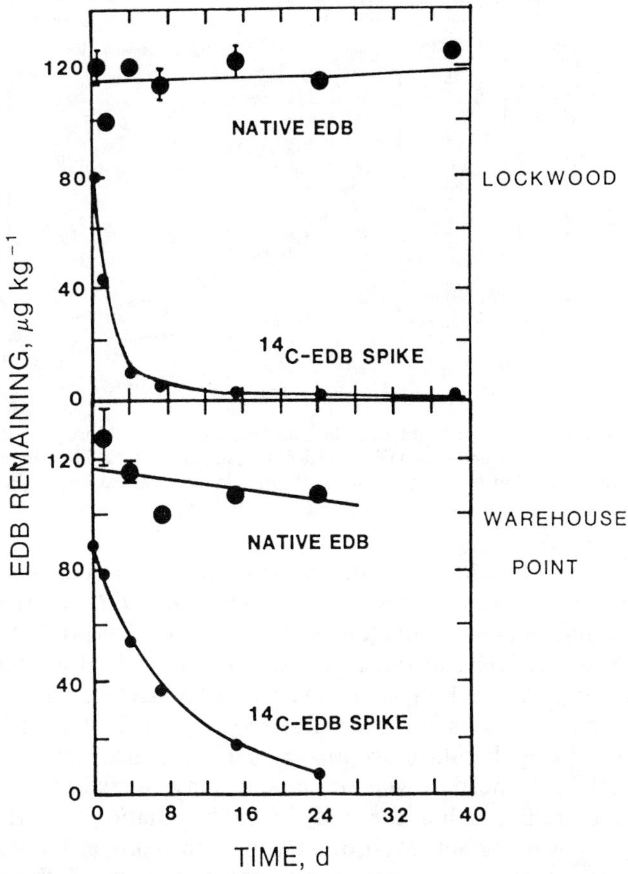

Fig. 6–1. Comparison of the degradation of native ethylene dibromide and a freshly added [^{14}C]ethylene dibromide spike in suspensions of Lockwood soil and Warehouse Point soil. (From Steinberg et al., 1987.)

in calcareous soil was found to be rate-limited by desorption of the pesticide (Rijnaarts et al., 1991) (Fig. 6–2). The rate of biodegradation of naphthalene in soil slurries under denitrification conditions was found to be rapid relative to the desorption rate of the chemical from soil aggregates and the maximum rate of biodegradation was proportional to the soil/water ratio (Mihelcic & Luthy, 1991). The rate and extent of degradation of dodecylbenzenesulfonates by a sewage inoculum was lowered by the addition of soil colloids (Inoue et al., 1978). Similarly, humic acids retarded the mineralization of benzylamine, and the extent of benzylamine mineralization declined as the percentage of the amine initially bound to the organic materials was increased (Amador & Alexander, 1988).

The use of soil slurries may confound biodegradation measurements by making soil organic C available for metabolism (see section entitled Effect of Available Carbon Associated with Organic Matter on Microbial Populations). The soil solution concentration or soil/solution volume ratio is an

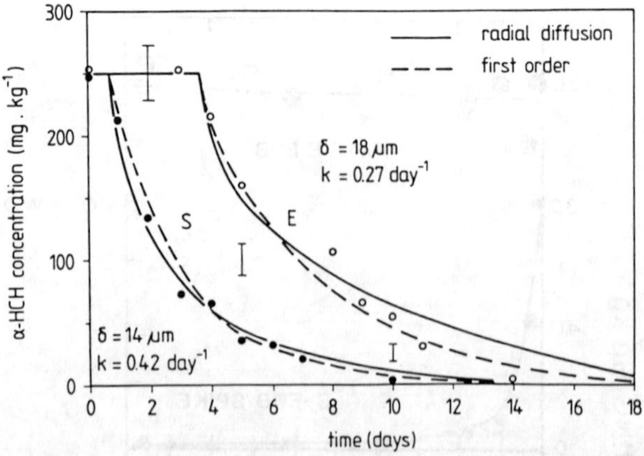

Fig. 6–2. Observed and predicted degradation kinetics of α-hexachlorocyclohexane in soil suspensions in end-over-end mixed (E) and stirred (S) batch reactors. Lines show the predictions of a radial diffusion (solid lines) and first-order model (dashed lines). (From Rijnaarts et al., 1991.)

important variable in the biodegradation of sorbed chemicals (Anderson, 1981; Mihelcic & Luthy, 1991; Scow & Hutson, 1992). Grover and Hance (1970) found a three- to fivefold increase in the adsorption of linuron and atrazine, respectively, in soil to solution ratios of 1:10 when compared to ratios of 4:1. The authors attributed the differences in sorption to the fact that the soil aggregates were more dispersed in slurries than at field moisture levels. It is likely that the desorption rate of a chemical from soil also would be strongly influenced by whether the soil is in a slurry or unsaturated state.

Another approach in the study of biodegradation of sorbed chemicals is to add known amounts of sorbents with defined properties of soil and measure the effect of the additions on the degradation rate. Although the added sorbents have different properties from soil organic matter, their sorptive properties are defined and characteristics such as sorbent quantity and affinity can be manipulated to test their importance in a soil system. Activated C was used as a model sorbent in several earlier studies. The rate of biodegradation of maleic hydrazide declined and the shapes of the degradation curves changed with an increase in the amount of activated C added to soil (Helweg, 1975). The addition of diallate to activated charcoal and soil mixtures giving different percentages of the initial fraction of herbicide sorbed demonstrated a direct relationship between the amount of chemical free in solution and the degradation rate (Anderson, 1981). Moyer et al. (1972) found that the decomposition rates of atrazine and chlorthiamid were five to six times lower in the presence of 2.5% charcoal than in its absence.

Chromatography beads are well-defined model sorbents that provide a wide choice of types of sorption affinities and a broad range of sorption partition coefficients. Organic chemicals mixed with hydrophobic SM-7 chromatography beads (BioRad) have sorption partition coefficients of two to three orders of magnitude greater than chemicals in soil (K.M. Scow, 1992, unpublished data). Addition of SM-7 beads to a silt loam soil decreased the

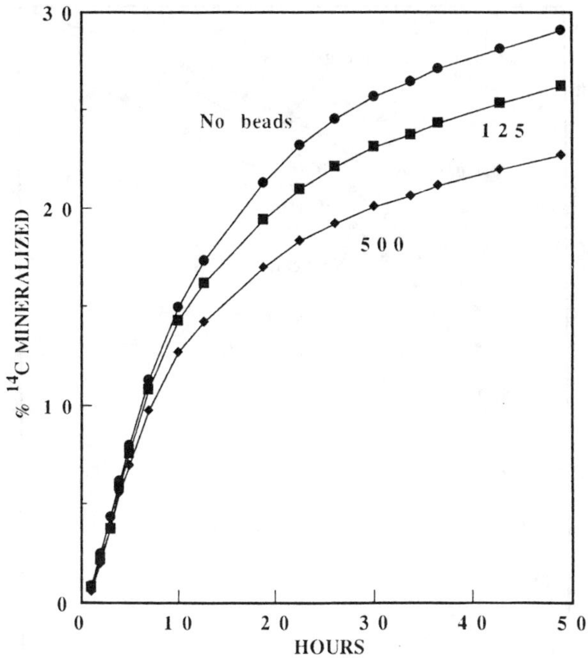

Fig. 6–3. Effect of addition of hydrophobic chromatography beads on the degradation of [^{14}C] *p*-nitrophenol in soil. Numbers refer to the concentration (mg kg^{-1} soil) of beads added.

percentage that had biodegraded of 50 υg *p*-nitrophenol kg^{-1} soil by 50 h (Fig. 6–3); however, the initial biodegradation rate was not affected. In contrast to the SM-7 beads, porous gel exclusion beads have little, if any, specific affinity for organic chemicals and separate molecules on the basis of size. The small diameter beads expand with water and hold a mass of water many times the dry weight of the beads. The amendment of soil with gel-exclusion beads with a molecular cut-off far exceeding the molecular weight of glutamate caused a slight increase in the rate and extent of biodegradation of 10 μg glutamate kg^{-1} soil (Fig. 6–4). One interpretation of the enhancement is that diffusion of glutamate into soil organic matter was retarded by movement into the gel-exclusion beads. Diffusion into and out of the gel-exclusion beads was quite rapid, compared to diffusion in soil organic matter, and thus the beads' presence permitted more degradation of the substrate before it became inaccessible in the soil matrix.

Studies in Defined Systems

Defined experimental systems permit the researcher control over the factors governing a process under study and often enable independent measurements of reactions or properties that are usually strongly coupled and difficult to separate from one another. Despite their possible lack of realism, defined systems sometimes allow the determination of a definitive relationship between a specific factor and an outcome. Scow and Alexander

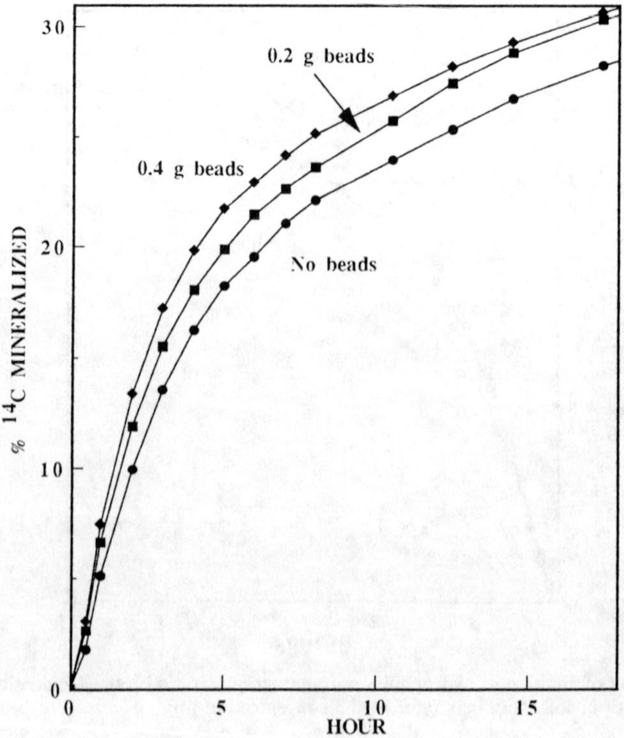

Fig. 6–4. Effect of addition of gel-exclusion chromatography beads on the degradation of [^{14}C] glutamate in soil. Numbers refer to the concentration (mg kg^{-1}) of beads added.

(1992) conducted experiments to determine how mass transfer processes affecting phenol in clay aggregates influenced the kinetics of its mineralization by bacteria. Porous kaolinite aggregates with radii of 0.55 or 0.75 cm were presoaked in solution containing 5 ng of radiolabeled phenol mL^{-1} and then placed in buffer-salts solution containing either 5.0 × 10^{6} or 5.0 × 10^{7} cells mL^{-1} of a *Pseudomonas* sp. capable of degrading phenol. Mineralization of the same concentration of phenol by the same inoculum sizes in solution with no aggregates was also measured and first-order rate constants of 3.15 and 3.84 d^{-1}, respectively, were calculated from the data. The mineralization rate of phenol that originated from the clay aggregates was lower than the rate in the absence of aggregates (Fig. 6–5).

A study was also conducted to measure the kinetics of mineralization of glutamate in an unsaturated matrix of gel-exclusion beads and buffer-salts solution. A buffer-salts solution with 50 ng of glutamate mL^{-1} and inoculated with a pure culture of *Pseudomonas* sp. was supplemented with 0, 0.5, 1.0, and 2.0 g of saturated beads that sequestered 0, 44, 61, and 76%, respectively, of the total solution. The shape of the biodegradation curve changed from first-order to a more sigmoidal shape with the addition of increased amounts of beads (Fig. 6–6). The time to inflection, the apparent rate at the inflection point, and the extent of mineralization were reduced with increas-

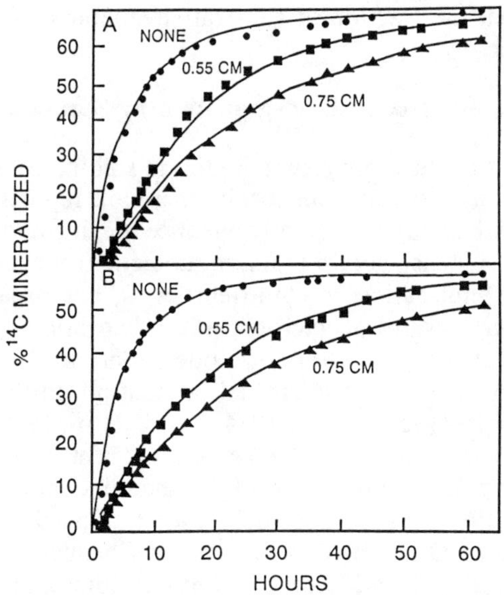

Fig. 6-5. Effect of varying the radius of clay aggregates on the biodegradation by different inoculum densities of *Pseudomonas* sp. of 5 ng [^{14}C] phenol mL^{-1} in solution. Numbers refer to the radii of the aggregates. Panels A and B show data for initial inoculum densities of 5 × 10^7 and 5 × 10^6, respectively. Solid lines show the simulations of the diffusion-sorption-biodegradation (DSB) model. (From Scow and Hutson, 1992.)

ing amounts of beads. The length of the period of deceleration also was extended with increasing amounts of beads. Rate constants obtained from fitting the first-order equation using nonlinear regression techniques were directly proportional to the percentage of total solution available to the bac-

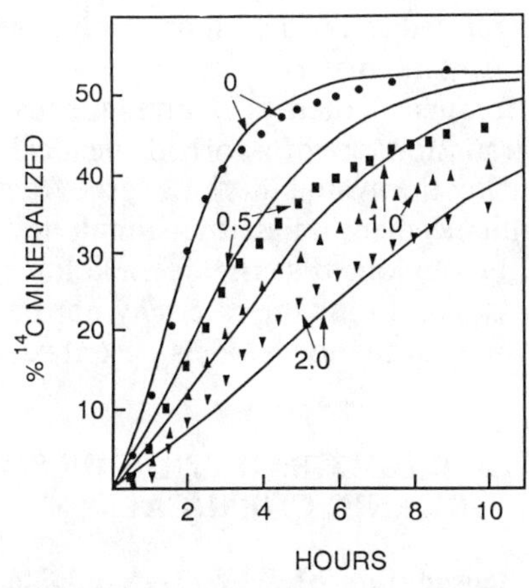

Fig. 6-6. Effect of different amounts of gel exclusion chromatography beads on the biodegradation by *Pseudomonas* sp. of 50 ng [^{14}C] glutamate mL^{-1} in solution. Numbers refer to the mass (*g*) of beads added. Solid lines show the simulations of the diffusion-sorption-biodegradation (DSB) model. (From Scow and Hutson, 1992.)

teria suggesting that the fraction of substrate available controlled the biodegradation rate.

Enhancement of Biodegradation by Sorption

Although most of the more recent studies in soil indicate that sorption impedes degradation of organic chemicals, there are reports that sorption of chemicals to particulates can actually enhance the degradation of chemicals. Most of this evidence comes from aquatic environments where there are low surface/volume ratios, low nutrient levels, low organic C content and low population densities of microorganisms. Sorption of chemicals to organic sediment enhanced their rate of biodegradation. The enhancement was attributed to growth of the microbial population on the C associated with the organic matter (Spain et al., 1984; Lee & Ryan, 1979). Low organic, mineral surfaces may support enhanced metabolism through concentration of organic compounds and nutrients from the surrounding medium (van Loosdrecht et al., 1990). These mechanisms are discussed in greater detail in the section entitled Other Effects of Sorbent-Chemical Interactions on Microbial Populations. The addition of montmorillonite, but not kaolinite or mica-vermiculite, clay to soils has been shown to enhance the decomposition of aldehydes (Kunc & Stotzky, 1974, 1977). Various explanations for the stimulation include effects on soil pH and buffering capacity, increased cation exchange capacity, enhanced induction of enzymes, and formation of a more favorable environment for microorganisms resulting in a stimulation of the total microbial population.

In contrast to what is usually observed for biological reactions, the adsorption of organic chemicals to clay surfaces appears to enhance their degradation rate by abiotic reactions, such as chemical hydrolysis, in soil. This has been observed for parathion (Mingelgrin & Yaron, 1974; Saltzman et al., 1974; Yaron, 1975) and atrazine (Armstrong & Chesters, 1969). The enhancement has been attributed to the action of catalysts in association with the surfaces of clay or organic matter.

It is possible that frequently there is a simultaneous enhancement and reduction in the biodegradation rate of a sorbed chemical. However, the effect of sorption in reducing the availability of a large fraction of the chemical may obscure any enhancement unless the stimulation is substantial. In such cases, it may not be obvious that the degradation rate is determined by two opposing processes and thus the rate may not be predictable using coupled mass-transfer-biodegradation models described below.

KINETICS OF BIODEGRADATION OF SORBED ORGANIC CHEMICALS

Kinetics of Biodegradation of a Freely Available Substrate

The kinetic equations most frequently used to describe microbial processes are the Michaelis-Menten and Monod equations (Simkins & Alexander,

1984; Alexander & Scow, 1989). The equations are similar in form but apply to conditions under which microbial populations remain constant and thus their substrate degradation can be described by enzyme kinetics (Michaelis-Menten) or population increases due to metabolism of substrate with a consequent increase in the degradation rate (Monod). The Michaelis-Menten equation can be derived as a variant of the Monod equation and both equations can be simplified to zero- or first-order kinetics (Simkins & Alexander, 1984). Other equations have been developed to describe microbial growth and have sometimes proven to provide better fits than the Monod equation to growth curves (Koch, 1982, 1985). The basic assumption that unifies all of these equations, however, is that mass transfer processes do not limit access of microorganisms to their substrates (Alexander & Scow, 1989).

Because an important variable in both Monod and Michaelis-Menten kinetics is the concentration of chemical available to the microorganisms that can degrade it, it is logical to assume that a process like sorption that significantly lowers the available concentration must influence the biodegradation rate. In addition, unless sorption simply lowers the solution concentration to some constant level (e.g., it is an equilibrium phenomenon), it seems unlikely that the Monod and Michaelis-Menten equations can describe, other than fortuitously, the microbial metabolism of sorbed chemicals. As will be discussed below, however, first-order kinetics sometimes appear to be adequate in describing these processes.

Kinetics of Biodegradation of Sorbed Chemicals

The interactions of the tightly coupled processes of mass transfer, sorption, and microbial metabolism determine the fate and distribution of a sorbed chemical in soil. These multiple processes are not only difficult to isolate experimentally, but also difficult to reconstruct under conditions when it is possible to independently measure the parameters describing each process. Kinetic models should, at least in principle, be particularly useful in examining a phenomenon as complex as the biodegradation of sorbed chemicals. A mechanistic, simulation model, especially if developed in consort with experimental studies, provides a means with which to predict, for example, the solution concentration of a chemical at a particular point in space and time, over a range of sorption partition coefficients, biodegradation rates, organic matter content, and physical geometry.

In the following section, kinetic models for describing the biodegradation of sorbed chemicals will be discussed. The approaches can be divided into three categories, with each group increasing in its physical complexity and demand for input parameters as more detail about geometry and spatial heterogeneity are included. The three groups include simple modifications to the first-order equation to correct for the concentration of substrate available for degradation, two-compartment models, and models with coupled reactions of diffusion-limited sorption/desorption and biodegradation. Interestingly, there are strong parallels between biodegradation models and models describing the kinetics of sorption-desorption. These models include the one-box model in which sorption is a first-order function of the concentration gradient between the sorbent and solution; the two compartment models that assume two classes of sorbing sites or a sorbent with an easily accessible and slowly exchanging site; and the radial diffusion model for sorption (Wu & Gschwend, 1986).

Simple Models

The first-order equation is the kinetic expression most commonly used to describe biodegradation of organic chemicals and the mineralization of plant residues in soils (Alexander & Scow, 1989). Theoretically, the equation should apply only to conditions where no net growth of the microbial population occurs and the initial substrate concentration is well below the half-saturation constant (Simkins & Alexander, 1984); however, not all sets of data fit by the equation meet these criteria (Hamaker, 1972a). The fact that the equation does fit data describing the degradation of high concentrations (calculated based on the total mass of chemical divided by the total mass of the soil sample) supports the idea that sorption reduces the available concentration to a value so low that the process falls within the first-order region. The first-order equation may be sufficient to describe biodegradation data when the curves do not consist of enough data points to permit a rigorous statistical comparison of different models. Certainly, the first-order equation is practical, and therefore favored, because it can be easily incorporated into more complex transport models. It has been argued that the multiple processes that simultaneously influence the fate of a chemical in the field counterbalance one another and the net result is a first-order disappearance curve. If a kinetic model, however, is to be used as a research tool, the lumping of the effects of multiple processes is not helpful in determining mechanisms of, and unraveling interactions between, processes.

An empirical approach that was popular 20 yr ago in describing the decomposition of pesticides in soil is the power rate law (Hamaker, 1972a) which in its integrated form is

$$C = [C_0^{(1-n)} + (n - 1)\,kt]^{1/1-n}$$

where C is the concentration of chemical remaining, C_0 is the initial concentration, k is the decomposition rate constant and n is the order of the reaction. When n is equal to 1, the equation is equivalent to the first-order equation. Often, though, disappearance curves follow greater than first-order kinetics ($n > 1$) and the half-life (time to reduce the concentration by 50%) increases as the concentration decreases with time. Far ahead of his time, Hamaker (1972a) hypothesized that long-term adsorption processes may be slowly reducing the solution concentration of the chemical thus making it less available for metabolism. The equation serves solely as a curve-fitting tool, does not provide information on mechanisms nor does it employ parameters that can be obtained from sorption measurements. It does, however, give an indication of the degree to which a reaction does deviate from first-order and may be useful for comparing chemicals with different sorption partition coefficients or the same chemical in soils with different organic matter contents.

If a chemical has reached an equilibrium with respect to sorption before significant biodegradation occurs, then the first-order rate constant may be modified to account for the fraction of a chemical that is available for degradation. Steen et al. (1980) proposed the following equation to describe the degradation of sorbed chemicals:

$$k_{app} = k_i/(1 + p \cdot K_{sw})$$

where k_{app} is the apparent rate constant, k_i is the rate constant in the absence of sorption, p is the ratio of solids to solution (M M^{-1}), and K_{sw} is the partition coefficient of the chemical between the soil solids and solution. Based on this relationship, as the solids/solution ratio increases or the sorption partition coefficient increases, the apparent rate constant will decline.

Helweg (1975) developed a relationship between activated C concentration and biodegradation rate constant to describe the degradation of maleic hydrazide in soil:

$$\log k = r \log (1/c_{ads})$$

where k is the degradation rate, c_{ads} is the concentration of sorbent, and r is the slope of the regression between activated C concentration and the degradation rate of the pesticide.

None of the preceding approaches accounts for the kinetics of sorption and desorption or distinguishes between available and unavailable pools of the chemical. The two-compartment model, a simple approach that introduces a modest amount of heterogeneity into a model of soil biodegradation, is frequently successful in fitting curves of biodegradation of sorbed chemicals.

Two-Compartment Models

Compartment models are widely used in pharmacokinetics and also have been applied to soil (Hamaker & Goring, 1976; Scow et al., 1986; Alexander & Scow, 1989). The models assume that a chemical exists in several, frequently two, phases or locations and the mass in one compartment is available for metabolism. In the other compartment, mass is not available or metabolized at a much slower rate. For a sorbed chemical in soil, one compartment could represent the free chemical in solution and available to microorganisms and the second compartment might be chemical that is not readily available because it is sorbed or sequestered in inaccessible micropores. After the supply of substrate in the available compartment is depleted at a relatively rapid rate, the subsequent rate of biodegradation would be governed by the rate of desorption or diffusion of the substrate from the inaccessible micropores to sites containing active microorganisms.

A two-compartment model, consisting of two first-order equations, fits significantly better than does a single first-order model to biodegradation data for numerous organic compounds in soil, many of which are subject to some degree of sorption (Scow et al., 1986; Hamaker & Goring, 1976; Hyzak & Zimdahl, 1974; Zimdahl & Gwynn, 1977). A more detailed discussion of applications of compartment models to soil can be found in Alexander and Scow (1989). Even though formulation of the two-compartment model as two first-order equations permits an analytical solution, the model need not be limited to simple rate expressions. It is possible to incorporate any kinetic expression, for example, the Monod equation to describe microbial growth and metabolism in the available compartment, into a two-compartment model that can be solved numerically (K.M. Scow, 1992, unpublished data).

There are striking parallels between the two-compartment model for biodegradation and the two-site sorption model where there is a rapid initial sorption onto external surfaces and slower, diffusion-controlled sorption inside the organic matrix or aggregates (Cameron & Klute, 1977) or the mobile-immobile model (van Genuchten & Wagenet, 1989). Often, these simple models approximate the predictions of more complex models that are more realistic in their geometry and explicitly include diffusion, both in the case of sorption (Rao et al., 1980) and the biodegradation of sorbed chemicals (Scow & Hutson, 1992) (see the following section).

Models Linking Biodegradation and Mass Transfer Processes

It is difficult to ignore the evidence that the rate-limiting step in sorption and desorption in soil is often a diffusional mass transfer process. Several groups have acknowledged the significance of this phenomenon in biodegradation by developing and testing kinetic models that integrate diffusion-limited processes of sorption and desorption occurring in porous spherical aggregates with a biodegradation reaction occurring in the solution phase (Scow & Hutson, 1992; Mihelcic & Luthy, 1991; Rijnaarts et al., 1991). Figure 6–7 (from Mihelcic and Luthy, 1991) depicts the processes and their spatial locations within the aggregate-solution system. In all three approaches, it is assumed that microorganisms are present only on the outside of aggregates, that an equilibrium relationship describes the distribution of chemical between the intraaggregate solution and solid phase, and that intraaggregate diffusion controls the rate of sorption and desorption.

A model to evaluate the effect of sorption/desorption and diffusion on the biodegradation of organic chemicals was developed by Scow and Hutson (1992) to aid in the interpretation of results obtained from well-defined experimental systems in which substrate was limited by diffusional mass transfer. The process-based deterministic model called the diffusion-sorption-biodegradation (DSB) model assumes diffusion is described by Fick's second law, sorption by a linear isotherm, and biodegradation by any of various rate equations that can be derived from the Monod equation, including first-order kinetics.

The DSB model was fit to the data depicted in Fig. 6–5 showing the biodegradation of phenol by a pure culture of bacterium in the absence and

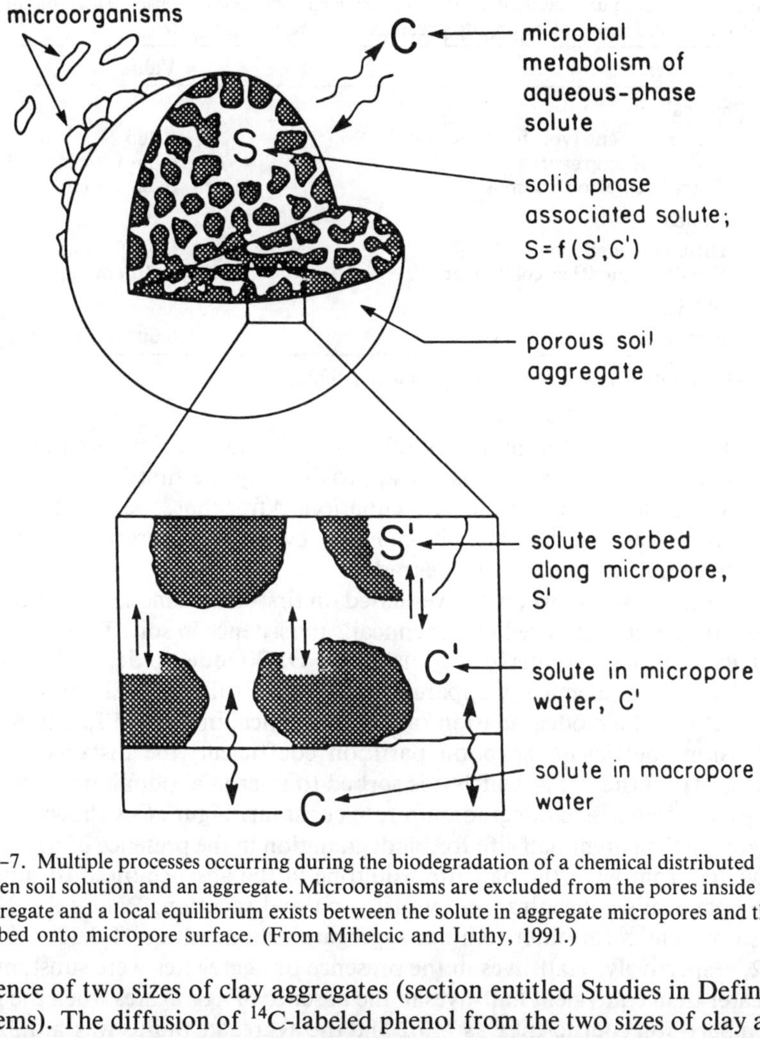

Fig. 6–7. Multiple processes occurring during the biodegradation of a chemical distributed between soil solution and an aggregate. Microorganisms are excluded from the pores inside the aggregate and a local equilibrium exists between the solute in aggregate micropores and that sorbed onto micropore surface. (From Mihelcic and Luthy, 1991.)

presence of two sizes of clay aggregates (section entitled Studies in Defined Systems). The diffusion of ^{14}C-labeled phenol from the two sizes of clay aggregates into sterile buffer was measured and a diffusion coefficient of 0.048 $cm^2\ d^{-1}$ was calculated data. Using the measurements for diffusion, the first-order biodegradation rate constants determined in the absence of aggregates, and measurements of the physical system, the DSB model provided a close fit to the experimental data for biodegradation of phenol in the presence of both sizes of clay aggregates and at both initial microbial population densities. The model simulations are shown as solid lines in Fig. 6–5.

The model was also fit to data measured in a defined system containing a pure culture of bacterium in the presence and absence of polyacrylamide gel-exclusion beads that sequestered part of the solution from bacteria (Fig. 6–6). Using the physical characteristics of the gel-exclusion system as input data (Table 6–1), simulations were performed with the DSB model. A diffusion coefficient of 0.048 $cm^2\ d^{-1}$ and a sorption partition coefficient to 0.3

Table 6–1. Input parameters used in diffusion-sorption-biodegradation (DSB) model simulations of biodegradation in the presence of porous aggregates.†

Parameter	Value
Physical	
Water content (vol. fraction)	0.365
Radius of aggregates	0.05–1.0 cm
Outside solution volume	0.32–3.2 cm^3
Chemical	
Diffusion coefficient	0.01–0.10 $cm^2\ d^{-1}$
Sorption partition coefficient	0–10 $cm^3\ kg^{-1}$
Biological	
First-order rate constant	0.693–7.74 d^{-1}

† Table adapted from Scow and Alexander (1992).

$dm^3\ kg^{-1}$ was used in model simulations. The model simulated biodegradation in 0.5 and 1.0 g beads for approximately the first 3.5 h and in 2.0 g beads for only the first 2 h of incubation. After that, the model deviated from the measured data and, in all cases, eventually overestimated the extent to which glutamata was degraded.

As discussed above, half-lives based on first-order kinetics are frequently used to describe and predict a chemical's persistence in soil. Therefore, sensitivity analyses were performed using the DSB model to determine the dependence of a chemical's apparent half-life on soil physical and chemical properties. The biodegradation rate of a chemical in soil is a function of its diffusion coefficient, sorption partition coefficient, the distance it must diffuse from the site at which it is sorbed to microbial populations that can degrade it, and its biodegradation rate constant. Figure 6–8 shows to what degree the apparent half-life for biodegradation in the presence of aggregates deviates from what the half-life would be in the absence of constraints on the chemical's concentration by diffusion and sorption. Panels A, B, and C show results for outer to inner aggregate volume ratios of 0.32, 1.0, and 3.2, respectively. Half-lives in the presence of aggregates were substantially greater than equivalent half-lives in the absence of aggregates when the sorption partition coefficient was ≥ 3.2 and the aggregate radius was ≥ 0.32 cm. Considering that many organic pollutants sorb strongly and numerous soils have some degree of aggregation and organic matter, taking into account the physical and chemical processes affecting a chemical's concentration may greatly improve kinetic models of biodegradation in soil.

Figure 6–9 shows simulations of the effect of varying the sorption partition coefficient on biodegradation (rate constant = 3.87 d^{-1}) over 120 h in the presence of aggregates of radius 0.05 or 0.25 cm (Scow & Hutson, 1992). Initially, solute diffuses rapidly from the outer solution into the 0.05-cm aggregates because of their high-surface area; whereas diffusion is initially slower into the larger aggregates and thus more degradation can occur in the outer solution. When the outer concentration of substrate is depleted, diffusion of substrate out of the aggregates limits the degradation rate. This is especially evident at higher partition coefficients. The shapes

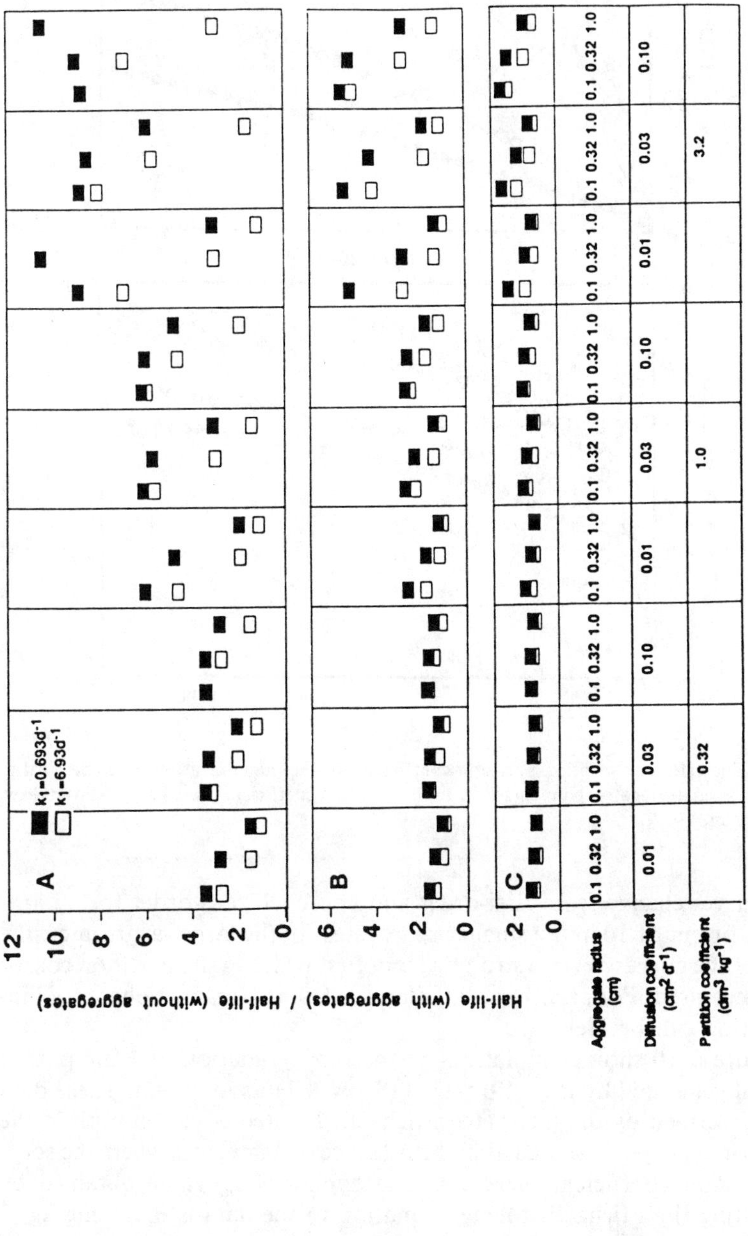

Fig. 6-8. The effect of outer solution volume, sorption partition coefficient, diffusion coefficient in the aggregate, and aggregate radius on the ratio of a chemical's half-life in the presence of aggregates to the half-life in the absence of aggregates. Solid and open squares show results from an intrinsic biodegradation rate constant of 0.693 and 6.93 d^{-1}, respectively. Panels A, B, and C show results for outer to inner aggregate volume ratios of 0.32, 1.0, and 3.2, respectively. (From Scow and Hutson, 1992.)

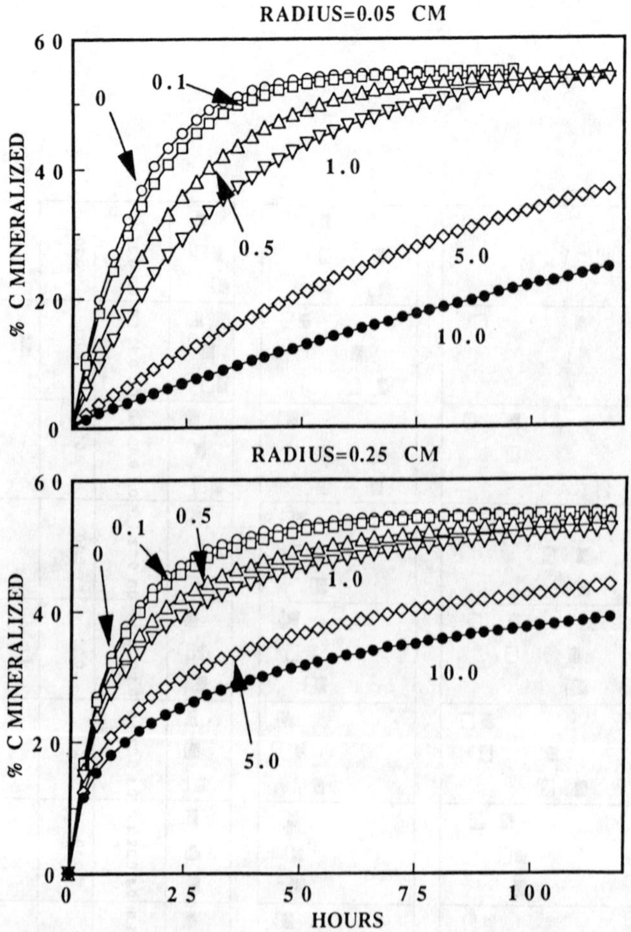

Fig. 6–9. The effect of sorption partition coefficient on the mineralization of a chemical in different sized aggregates. Numbers refer to the partition coefficient (dm^3 kg^{-1}). (From Scow and Hutson, 1992.)

of the curves change from first-order and approach zero-order for a partition coefficient of 10 in the smaller aggregates. In the larger aggregates, the shapes of the curves deviate strongly from first order as the partition coefficient is increased. For partition coefficients ≥5.0, two distinct phases of biodegradation can be identified.

Figure 6–10 shows simulations of the biodegradation of 1 mg phenol kg^{-1} soil generated by the DSB model (Scow & Hutson, 1992). These data can be described by the integrated Michaelis-Menten equation both in the absence of aggregates, as well as in the presence of aggregates where the sorption partition coefficient increases. The apparent K_m value obtained by curve-fitting the Michaelis-Menten equation to the data is 0.378 mg kg^{-1} in the absence of spherical aggregates and, in the presence of aggregates,

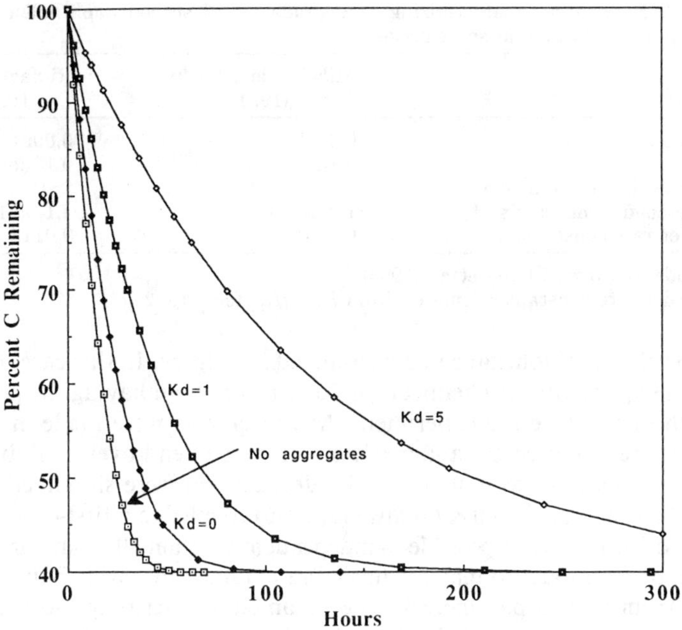

Fig. 6–10. Comparison of the apparent half-saturation constant for biodegradation of phenol in the absence and presence of aggregates and for different sorption partition coefficients.

increases to 0.835, 1.99, and 6.36 μg g^{-1} with an equilibrium distribution coefficient (dm kg^{-1}), K_d, of 0, 1, and 5, respectively.

To describe the biodegradation of naphthalene under denitrification conditions in soil-water suspensions, Mihelcic and Luthy (1991) used a microbial degradation kinetic equation coupled with the solute-soil radial diffusion model of Wu and Gschwend (1986). The model assumed inaccessibility of sorbate to microorganisms, a rapid rate of solute release from the aggregates compared to the diffusion rate inside the aggregates, and Michaelis-Menten kinetics for degradation. The coupled desorption-degradation model was compared to a simple desorption-equilibrium model described by the equation:

$$dC/dt = -(K_{max}C/K_s + C)(1 + \rho K_p)^{-1}$$

where C is the chemical concentration, K_{max} is the maximum rate of the reaction, K_s is the half-saturation constant, ρ is the soil/water ratio, and K_p is a linear sorption partition coefficient. Model calculations used measured data as input parameters and the values of the parameters (from Mihelcic and Luthy, 1991, Exp. 2) are summarized in Table 6–2. The solutions to both models were identical and thus indicated that intraparticle diffusion was rapid compared to the rate of microbial degradation, and sorption could be treated as an equilibrium relationship.

A similar approach was used by Rijnaarts et al. (1991) to evaluate the effect of desorption and intraparticle mass transfer on the biodegradation

Table 6–2. Parameters characterizing biodegradation of sorbed naphthalene and α-hexachlorocyclohexane in soil slurries.†

Parameter	Mihelcic and Luthy (1991)	Rijnaarts et al. (1991)
Aggregate radius, cm	0.0017	0.0091
Porosity	0.13	0.13 (assumed)
Partition coefficient, $cm^3\ cm^{-3}$)	12	251
Diffusion coefficient, $cm^2\ s^{-1}$	1.2×10^{-7}	4.1×10^{-13}
First-order rate constant, h^{-1}	0.049‡	0.0113

† Table adapted from Chung et al. (1993)
‡ First-order rate constant calculated from K_{max}/K_m for Exp. 2.

of α-hexachlorocyclohexane in a well-mixed soil slurry. In this case, the soil had been exposed to the chemical for 20 yr rather than having the pesticide added shortly before the experiment. Measurements were made in a well-mixed "stirred" bioreactor and in a less well-mixed "end-over-end" bioreactor. Experimental measurements of biodegradation were simulated with a radial diffusion model (Wu & Gschwend, 1986) coupled to a first-order degradation model. It was not possible to independently obtain all input parameters required by the model so many values were obtained by curve-fitting or calculated from known parameters. The parameters describing the end-over-end bioreactor are presented in Table 6–2. By setting the intraparticle diffusion distance as an independent model parameter rather than assuming the volumetric-mean aggregate size used in the diffusion estimates, the radial diffusion-biodegradation model was found to provide a good fit to biodegradation data in both systems. However, a first-order equation fit the biodegradation data equally well. This result was surprising because desorption kinetics in the less well-mixed, end-over-end system were described by the radial diffusion model suggesting that the microbial population was not dependent on the apparent solution concentration. The authors suggested that penetration of bacteria into the interiors of the soil aggregates was partly responsible for the deviation of biodegradation kinetics from the radial-diffusion model.

The Thiele modulus is a dimensionless group used in chemical engineering to estimate whether a reaction occurring *inside* of aggregates is controlled by the rate of diffusion of the substrate to the reaction site. Myrold and Tiedje (1985) used the Thiele modulus to ascertain specific conditions (e.g., aggregate radius and reaction rate) under which denitrification would be important in soil. Based on a modified version of the model of Scow and Hutson (1992), Chung et al. (1993) derived a dimensionless group analagous to the Thiele modulus differing in that it described a reaction occurring external to rather than inside the aggregates. The dimensionless group, ϕ, serves as a criterion for determining when mass transfer processes can be ignored in biodegradation studies:

$$\phi = R \sqrt{\{\epsilon + K_d\,(1 - \epsilon)\}\ k/D_e}$$

where R is the radius of aggregate (cm), ϵ = aggregate porosity, K_d is the sorption partition coefficient (cm^3 g solid^{-3}), k is the first-order rate constant for biodegradation (h^{-1}), and D_e is the effective diffusion coefficient for the chemical in the pores of the aggregates ($cm^2\ m^{-1}$). Intraparticle diffusion is not significant for values of $\phi < 1$.

Using data reported in the studies of Mihelcic and Luthy (1991) and Rijnaarts et al. (1991), ϕ was calculated to test whether the magnitude of the calculated values were in agreement with the different conclusions of the two studies. Table 6–2 summarizes the input parameters obtained from the two studies used in the calculation of ϕ. The first-order rate constant for Exp. 2 of the Mihelcic and Luthy (1991) study was derived from the Michaelis-Menten parameters K_{max} and K_m. The parameter values chosen from the Rijnaarts et al. (1991) study were those describing the end-over-end reaction system. Some of these parameters were estimated from other parameters, and thus do not represent independent measurements. For the Mihelcic and Luthy (1991) study, in which biodegradation of naphthalene was rapid relative to the rate of desorption of the chemical, ϕ was calculated to be 3.5. On the other hand, in the Rijnaarts et al. (1991) study in which biodegradation of α-hexachlorocyclohexane was indeed rate-limited by desorption of the pesticide, a high value of 22×10^4 was calculated for ϕ. Even if the first-order biodegradation rate constant, which is not explicitly known in the Rijnaarts et al. (1991) study, is assumed to be an order of magnitude greater or smaller than the value used in these calculations, the value of ϕ is still very high, ranging from 6.6×10^4 and 6.6×10^3, respectively. These results illustrate the potential utility of the criterion ϕ in assessing the conditions under which the biodegradation of a sorbed chemical is likely to be rate-limited by mass transfer from soil to microorganisms.

It may appear that the biodegradation rate exceeds the desorption rate if microorganisms at the surface of the sorbent immediately use the chemical as it becomes available and before it diffuses into the bulk solution where it can be analyzed using conventional methods. Rijnaarts et al. (1991) found that in their soil slurries the effective diffusion pathlength estimates from curve-fitting was significantly shorter than the measured value (the volumetric mean aggregate size) leading the authors to suggest that bacteria had penetrated into the soil aggregates and were actively degrading α-hexachlorocyclohexane.

To test if the two-compartment model fits biodegradation data obtained under known mass-transfer limitations, Scow and Hutson (1992) compared fits of the first-order and two-compartment models to data simulating the biodegradation of a chemical limited in its availability by diffusion and desorption. The two-compartment model provided a better fit than did a simple first-order model to simulated data showing the biodegradation of a chemical with a sorption partition coefficient of 10 $dm^3\ kg^{-1}$ in the presence of large (0.25-cm radii) aggregates. There was no difference between the two models when fit to simulations of biodegradation of the same chemical in the presence of aggregates of radii of 0.05 cm. Biodegradation was overestimated when the chemical's long-term persistence in 0.25-cm aggregates was extrapolated from a half-life obtained from fitting the first-order model to a curve showing the degradation of 50% of the initial mass of the chemical

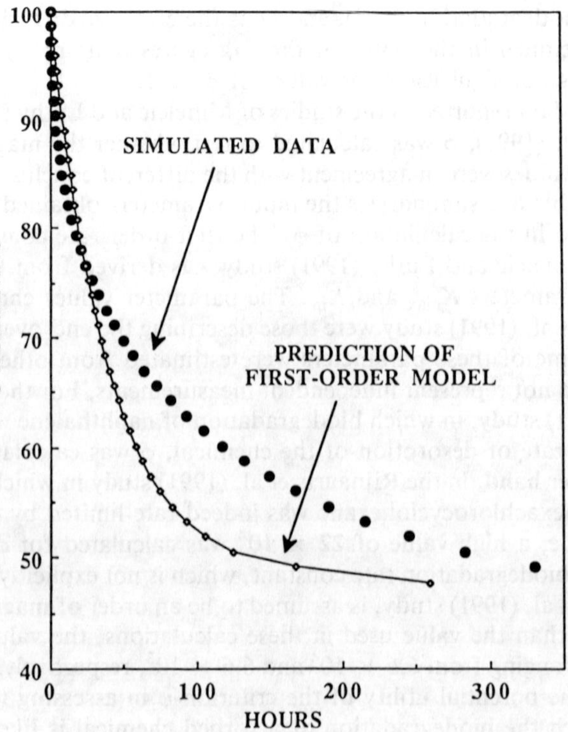

Fig. 6–11. Comparison of simulated data and prediction by first-order model of the biodegradation of a chemical in the presence of aggregates of 0.25-cm radius over a duration of four half-lives. (From Scow and Hutson, 1992.)

(Fig. 6–11). This finding is consistent with Hamaker's (1972a) rule-of-thumb that the rate constant from which extrapolations of a chemical's persistence are made should be obtained from data showing the biodegradation of the chemical over a minimum of 1.5 half-lives.

Zero-order, or linear, kinetics are generally assumed to occur at high substrate concentrations and population densities when the reaction is saturated with respect to substrate and population, and there is no change in the biodegradation rate (Simkins & Alexander, 1984). However, such conditions are uncommon in nature where low nutrient conditions are prevalent. Interestingly, linear kinetics may occur in nature under conditions where biodegradation is severely limited by diffusion and desorption of a substrate to locations where it can be degraded. Although biodegradation in these cases will not be strictly linear over the entire period of activity, it may appear to be linear if the rate declines slowly during the incubation period, as can be observed in the biodegradation curves for K_d values of 5 and 10 shown in Fig. 6–9. The appearance of linearity may result from a high sorption partition coefficient, a small diffusion coefficient, or a long diffusion path. In Helweg's (1975) study of the effect of adding differing amounts of activated C to soil on the biodegradation of maleic hydrazide, zero-order kinetics described the degradation of the chemical sorbed onto the largest amount, 5000 mg kg^{-1}, of activated C added. It is possible that adsorption or diffusion may contribute to linear kinetics sometimes reported for the biodegradation of low concentrations of substrates in the environment, as in Subba-Rao and Alexander (1982) or Spain et al. (1984).

OTHER EFFECTS OF SORBENT-CHEMICAL INTERACTIONS ON MICROBIAL POPULATIONS

It almost seems naive to assume that the only impact a sorbent such as organic matter has on a microbial population in soil is to lower the solution concentration of its substrate. Although the sorbent-sorbate interaction may be the most important determinant of degradation rates, it is important to recognize other impacts that organic matter may have on microbial processes.

Alternate effects of sorbent-chemical interactions may have a significant impact on the kinetic models described in the section entitled Kinetics of Biodegradation of Sorbed Organic Chemicals. Even if it is possible to realistically model the rate of mass transfer of a sorbed chemical to the microbial populations able to degrade it, the assumption remains that the microbial population behaves the same in the presence and absence of soil. The assumption of equivalence, however, between populations in liquid and porous media may not always be reasonable, as will be discussed in the following section.

Relationship between Organic Matter Content and Microbial Population Density

A positive correlation between organic matter content and microbial biomass density can be found in soils that have been subject to similar management practices (Anderson & Domsch, 1989; Schnürer et al., 1985). Because the metabolism rate is a function of the population density of the microorganisms carrying out the reaction (Simkins & Alexander, 1984), the potential biodegradation rate of a specific chemical may be greater in soils with a higher organic matter content than in, for example, sandy, low organic matter soils. Of course, the size of the specific population responsible for the degradation of the chemical of interest may not necessarily be correlated with the total microbial biomass. However, Anderson (1981) found that the microbial biomass in a soil was directly related to the degradation rate of diallate. Dubey et al. (1966) found a positive correlation between the apparent disappearance of linuron and diphenamid and the organic matter content of several soils. Correlations between organic matter content and rates of apparent disappearance have also been reported for numerous other compounds (Hamaker, 1972a). The interpretation of results from some of the early studies should be viewed with caution because plant bioassays were frequently used to determine the concentration of pesticides remaining. No control experiments were performed to ensure that the "loss" of the pesticide from higher organic soils was due to microbial activity and not to a lower availability of the pesticide to the bioassay plant.

The initial microbial population density may be important only in the initial phase of degradation. A *Flavobacterium* sp. was inoculated at four different population densities into a sandy soil contaminated with pentachlorophenol (Crawford & Mohn, 1985). During the first 2 d of incubation, the biodegradation rate of pentachlorophenol was directly proportional to population density after which the rates leveled off and there was no difference among treatments. The final percentage of pentachlorophenol degraded was also proportional to the initial population density in most cases. These results suggest that population density may influence the biodegradation rate of a chemical initially present in the soil solution. Once the chemical has reached an equilibrium with respect to sorption, and its degradation rate is controlled by desorption, population density may no longer be an important factor. A study of the biodegradation of dichloroaniline that had formed irreversibly bound residues with soil organic matter contradicted this conclusion. In this study, the addition of aniline to soil stimulated the degradation of the dichloraniline and the stimulation was attributed to an increase in the population size, or enzyme activity, of microorganisms able to degrade dichloroaniline (You & Bartha, 1982).

Sorption of Microorganisms and Extracellular Enzymes

Microorganisms themselves can be sorbed and, in fact, the majority of microorganisms in natural ecosystems are found in association with organic, clay, or other mineral surfaces rather than free in the solution phase. Several reviews have considered the interactions of microorganisms and surfaces (Stotzky, 1967, 1972; Filip, 1973; Hattori, 1973; Marshall, 1971, 1976). Once microorganisms have attached to surfaces, their physiology may differ from those cells suspended in solution, due to either direct or indirect effects of the surface. Stotzky (1986) concluded that surfaces indirectly influence microorganisms by "modifying the physicochemical characteristics of microbial habitats, and this either enhances or attenuates the growth and metabolic activities of individual microbial populations." McLaren (1978), on the other hand, concluded that the effect of surfaces is more direct, resulting in changes in the physiology of the cell. Sorption of microorganisms to surfaces may aid in their use of substrates bound to surfaces (Marshall, 1976; Hermansson & Marshall, 1985). Under optimum nutrient conditions, *Vibrio proteolytica* was found to be less active, based on ^{3}H-thymidine incorporation, *p*-iodonitrotetrazolium (INT) reduction, and the ratio of ATP/DNA, when attached to polystyrene than when free in solution (Jeffrey & Paul, 1986). Under low nutrient conditions, however, the activity of attached cells exceeded that of the unattached cells. The authors suggested that in the presence of ample nutrients, the lower exposed surface area of attached cells resulted in rates of absorption or transport of nutrients that were slower than for unattached cells. Extracellular enzymes, often involved in the breaking up of polymers and complex organic molecules, are themselves subject to sorption and deactivation by soil colloids (Skujins, 1967).

Most studies that have evaluated the effect of surfaces on microbial activity have been conducted in aquatic systems. Comparable studies are difficult, if even possible, to conduct in soil. In soil, Orgram et al. (1985) found that the model that best described the trends in first-order rate constants of biodegradation of 2,4-D in soil slurries assumed that sorbed and solution-phase bacteria degraded 2,4-D in solution at equal rates. Alternate models that were rejected assumed that only microbial populations that were sorbed or in the soil solution could metabolize sorbed chemicals, or that the sorbed chemical could be metabolized. Van Loosdrecht et al. (1990) reviewed the literature demonstrating the direct effects of surfaces on microorganisms and concluded that the reported evidence of direct effects could just as well be explained by indirect effects.

Effects of Available Carbon Associated with Organic Matter on Microbial Populations

It is well known that the kinetics of biodegradation of a particular chemical can be influenced by the presence of other compounds (Schmidt et al., 1985, 1987; Scow et al., 1989). The effects of secondary compounds are varied and may result in either an increase or reduction in the lag period and degradation rate of the compound of interest. Microbial populations in soil rely on organic matter as a source of C, other nutrients, and growth factors. Thus, it is reasonable to question whether an increase in the organic matter content of a soil, while it may decrease the solution concentration of a sorbed chemical, may also stimulate the degradation rate of the chemical in solution by providing additional sources of C and energy. Hance (1974) found that decomposition rates of atrazine and linuron [3-(3,4-dichlorophenyl)-1-methoxy-1-methylurea] were substantially lower in a reinoculated soil that had been subject to peroxidation and removal of organic C than in untreated soil. He suggested that C and N from the organic matter was required to support degradation of the pesticides. The drastic measures used, however, to remove C from the soil may have resulted in other adverse effects on the reintroduced organisms. McCormick and Hiltbold (1966) found that degradation of atrazine and diuron was directly linked to metabolism of soil organic C. To measure the effect of indigenous C associated with sand on biodegradation of an added substrate, experiments were conducted with a pure culture of bacteria in: (i) solution culture, (ii) moist sand that had been combusted to remove all C, and (iii) moist sand that had not been combusted (Scow, 1989). The rate and extent of biodegradation of a low concentration of phenol was lower in the combusted sand than in solution, whereas it was higher in the uncombusted sand because of growth of the bacterium on the contaminating C (Fig. 6–12).

Fulvic and humic acids, which are major components of soil organic matter, have been shown to have stimulatory effects on microbial metabolism (McLoughlin & Küster, 1972; Visser, 1985). Fulvic acid appeared to be in-

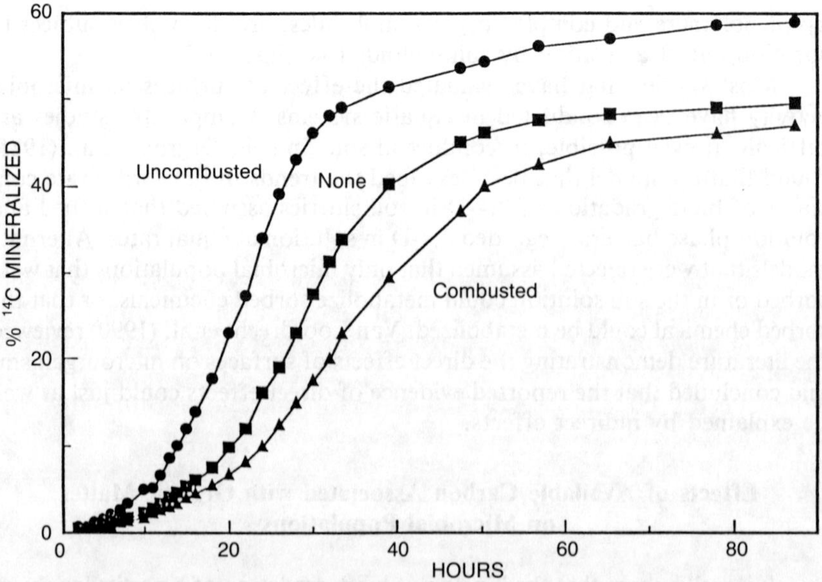

Fig. 6-12. Effect of combusted or uncombusted sand on the biodegradation of 50 ng of phenol mL^{-1} by *Pseudomonas* sp.

volved in the induction of the enzyme catechol-1,2-oxygenase, thus stimulating the degradation of benzoate by an *Arthrobacter* sp. (De Haan, 1974). Other microbial species including pseudomonads (DeHaan, 1974; Paul & Mathur, 1967), fungi (Mathur & Paul, 1966, 1967), and a variety of other bacterial species (Taha et al., 1973) have been found to either cometabolize or metabolize as sole C and energy sources, humic and fulvic acids.

The way in which indigenous or other sources of C affect the kinetics of biodegradation of a pollutant will depend on whether the pollutant is used simultaneously with the secondary C source or if it is metabolized independently. An important consideration is the similarity in chemical structure between the added chemical and indigenous C and whether both substances are degraded by a related, or the same, metabolic pathway. For example, addition of glucose or glutamate had no effect on the rates of mineralization of *p*-nitrophenol in soil, whereas addition of phenol increased the lag period before mineralization of *p*-nitrophenol began (Scow et al., 1989). Tranvik and Höfle (1987) found that lake populations of bacteria required the presence of humic substances to degrade a low concentration of phenol. The observation that the microbial populations grew to a high density, presumably on available C in the humic material, before starting to metabolize phenol, was also noted.

Mixing a soil in a slurry causes a sudden release of organic C that was previously protected from degradation and also releases smaller, more degradable molecules from the organic matrix. The presence of freshwater or estuarine sediments has been shown to enhance the degradation of many organic chemicals, of which some are substantially sorbed to organic matter (Herbes

& Schwall, 1981; Walker et al., 1984; Jones et al., 1982; Lee & Ryan, 1979). While it may be argued that differences between sediments and soil in slurries are substantial, the possible effect of available C associated with organic matter should at least be considered in experiments conducted with soil slurries.

Threshold Concentrations for Growth or Metabolism

Another assumption of the Monod model and its variants is that the rate constants for biodegradation are truly constants and independent of substrate concentration. However, threshold concentrations for growth on, and metabolism of, organic chemicals by microorganisms have been reported in aquatic environments (Alexander, 1985; Boethling & Alexander, 1979). Schmidt et al. (1985) proposed an equation for predicting the threshold concentration for growth of a bacterial population in solution culture that includes in its denominator the coefficient for diffusion of the chemical in the surrounding medium:

$$\tau = [1/Y_{max}\ (R_d^2 - R_b^2)/2]/\{(D_{AB}\ C_B)/[\rho - (m/\ln 2)(R_d^2 - R_b^2)]/2\}$$

where

τ = cell doubling time,
Y_{max} = yield coefficient,
R_d = radius of cell at division,
R_b = radius of cell at first appearance,
D_{AB} = diffusivity of chemical in solution,
C_B = concentration of substrate in solution,
ρ = dry weight density of cell, and
m = maintenance coefficient.

Using a diffusion coefficient of 1×10^{-5} cm^2 s^{-1} and values from the literature for the biological parameters, the authors predicted that a threshold for growth would occur at a chemical concentration of 0.20 ng mL^{-1}. As long as other organic compounds present in soil do not support growth of the microbial population, the threshold concentration for growth would be at least two or more orders of magnitude higher in soil than that calculated above. This is because the effective diffusion coefficient, which would replace D_{AB} in the equation, is roughly two to three orders of magnitude lower for organic chemicals in soil than in bulk solution (Hamaker, 1972b).

In certain bacterial species, it has been shown that there are threshold concentrations for specific organic chemicals below which the organism is unable to biodegrade the chemical (Schmidt et al., 1986). Threshold concentrations have been observed for bacteria degrading phenylamide herbicides (Lechner & Straube, 1984) and chlorobenzoates (Reber, 1982). A threshold concentration of 10 ng *p*-nitrophenol mL^{-1} has been demonstrated in a strain of *Pseudomonas* in liquid culture (Schmidt et al., 1987). Sorption to SM-7 chromatography beads reduced the concentration in solution of *p*-nitrophenol to below its threshold concentration for degradation for the same strain of *Pseudomonas* (Araujo and K.M. Scow, 1989, unpublished data) resulting in no biodegradation. In the absence of sorbent, the same mass of chemical was rapidly degraded. A coupled mass transfer-biodegradation model would not predict an absence of degradation of *p*-nitrophenol because the model assumes that some biodegradation activity occurs at all substrate concentrations and mass transfer processes limit only the rate, not the induction of, enzymes responsible for the particular reaction.

The biodegradation of sorbed chemicals raises many interesting questions regarding how those microorganisms that degrade these chemicals can survive and stay active when all they have access to are the low solution concentrations associated with strongly sorbed chemicals. These organisms may be efficient at scavenging dissolved organic C associated with organic matter or may be capable of metabolic dormancy in the absence of substrate. Research on the physiology of microorganisms that metabolize sorbed chemicals could have immediate benefits in the development of bioremediation techniques that are based on biological, rather than engineering, principles.

Sorption of Metabolites or Nutrients

As a rule, the products of microbial metabolism tend to be more polar and thus less subject to hydrophobic sorption than parent compounds from which they originate. This is not always the case, however. Brouwer et al. (1990) found that hydroxyatrazine, a transformation product of atrazine, had a distribution coefficient between organic matter and solution two to six times greater than that of atrazine. Also, not all microbial reactions result in the degradation of a compound. Alexander (1981) summarizes and gives examples of several conjugative and other reactions involving environmental pollutants. Some reactions, such as dimerization, result in the formation of bound residues with soil organic matter and the biodegradation rate of the bound chemical is substantially decreased from its rate in solution.

Sorption is not exclusively limited to the particular organic chemical undergoing biodegradation, but can also affect the availability of other substances required by the organism such as inorganic nutrients. The concentration of nutrients at surfaces from a dilute solution can enhance the growth of organisms at the solid-liquid interface (Marshall, 1976). This phenomenon appears to be important in aquatic systems (van Loosdrecht et al., 1990), but is either not as obvious or confounded by other processes in soils. Again, it is not surprising that soil and aquatic systems demonstrate different behaviors considering that soils have significantly higher surface/volume ratios, organic contents, and solids/solution ratios than do aquatic environments.

The presence of clay in soil has been shown to increase the retention of metabolites resulting from the decomposition of organic residues and increase the stability of soil organic matter (van Veen et al., 1985; Jenkinson, 1977). Sörensen (1972) found that the effect of adding clay to a sandy soil resulted in a two- to threefold accumulation of amino acid metabolites from microbial populations degrading carbohydrates. In different soils, the accumulation of amino acid metabolites from ^{14}C-cellulose was proportional to the amount of clay present (Sörenson, 1975). A greater percentage of ^{14}C from added glucose was incorporated into microbial biomass in a clay soil than in a sandy soil (van Veen et al., 1985). Three hypotheses proposed to explain the higher incorporation in clay were: (i) biomass C was more protected from decomposition, (ii) a closer interaction between microbial populations and products of their decay, and (iii) C was used more efficiently for biosynthesis with less lost through respiration. Martin et al. (1976) found the presence of clay to increase the efficiency of incorporation of C from organic compounds in actinomycetes. If $^{14}CO_2$-evolution is used as the measure of biodegradation of an organic chemical, then the presence of different amounts of clay may influence the results not just because of its ability to sorb the chemical but also through its influence on the fraction of the total C in the molecule that will be evolved as CO_2.

Variations in Sorbent Characteristics and Multiple Sorption Reactions

Sorption of nonpolar organic compounds is proportional to the organic matter content of a soil and assumed to be independent of the origin of the organic matter (Chiou, 1989). Use of the octanol-water partition coefficient for a chemical, normalized for soil organic C, to estimate the sorption of the chemical in a soil (Karickhoff, 1981) implicitly assumes that all organic matter behaves the same with regard to sorption.

In actuality, organic matter is highly variable in its structure and chemical properties depending on its age, source, management practices to which it has been subjected, and local climatic conditions (Whitehead et al., 1975). Nuclear magnetic spectroscopy has shown that humic acids from tropical soils are highly aliphatic with a large portion of C in carbohydrates and polysaccharides, whereas temperate soils contained more carboxyl and aromatic C (Lobartini & Tan, 1988). The way in which a humic substance is associated with a mineral surface may affect the sorptivity of the complex (Murphy et al., 1990; Pignatello, 1990a,b). As organic matter ages, there is a decrease in hydrophilic groups, an increase in C, a decrease in H and O, and an increase in molecular weight, polymerization, aromatization, and polycondensation. Sorptivity tends to decrease with an increase in oxygen-containing functional groups (Grathwohl, 1990). Garbarini and Lion (1986) found that the ratio of C to O strongly influenced the partition coefficient for toluene and trichloroethylene in organic matter of different origins.

Organic matter characteristics also influence the size and activity of microbial populations in soil. New organic matter supports a higher and more active biomass than does more aged organic matter (Doran & Smith, 1987). A high correlation was found between the aliphatic fraction of organic matter and microbial biomass in temperate, agricultural soils that had been subjected to different management systems (Capriel et al., 1990).

METHODOLOGICAL PROBLEMS IN COMPARISON OF SORPTION-DESORPTION AND BIODEGRADATION EXPERIMENTS

Measurements of sorption partition coefficients and other parameters can be dramatically influenced by the presence of microbial populations (Scott et al., 1982). Ideally, measurements of the sorption and desorption of biodegradable chemicals should be conducted in sterile soil so that removal of the chemical from solution will be due only to sorption and not biodegradation. The ideal method for sterilizing soil is one that deactivates the microbial population without lysing the cells (Skujins, 1967). Unfortunately, no single method meets these criteria and each one has impacts in addition to inhibiting microbial activity. Identifying the alternative impacts is crucial before assuming that measurements (e.g., desorption kinetics) in sterile soil are representative of the same process in nonsterile soil.

Gamma-irradiation is commonly believed to have the least impact on soil properties; however, increases in dissolved organic C, extractable Mn, Ca, Mg, B, Cu, NH_4, and conductivity have been observed (Skipper & Westermann, 1973; Eno & Popenoe, 1964). The use of toluene as a sterilant requires incorporation of the chemical at levels far exceeding its solubility and thus introducing an additional organic phase. Ethylene oxide increases soluble organic matter, available P, and Cu and Mn (Skipper & Westermann, 1973). Propylene oxide has been shown to reduce the surface area of soils, increase the pH, increase extractable Mn, and may leave behind residual epoxides (Wolf et al., 1989; Skipper & Westermann, 1973). Methyl bromide may increase extractable N, P, and S (Eno & Popenoe, 1964). Sodium azide, a general enzyme inhibitor, cause changes in soil pH (Skipper & Westermann, 1973; Wolf et al., 1989). Mercuric chloride increases extractable Mn (Wolf et al., 1989). Autoclaving, generally believed to be the most destructive of soil sterilization methods, disrupts soil structure, increases the available C concentration, increases the concentration of extractable N, P, and K, and changes the oxidation state of certain minerals (Eno & Popenoe, 1964; Salonius et al., 1967). Because no single method of sterilization is ideal, it may be desireable to compare results from soils sterilized in more than one way, conduct additional experiments to test whether the sterilization method influenced the observed result, and, at the very least, be cognizant of possible contributions of the sterilization method to any conclusions that are drawn from an experiment.

BENEFICIAL EFFECTS OF SORPTION ON THE BIODEGRADATION OF ENVIRONMENTAL POLLUTANTS

Under certain conditions, sorption may actually benefit bioremediation efforts in soil by reducing the solution concentration of a pollutant from a level toxic to microorganisms to a level at which it can be biodegraded. Adding bark chips to solution containing pentachlorophenol lowered the solution concentration enough to substantially reduce its toxicity (Apajalahti & Salkinoja-Salonen, 1984). The presence of chips significantly reduced the lag phase and decreased the half-life from 10 to 3 mo (Apajalahti & Salkinoja-Salonen, 1984). Carbon associated with the wood chips did not serve as a substrate for the microbial populations and contact between the organisms and the wood chips was not necessary for degradation. An important property of wood chips as a sorbent is that sorption is reversible and the desorption rate is more rapid compared to the potential biodegradation rate of the chemical in solution. Enhancement of biodegradation rate due to a reduction in toxicity to microbial degraders also has been observed for high concentrations of phenol in the presence of activated C (Erhardt & Rehm, 1985; Mörsen & Rehm, 1987).

CONCLUSION

One simply has to note the multitude of definitions of sorption, the numerous types of chemical-surface interactions that exist, and the lack of agreement on how to describe the kinetics of the sorption-desorption process to realize that investigation of the biodegradation of sorbed chemicals only adds another layer of complexity onto a topic that is already complicated and riddled with uncertainties. Even if soil chemists clarify the structure of organic matter, elucidate the mechanism(s) of hydrophobic sorption, and identify the parameters controlling sorption and desorption of organic chemicals in soil, there may still not be sufficient information to ascertain what controls the biodegradation rates of sorbed chemicals. Many unknowns remain concerning the spatial location of microorganisms—particularly those that can degrade the chemical of interest—the physiology and metabolic activity of organisms that must survive on the low solution concentrations of sorbed chemicals, and the significant indirect and direct effects of organic matter on microbial populations.

Many investigations of biodegradation of sorbed chemicals in soils are descriptive, and from these results it is difficult to isolate individual factors out of the large number of factors that may be influencing biodegradation. More controlled studies are generally confined to simplified and artificial systems and how applicable these findings are to soil is difficult to demonstrate. Although previous work has been crucial in the development of our understanding of the problem, further studies of the interaction between sorption and biodegradation must be pursued in more realistic soil matrics under conditions likely to be encountered in the environment. In addition, more quantitative approaches are needed to understand the interactions and feedback between the simultaneously occurring processes that together determine the behavior of a sorbed chemical.

The development of predictive models will advance only if parameter values can be independently measured as opposed to being obtained from calibration of models with measured data (Brusseau & Rao, 1989b). Unfortunately, methods for obtaining critical parameters in soil are often poor or nonexistent and this especially holds true for parameters describing microbial processes. There is a strong need for the development of sensitive and nondestructive measures of microbial populations, activity, and quantification of microscale solution concentrations of chemicals in soil. Finally, experiments with different combinations of sorbents, sorbates, and microorganisms are needed to determine what factors are important in controlling degradation rates of sorbed chemicals and what mechanisms are involved in their biodegradation.

ACKNOWLEDGMENT

This work was supported by the National Institute of Environmental Health Science's Superfund Basic Research program P42 ESO 4699; the U.S. Department of Energy, Office of Environmental Restoration and Waste Management Young Faculty Awards Program administered by Oak Ridge Associated Universities for DOE; the U.S. Department of Interior, Geological Survey, through the State Water Resources Research Institute of California, Water Resources Project UCAL-WRC-W-759; and a Kearney Foundation of Soil Science Project, No. 3958-RR.

REFERENCES

Alexander, M. 1981. Biodegradation of chemicals of environmental concern. Science 211:132–138.

Alexander, M. 1985. Biodegradation of organic chemicals. Environ. Sci. Technol. 19:106–111.

Alexander, M., and K.M. Scow. 1989. Kinetics of biodegradation in soil. p. 243–269. *In* B.L. Sawhney and K. Brown (ed.) Reactions and movement of organic chemicals in soil. SSSA Spec. Publ. 22. SSSA and ASA, Madison, WI.

Alison, F.E. 1968. Soil aggregation—Some facts and fallacies as seen by a microbiologist. Soil Sci. 106:136–143.

Amador, J., and M. Alexander. 1988. Effect of humic acids on the mineralization of low concentrations of organic compounds. Soil Biol. Biochem. 20:185–191.

Anderson, J.P.E. 1981. Soil moisture and the rates of biodegradation of diallate and triallate. Soil Biol. Biochem. 13:155–161.

Anderson, T.-H., and K.H. Domsch. 1989. Ratios of microbial biomass carbon to total organic carbon in arable soils. Soil Biol. Biochem. 21:471–479.

Apajalahti, J.H.A., and M.S. Salkinoja-Salonen. 1984. Absorption of pentachlorophenol (PCP) by bark chips and its role in microbial PCP degradation. Microb. Ecol. 10:359–367.

Armstrong, D.E., and G. Chesters. 1969. Adsorption catalyzed chemical hydrolysis of atrazine. Environ. Sci. Technol. 2:683–689.

Balesdent, J., G.H. Wagner, and A. Mariott. 1988. Soil organic matter turnover in longterm field experiments as revealed by carbon-13 natural abundance. Soil Sci. Soc. Am. J. 52:118–124.

Ball, W.P., M.N. Goltz, and P.V. Roberts. 1991. Comment on "Modeling the transport of solutes influenced by multiprocess nonequilibrium" by M.L. Brusseau, R.E. Jessup, and P.S.C. Rao. Water Resour. Res. 27:653–656.

Bettmann, H., and H.J. Rehm. 1984. Degradation of phenol by polymer entrapped microorganisms. Appl. Microbiol. Biotechnol. 20:285–290.

Boethling, R.S., and M. Alexander. 1979. Effect of concentration of organic chemicals on their biodegradation by natural microbial communities. Appl. Environ. Microbiol. 37:1211–1216.

Brouwer, W.W.M., J.J.T.I. Boesten, and W.G. Siegers. 1990. Adsorption of transformation products of atrazine by soil. Weed Res. 30:123–128.

Brusseau, M.L., and P.S.C. Rao. 1989a. Sorption nonideality during organic contaminant transport in porous media. Crit. Rev. Environ. Control 19:33–99.

Brusseau, M.L., and P.S.C. Rao. 1989b. The influence of sorbate-organic matter interactions on sorption nonequilibrium. Chemosphere 18:1691–1706.

Brusseau, M.L., R.E. Jessup, and P.S.C. Rao. 1991. Nonequilibrium sorption of organic chemicals: Elucidation of rate-limiting processes. Environ. Sci. Technol. 25:134–142.

Calvet, R. 1989. Adsorption of organic chemicals in soils. Environ. Health Perspect. 83:145–177.

Cameron, D.A., and A. Klute. 1977. Convective-dispersive solute transport with a combined equilibrium and kinetic adsorption model. Water Resour. Res. 13:183–188.

Capriel, P., T. Beck, H. Borchert, and P. Härter. 1990. Relationship between soil aliphatic fraction extracted with supercritical hexane, soil microbial biomass, and soil aggregate stability. Soil Sci. Soc. Am. J. 54:415–420.

Casida, L.E. 1971. Microorganisms in unamended soil as observed by various forms of microscopy and staining. Appl. Environ. Microbiol. 21:1040–1045.

Chiou, C.T. 1989. Theoretical considerations of the partition uptake of nonionic organic compounds by soil organic matter. p. 1–29. *In* B.L. Sawhney and K. Brown (ed.) Reactions and movement of organic chemicals in soil. SSSA Spec. Publ. 22. SSSA and ASA, Madison, WI.

Chung, G.-Y., B.J. McCoy, and K.M. Scow. 1993. Criteria to assess when biodegradation is kinetically limited by intraparticle diffusion and sorption. Biotechnol. Bioeng. 41:625–636.

Crawford, R.L., and W.W. Mohn. 1985. Microbiological removal of pentachlorophenol from soil using a *Flavobacterium*. Enzyme Microbiol. Technol. 7:617–620.

Davidson, J.M., and R.K. Chang. 1972. Transport of picloram in relation to soil physical conditions and pore-water velocity. Soil Sci. Soc. Am. Proc. 36:257–261.

De Haan, H. 1974. Effect of a fulvic acid fraction on the growth of a *Pseudomonas* from Tjeukemeer (The Netherlands). Freshwater Biol. 4:301–310.

Dommergues, Y., and R. Mangenot. 1970. Ecologie microbienne du sol. Masson et Cie, Paris.

Doran, J.W., and M.S. Smith. 1987. Organic matter management and utilization of soil and fertilizer nutrients. p. 53–72. *In* R.F. Follett (ed.) Soil fertility and organic matter as critical components of production systems. SSSA Spec. Publ. 19. SSSA and ASA, Madison, WI.

Dubey, H.D., R.E. Sigafus, and J.R. Freeman. 1966. Effect of soil properties on the persistence of linuron and diphenamid in soils. Agron. J. 58:228–231.

Elder, V.A., B.L. Proctor, and R.A. Hites. 1981. Organic compounds found near dump sites in Niagara Falls, New York. Environ. Sci. Technol. 15:1237–1243.

Eno, C.F., and H. Popenoe. 1964. Gamma-radiation compared with steam and methyl bromide as a soil sterilizing agent. Soil Sci. Soc. Am. Proc. 28:533–535.

Erhardt, H.M., and H.J. Rehm. 1985. Phenol degradation by microorganisms adsorbed on activated carbon. Appl. Microbiol. Technol. 21:32–36.

Filip, Z. 1973. Clay minerals as factors influencing biochemical activity of soil microorganisms. Folia Microbiol. (Prague) 18:56–74.

Firestone, M.K. 1982. Biological denitrification. p. 289–326. *In* F.J. Stevenson (ed.) Nitrogen in agricultural soils. ASA, CSSA, and SSSA, Madison, WI.

Focht, D.D., and D. Shelton. 1987. Growth kinetics of *Pseudomonas alcaligenes* C-O relative to inoculation and 3-chlorobenzoate metabolism in soil. Appl. Environ. Microbiol. 53:1846–1849.

Foster, R.C. 1988. Microenvironments of soil microorganisms. Biol. Fertil. Soils 6:189–203.

Garbarini, D.R., and L.W. Lion. 1986. Influence of the nature of soil organics on the sorption of toluene and trichloroethylene. Environ. Sci. Technol. 20:1263–1269.

Geissbühler, H., C. Haselbach, and H. Aebi. 1963. The fate of N′(4-chlorophenoxy)-phenyl-N,N dimethylurea (C-1983) in soils and plants. III. Breakdown in soils and plants. Weed Res. 3:277–297.

Grathwohl, P. 1990. Influence of organic matter from soils and sediments from various origins on the sorption of some chlorinated aliphatic hydrocarbons: Implications of K_{oc} correlations. Environ. Sci. Technol. 24:1687–1693.

Gray, T.R.G. 1967. Stereoscan electron microscropy of soil microorganisms. Science 155:1668–1670.

Gray, T.R.G., P. Baxby, I.R. Hill, and M. Goodfellow. 1968. Direct observation of bacteria in soil. p. 171–192. *In* T.R.G. Gray and D. Parkinson (ed.) The ecology of soil bacteria. Liverpool Univ. Press, Liverpool.

Grover, R., and R.J. Hance. 1970. Effect of ratio of soil to water on adsorption of linuron and atrazine. Soil Sci. 109:136–138.

Haider, K. 1992. Problems related to the humification processes in soils of temperate climates. p. 55–94. *In* G. Stotzky and J.M. Bollag (ed.) Soil biochemistry. Vol. 7. Marcel Dekker, New York.

Hamaker, J.W. 1972a. Decomposition: Quantitative aspects. p. 153–240. *In* C.A.I. Goring and J.W. Hamaker (ed.) Organic chemicals in the soil environment. Marcel Dekker, New York.

Hamaker, J.W. 1972b. Diffusion and volatilization. p. 341–397. *In* C.A.I. Goring and J.W. Hamaker (ed.) Organic chemicals in the soil environment. Marcel Dekker, New York.

Hamaker, J.W., and C.A.I. Goring. 1976. Turnover of pesticide residues in the soil. p. 219–243. *In* D.D. Kaufman et al. (ed.) Bound and conjugated pesticide residues. Am. Chem. Soc., Washington, DC.

Hance, R.J. 1974. Soil organic matter and the adsorption and decomposition of the herbicides atrazine and linuron. Soil Biol. Biochem. 6:39–42.

Harmon, T.C., W.P. Ball, and P.V. Roberts. 1989. Nonequilibrium transport of organic contaminants in groundwater. p. 405–437. *In* B.L. Sawhney and K. Brown (ed.) Reactions and movement of organic chemicals in soil. SSSA Spec. Publ. 22. SSSA and ASA, Madison, WI.

Hassett, J.J., and W.L. Banwart. 1989. The sorption of nonpolar organics by soils and sediments. p. 31–44. *In* B.L. Sawhney and K. Brown (ed.) Reactions and movement of organic chemicals in soil. SSSA Spec. Publ. 22. SSSA and ASA, Madison, WI.

Hattori, T. 1966. Bacterial distribution and variation in soil aggregates. J. Sci. Soil Manure (Jpn) 37:32.

Hattori, T. 1973. Microbial life in the soil: An introduction. Marcel Dekker, New York.

Hattori, T., and R. Hattori. 1976. The physical environment in soil microbiology: An attempt to extend principles of microbiology to soil. Crit. Rev. Microbiol. 4:423–461.

Hayes, M.H.B. 1985. Extraction of humic substances from soil. p. 329–362. *In* G.A. Aiken et al. (ed.) Humic substances in soil, sediment, and water: Geochemistry, isolation, and characterization. John Wiley and Sons, New York.

Hayes, M.H.B., P. MacCarthy, R.L. Malcolm, and R.S. Swift (ed.). 1989. Humic substances II—In search of structure. p. 1–764. John Wiley and Sons, New York.

Helweg, A. 1975. Degradation of ^{14}C-maleic hydrazide in soil as influenced by adsorption on activated carbon. Weed Res. 15:129–133.

Herbes, S.E., and L.R. Schwall. 1981. Rate of microbial transformation of polycyclic aromatic hydrocarbons in water and sediments in the vicinity of a coal-coking waste water discharge. Appl. Environ. Microbiol. 41:20–28.

Hermansson, M., and K.C. Marshall. 1985. Utilization of surface localized substrate by nonadhesive marine bacteria. Microb. Ecol. 11:91–105.

Hissett, R., and T.R.G. Gray. 1976. Microsites and time changes in soil microbe ecology. p. 23–39. *In* J.M. Anderson and A. MacFadyen (ed.) The role of terrestrial and aquatic organisms in decomposition processes. Blackwell Sci. Publ., Oxford.

Hyzak, D.L., and R.L. Zimdahl. 1974. Rate of degradation of metribuzin and two analogs in soil. Weed Sci. 22:75–79.

Inoue, K., K. Kaneko, and M. Yoshida. 1978. Adsorption of dodecylbenzenesulfonates by soil colloids and influence of soil colloids on their degradation. Soil Sci. Plant Nutr. 24:91–102.

Jeffrey, W.H., and J.H. Paul. 1986. Activity of an attached and free-living *Vibrio* sp. as measured by thymidine incorporation, *p*-iodonitrotetrazolium reduction, and ATP/DNA ratios. Appl. Environ. Microbiol. 51:150–156.

Jenkinson, D.S. 1977. Studies on the decomposition of plant material in soil. V. The effects of plant cover and soil type on the loss of carbon from ^{14}C-labelled rye grass decomposing under field conditions. J. Soil Sci. 28:424–434.

Jones, T.W., W.M. Kemp, J.C. Stevenson, and J.C. Means. 1982. Degradation of atrazine in estuarine water/sediment systems and soils. J. Environ. Qual. 11:632–638.

Karickhoff, S.W. 1980. Sorption kinetics of hydrophobic pollutants in natural sediments. p. 193–205. *In* R.A. Baker (ed.) Contaminants and sediments. Vol. 2. Ann Arbor Sci., Ann Arbor, MI.

Karickhoff, S.W. 1981. Semi-empirical estimation of sorption of hydrophobic pollutants on natural sediments and soils. Chemosphere 10:833–846.

Karickhoff, S.W., and K.R. Morris. 1985. Sorption dynamics of hydrophobic pollutants in sediment suspensions. Environ. Toxicol. Chem. 4:469–479.

Kilbertus, G. 1980. Etudé des microhabitats contenus dans les aggrégats du sol. Leur relation avec la biomass bacteriénne et la taille des procaryotes présents. Rev. Ecol. Biol. Sol. 17:543–557.

Koch, A.L. 1982. Multistep kinetics: Choice of models for the growth of bacteria. J. Theor. Biol. 98:401–417.

Koch, A.L. 1985. The macroeconomics of bacterial growth. p. 1–42. *In* M. Fletcher and G.D. Floodgate (ed.) Bacteria in their natural environments. Soc. for Gen. Microbiol., Cambridge, England.

Kunc, F., and G. Stotzky. 1974. Effect of clay minerals on heterotrophic microbial activity in soil. Soil Sci. 118:186–195.

Kunc, F., and G. Stotzky. 1977. Acceleration of aldehyde decomposition in soil by montmorillonite. Soil Sci. 124:167–172.

Lechner, U., and G. Straube. 1984. Influence of substrate concentration on the induction of amidases in herbicide degradation. Z. Allg. Mikrobiol. 24:581–584.

Lee, R.F., and C. Ryan. 1979. Microbial degradation of organochlorine compounds in estuarine water and sediments. p. 443–450. *In* A.W. Bourquin and P.H. Pritchard (ed.) Microbial degradation of pollutants in marine environments. EPA-600/9-79-012. U.S. Environmental Protection Agency, Washington, DC.

Lobartini, J.C., and K.H. Tan. 1988. Differences in humic acid characteristics as determined by carbon-13 nuclear magnetic resonance, scanning electron microscopy, and infrared analysis. Soil Sci. Soc. Am. J. 52:125–130.

Marshall, K.C. 1971. Sorptive interactions between soil particles and microorganisms. p. 409–445. *In* A.D. McLaren and J.J. Skujins (ed.) Soil biochemistry. Vol. 2. Marcel Dekker, New York.

Marshall, K.C. 1976. Interfaces in microbial ecology. Harvard Univ. Press, Cambridge.

Martin, J.P., Z. Filip, and K. Haider. 1976. Effect of montmorillonite and humate on growth and metabolic activity of some actinomycetes. Soil Biol. Biochem. 8:409–413.

Mathur, S.P., and E.A. Paul. 1966. A microbiological approach to the problem of soil humic acid structures. Nature (London) 212:646–647.

Mathur, S.P., and E.A. Paul. 1967. Microbial utilization of soil humic acids. Can. J. Microbiol. 13:573–580.

McCall, P.J., and G.L. Agin. 1985. Desorption kinetics of picloram as affected by residence time in the soil. Environ. Toxic. Chem. 4:37–44.

McCormick, L.L., and A.E. Hiltbold. 1966. Microbiological decomposition of atrazine and diuron in soil. Weeds 14:77–82.

McLaren, A.D. 1978. Kinetics and consecutive reactions of soil enzymes. p. 97–116. *In* R.G. Burns (ed.) Soil enzymes. Academic Press, New York.

McLoughlin, A., and E. Küster. 1972. The effect of humic substances on the respiration and growth of microorganisms. I. Effect on the yeast *Candida utilis.* Plant Soil 37:17–25.

Mihelcic, J.R., and R.G. Luthy. 1991. Sorption and microbial degradation of naphthalene in soil-water suspensions under denitrification conditions. Environ. Sci. Technol. 25:169–177.

Mingelgrin, U., and B. Yaron. 1974. The effect of calcium salts on the degradation of parathion in sand and soil. Proc. Soil Sci. Soc. Am. 38:914–917.

Mörsen, A., and H.J. Rehm. 1987. Degradation of phenol by a mixed culture of *Pseudomonas putida* and *Cryptococcus elinovii* adsorbed on activated carbon. Appl. Microbiol. Biotechnol. 26:283–288.

Mortland, M.M., and L.J. Halloran. 1976. Polymerization of aromatic molecules on smectite. Soil Sci. Soc. Am. J. 40:367–370.

Moyer, J.R., RJ. Hance, and C.E. McKone. 1972. The effect of adsorbents on the rate of degradation of herbicides incubated with soil. Soil Biol. Biochem. 4:307–311.

Murphy, E.M., J.M. Zachara, and S.C. Smith. 1990. Influence of mineral-bound humic substances on the sorption of hydrophobic organic compounds. Environ. Sci. Technol. 24:1507–1516.

Myrold, D.D., and J.M. Tiedje. 1985. Diffusional constraints on denitrification in soil. Soil Sci. Soc. Am. J. 49:651–657.

Nedwell, D.B., and T.R.G. Gray. 1987. Soils and sediments as matrices for microbial growth. p. 21–54. *In* M. Fletcher et al. (ed.) Ecology of microbial communities. Cambridge Univ. Press, Cambridge, England.

Nomura, N.S., and H.W. Hilton. 1977. The adsorption and degradation of glyphosate in five Hawaiian sugarcane soils. Weed Res. 17:113–121.

Ogram, A.W., R.E. Jessup, L.T. Ou, and P.S.C. Rao. 1985. Effects of sorption on biological degradation rates of (2,4-dichlorophenoxy) acetic acid in soil. Appl. Environ. Microbiol. 49:582–587.

Papendick, R.I., and G.S. Campbell. 1981. Theory and measurement of water potential. p. 1–22. *In* J.F. Parr et al. (ed.) Water potential relations in soil microbiology. SSSA Spec. Publ. 9. SSSA and ASA, Madison, WI.

Paul, E.A., and S.P. Mathur. 1967. Cleavage of humic acid by *Penicillium frequentans*. Plant Soil 27:297–299.

Pignatello, J.J. 1989. Sorption dynamics of organic compounds in soils and sediments. p. 45–80. *In* B.L. Sawhney and K. Brown (ed.) Reactions and movement of organic chemicals in soil. SSSA Spec. Publ. 22. SSSA and ASA, Madison, WI.

Pignatello, J.J., B.L. Sawhney, and C.R. Frink. 1987. EDB: Persistence in soil. Science 236:898.

Pignatello, J.J. 1990a. Slowly reversible sorption of aliphatic halocarbons in soils. I. Formation of residual fractions. Environ. Toxic. Chem. 9:1107–1115.

Pignatello, J.J. 1990b. Slowly reversible sorption of aliphatic halocarbons in soils. II. Mechanistic aspects. Environ. Toxicol. Chem. 9:1117–1126.

Rao, P.S.C., J.M. Davidson, R.E. Jessup, D.E. Rolston, J.M. Davidson, and D.P. Kilcrease. 1980. Experimental and mathematical description of nonadsorbed solute transfer by diffusion in spherical aggregates. Soil Sci. Soc. Am. J. 44:684–688.

Reber, H.H. 1982. Inducibility of benzoate oxidizing cell activities in *Acinetobacter calcoaceticus* strain Bs 5 by chlorobenzoates as influenced by the position of chlorine atoms and the inducer concentration. Eur. J. Appl. Microbiol. Biotechnol. 15:138–140.

Rijnaarts, H.H.M., A. Bachmann, J.C. Jumelet, and A.J. Zehnder. 1991. Effect of desorption and intraparticle mass transfer on the aerobic biomineralization of α-hexachlorocyclohexane in a contaminated calcareous soil. Environ. Sci. Technol. 24:1349–1354.

Salonius, P.O., J.B. Robinson, and F.E. Chase. 1967. A comparison of autoclaved and gamma-irradiated soils as media for microbial colonization experiments. Plant Soil 27:239–248.

Saltzman, S., B. Yaron, and U. Mingelgrin. 1974. The surface catalyzed hydrolysis of parathion on kaolinite. Soil Sci. Soc. Am. Proc. 38:231–234.

Sawhney, B.L. 1989. Movement of organic chemicals through landfill and hazardous waste disposal sites. p. 447–474. *In* B.L. Sawhney and K. Brown (ed.) Reactions and movement of organic chemicals in soil. SSSA Spec. Publ. 22. SSSA and ASA, Madison, WI.

Schmidt, S.K., M. Schuler, and M. Alexander. 1986. Predicting threshold concentrations of organic substrates for bacterial growth. J. Theor. Biol. 114:1–8.

Schmidt, S.K., K. Scow, and M. Alexander. 1987. Kinetics of *p*-nitrophenol mineralization by a *Pseudomonas* sp.: Effects of second substrates. Appl. Environ. Microbiol. 53:2617–2623.

Schmidt, S.K., S. Simkins, and M. Alexander. 1985. Models for the kinetics of biodegradation of organic compounds not supporting growth. Appl. Environ. Microbiol. 50:323–331.

Schnitzer, M., and S.U. Khan. 1978. Soil organic matter. Elsevier Sci. Publ. Co., New York.

Schnürer, J., M. Clarholm, and T. Rosswall. 1985. Microbial biomass and activity in an agricultural soil with different organic matter contents. Soil Biol. Biochem. 17:611–618.

Scott, A.D., D.C. Wolf, and T.L. Lavy. 1982. Apparent adsorption and microbial degradation of phenol by soil. J. Environ. Qual. 11:107–112.

Scott, A.D., D.C. Wolf, and T.L. Lavy. 1983. Adsorption and degradation of phenol at low concentrations in soil. J. Environ. Qual. 12:91–95.

Scow, K.M. 1989. The effect of the soil matrix on the kinetics of biodegradation. Ph.D. diss. Cornell Univ., New York (Diss. Abstr. 8915112).

Scow, K.M., and M. Alexander. 1992. Effect of diffusion on the kinetics of biodegradation: Experimental results with synthetic aggregates. Soil Sci. Soc. Am. J. 56:128–134.

Scow, K.M., and J. Hutson. 1992. Effect of diffusion and sorption on the kinetics of biodegradation: Theoretical considerations. Soil Sci. Soc. Am. J. 56:119–127.

Scow, K.M., S.K. Schmidt, and M. Alexander. 1989. Kinetics of biodegradation of mixtures of substrates in soil. Soil Biol. Biochem. 21:703–708.

Scow, K.M., S. Simkins, and M. Alexander. 1986. Kinetics of mineralization of organic compounds at low concentrations in soil. Appl. Environ. Microbiol. 51:1028–1035.

Seech, A.G., and E.G. Beauchamp. 1988. Denitrification in soil aggregates of different sizes. Soil Sci. Soc. Am. J. 52:1616–1621.

Siala, A., I.R. Hill, and T.R.G. Gray. 1974. Populations of spore-forming bacteria in an acid forest soil, with special reference to *Bacillus subtilis*. J. Gen. Microbiol. 81:183–190.

Simkins, S., and M. Alexander. 1984. Models for mineralization kinetics with the variables of substrate concentration and population density. Appl. Environ. Microbiol. 47:1299–1306.

Skipper, H.D., and D.T. Westermann. 1973. Comparative effects of propylene oxide, sodium azide, and autoclaving on selected soil properties. Soil Biol. Biochem. 5:409–414.

Skujins, J.J. 1967. Enzymes in soil. p. 371–414. *In* A.D. McLaren and G.H. Peterson (ed.) Soil biochemistry. Marcel Dekker, New York.

Smith, D.T., and W.F. Meggitt. 1970. Persistence and degradation of pyrazon in soil. Weed Sci. 18:260–264.

Sörensen, L.H. 1972. Stabilization of newly formed amino acid metabolites in soil by clay minerals. Soil Sci. 114:5–11.

Sörensen, L.H. 1975. The influence of clay on the rate of decay of amino acid metabolites synthesized in soils during decomposition of cellulose. Soil Biol. Biochem. 7:171–177.

Spain, J.C., P. Van Veld, C.A. Monti, P.H. Pritchard, and C.R. Cripe. 1984. Comparison of *p*-nitrophenol biodegradation in field and laboratory test systems. Appl. Environ. Microbiol. 48:944–950.

Steen, W.C., D.F. Paris, and G.L. Baughman. 1980. Effects of sediment sorption on microbial degradation of toxic substances. p. 477–482. *In* R.A. Barker (ed.) Contaminants and sediments. Ann Arbor Sci. Publ., Ann Arbor, MI.

Steinberg, S.M., J.J. Pignatello, and B.L. Sawhney. 1987. Persistence of 1,2-dibromoethane in soils: Entrapment in intraparticle micropores. Environ. Sci. Technol. 21:1201–1208.

Stevenson, F.J. 1982. Humus chemistry: Genesis, composition, and reactions. John Wiley and Sons, New York.

Stotzky, G. 1967. Clay minerals and microbial ecology. Trans. N.Y. Acad. Sci. 30:11–21.

Stotzky, G. 1972. Activity, ecology and population dynamics of microorganisms in soil. Crit. Rev. Microbiol. 2:59–137.

Stotzky, G. 1986. Influence of soil mineral colloids on metabolic processes, growth, adhesion, and ecology of microbes and viruses. p. 305–428. *In* P.M. Huang and M. Schnitzer (ed.) Interactions of soil minerals with natural organics and microbes. SSSA Spec. Publ. 17. SSSA, Madison, WI.

Subba-Rao, R.V., and M. Alexander. 1982. Effect of sorption on mineralization of low concentrations of aromatic compounds in lake water samples. Appl. Environ. Microbiol. 44:659–688.

Taha, S.M., M.N. Zayed, and L. Zohdy. 1973. Studies in humic acids decomposing bacteria in soil. II. Isolation and identification of micro-organisms. Zentralbl. Bakteriol. Abt. 136:189–197.

Tranvik, L.J., and M.G. Höfle. 1987. Bacterial growth in mixed cultures on dissolved organic carbon from humic and clear waters. Appl. Environ. Microbiol. 53:482–488.

Tsernoglou, D., and E.H. Anthony. 1971. Particle size, water-stable aggregates, and bacterial populations in lake sediments. Can. J. Microbiol. 17:217–227.

van Genuchten, M.T., J.M. Davidson, and P.J. Wierenga. 1974. An evaluation of kinetic and equilibrium equations for the prediction of pesticide movement in porous media. Soil Sci. Soc. Am. Proc. 38:29–35.

van Genuchten, M.T., and R.J. Wagenet. 1989. Two-site/two-region models for pesticide transport and degradation: Theoretical development and analytical solutions. Soil Sci. Soc. Am. J. 53:1303–1310.

van Loosdrecht, M.C.M., J. Lyklema, W. Norde, and A.J.B. Zehnder. 1990. Influences of interfaces on microbial activity. Microbiol. Rev. 54:75–87.

van Veen, J.A., J.N. Ladd, and M. Amato. 1985. Turnover of carbon and nitrogen through the microbial biomass in a sandy loam and a clay soil incubated with [^{14}C(U)]glucose and [^{15}N]$(NH_4)_2SO_4$ under different moisture regimes. Soil Biol. Biochem. 17:747–756.

Visser, S.A. 1985. Effect of humic acids on numbers and activities of microorganisms within physiological groups. Org. Geochem. 8:81–85.

Waid, J.S. 1973. A method to study microorganisms from soil particles with the aid of the transmission electron microscope. Bull. Ecol. Res. Comm. (NFR) 17:103–108.

Waid, F.S., and T.B. Suan. 1974. Transmission electron microscopy of bacteria in soils during acetate decomposition. Soil Biol. Biochem. NFR 6:63–68.

Walker, W.W., C.R. Cripe, P.H. Pritchard, and A.W. Bourquin. 1984. Biotic and abiotic degradation of dibutylphthalate in estuarine and freshwater environments. Chemosphere 13:1283–1294.

Whitehead, D.C., H. Buchan, and R.D. Hartley. 1975. Components of soil organic matter under grass and arable cropping. Soil Biol. Biochem. 7:65–71.

Wolf, D.C., T.H. Dao, H.D. Scott, and T.L. Levy. 1989. Influence of sterilization methods on selected soil microbiological, physical, and chemical properties. J. Environ. Qual. 18:39–44.

Wu, S., and P.M. Gschwend. 1986. Sorption kinetics of hydrophobic compounds to natural sediments and soils. Environ. Sci. Technol. 20:725–730.

Yaron, B. 1975. Chemical conversion of parathion on soil surfaces. Soil Sci. Soc. Am. J. 39:639–643.

You, I.-S., and R. Bartha. 1982. Stimulation of 3,4-dichloroaniline mineralization by aniline. Appl. Environ. Microbiol. 44:678–681.

Zimdahl, R.L., and S.M. Gwynn. 1977. Soil degradation of three dinitroanilines. Weed Sci. 25:247–251.

Zoeteman, B.C.J., K. Harmsen, J.B.H.J. Linders, C.F.H. Morra, and W. Slooff. 1980. Persistent organic pollutants in river water and ground water of the Netherlands. Chemosphere 9:231–249.

7 Miscible Displacement and Theoretical Techniques for Simultaneous Study of Pesticide Sorption and Degradation During Transport

A. P. Gamerdinger
University of Rhode Island
Kingston, Rhode Island

Kathryn C. Dowling and A. T. Lemley
Cornell University
Ithaca, New York

Our current understanding of the mechanisms that control sorption and transformation of pesticides during transport is not complete. However, quantitative estimates of sorption and degradation parameters are required in management and research models for predicting the fate of pesticides in the environment. Solutions to transport equations that consider both sorption and transformation, when coupled with miscible displacement techniques, offer substantial promise for estimating parameters which are required in environmental fate models. This research was undertaken to develop and improve reliable methods for estimating these parameters. The term *degradation* is used to describe the transformation of parent compound and is not meant to imply mineralization or total degradation.

Model compounds including aldicarb [2,methyl-2(methylthio)propionaldehyde *O*(methylcarbamoyl)oxime], aldicarb sulfoxide [2-methyl-2-(methylsulfinyl)propionaldehyde *O*-(methylcarbamoyl)oxime], methomyl [*S*-Methyl-*N*-((methylcarbamoyl)oxy)-thioacetimidate] (carbamates); atrazine [2-chloro-4-ethylamino-6-isopropylamino-*s*-triazine], simazine [2-chloro-4,6-bis-(ethylamino)-*s*-triazine], cyanazine {2-[[4-chloro-6-(ethylamino)-1,3,5-triazin-2-yl]amino]-2-methylpropionitrile} (*s*-triazines); and ethoprop (O-ethyl S,S-dipropyl phosphorodithioate) (organophosphate) were selected based on their use, documented presence in groundwater, and potential toxicity. A combination of batch incubation, miscible displacement, and theoretical modeling techniques were employed. This chapter reviews the application and

 Sorption and Degradation of Pesticides and Organic Chemicals in Soil. SSSA Special Publication no. 32.

modification of these techniques in the context of work performed in our laboratory for the simultaneous study of sorption and transformation processes during transport. We also present an inverse relationship between equilibrium and nonequilibrium sorption parameters. Other significant contributions in this area are reviewed elsewhere in this publication.

Lemley, Wagenet, and co-workers first applied miscible displacement and theoretical modeling techniques for the systematic study of aldicarb degradation during transport (Zhong et al., 1986; Lemley et al., 1988). Although the experiments were controlled laboratory studies, the dynamic laboratory soil-column was more representative of the field than a batch incubation system. The application of these experimental methods for characterizing the transport of more sorptive solutes led to further model development, evaluation, and modifications of the experimental and theoretical procedures (van Genuchten & Wagenet, 1989; Gamerdinger et al., 1990, 1991; Dowling, 1990).

MATERIALS AND METHODS

Detailed methods for each experiment are contained in the above references; general procedures are summarized here. Significant developments and modifications are noted in the discussion.

In all cases, analytical grade (>99% pure) pesticides and tracers (Cl^-, ^3H_2O) were prepared in 0.005 *M* $CaSO_4$ solutions. All surface soils were obtained from the 5 to 30 cm depth, air dried, and passed through a 2-mm sieve prior to use. The effect of air drying on microbial processes was not of concern as chemical hydrolysis was the primary degradation pathway. Relevant soil properties are represented in Table 7-1.

Parent compounds and degradation products were analyzed using either gas or liquid chromatographic techniques. Liquid scintillation and coulometric titration techniques were used for quantifying ^3H_2O and Cl^-, respectively.

Batch incubation techniques (e.g., Rao & Davidson, 1979) were used to determine sorption isotherms for methomyl, atrazine, simazine, cyanazine, and ethoprop. Miscible displacement experiments were conducted following the method detailed by Zhong et al. (1986). Columns were saturated (90–99%) with $CaSO_4$ solution and equilibrated until constant flow-rate and temperature were established. A pulse of the pesticide solution and tracer

Table 7-1. Soil chemical and physical properties.

	Organic C	Sand	Silt	Clay	pH (1:1 H_2O)
	%				
Palmyra	1.17	53	37	10	4.9
Riverhead 1	2.32	75	21	4	4.5
Riverhead 2	1.07	51	36	12	4.4
Valois	1.64	30	55	15	5.9
Rhinebeck	3.13	12	52	36	6.7

was applied; effluent was collected and analyzed for the solute of interest. Breakthrough curves (BTCs), or temporal concentration distributions, were generated and then interpreted with the models described below.

Detailed derivations of the two models applied here are presented elsewhere (van Genuchten & Wagenet, 1989). Both are based on the convection-dispersion equation (CDE) and include a description of first-order degradation. The simpler model considers sorption to attain a rapid equilibrium which is described with a linear isotherm and the equilibrium sorption constant, K_d. The more complex considers sorption in two domains where one fraction attains a rapid equilibrium and is also described by K_d. Sorption in the other domain is time dependent and characterized by the reverse sorption rate coefficient, k_2, and (l–f) where f is the fraction of sorbent for which sorption attains a rapid equilibrium. Analytic solutions to the equilibrium and nonequilibrium forms of the models are the basis for parameter optimization codes CXTFIT and CXT4, respectively. The governing equations follow.

Linear Equilibrium Sorption Formulation

$$R\,(\partial c/\partial t) = D\,(\partial^2 c/\partial x^2) - v\,(\partial c/\partial x) - \mu c \qquad [1]$$

$$R = 1 + (\rho\, K_d/\theta) \qquad [2]$$

where K_d is the distribution coefficient for linear solute partitioning between the solution (c) and sorbed (s) phases; R is the dimensionless retardation factor; ρ is the soil bulk density; θ is the volumetric water content; D is the apparent diffusion coefficient and accounts for hydrodynamic dispersion and diffusion; v is the average pore water velocity; μ is the first-order degradation rate constant; t and x represent time and distance, respectively.

Nonequilibrium Sorption (Two-Site) Formulation

$$1 + \frac{\rho f K_d}{\theta}\frac{\partial c}{\partial t} = D\frac{\partial^2 c}{\partial x^2} - v\frac{\partial c}{\partial x} - \frac{k_2 \rho}{\theta}[(1 - f)K_d\, c - s_2] - \mu_l c - \frac{\mu_{sl}\, f\, \rho\, K_d\, c}{\theta} \qquad [3]$$

$$\frac{\partial s_2}{\partial t} = k_2[(1 - f)K_d c - s_2] - \mu_{s2}\, s_2 \qquad [4]$$

where all terms are as defined previously and the subscripts 1 and 2 represent the equilibrium and time-dependent sorption domains, respectively, and l represents the liquid phase.

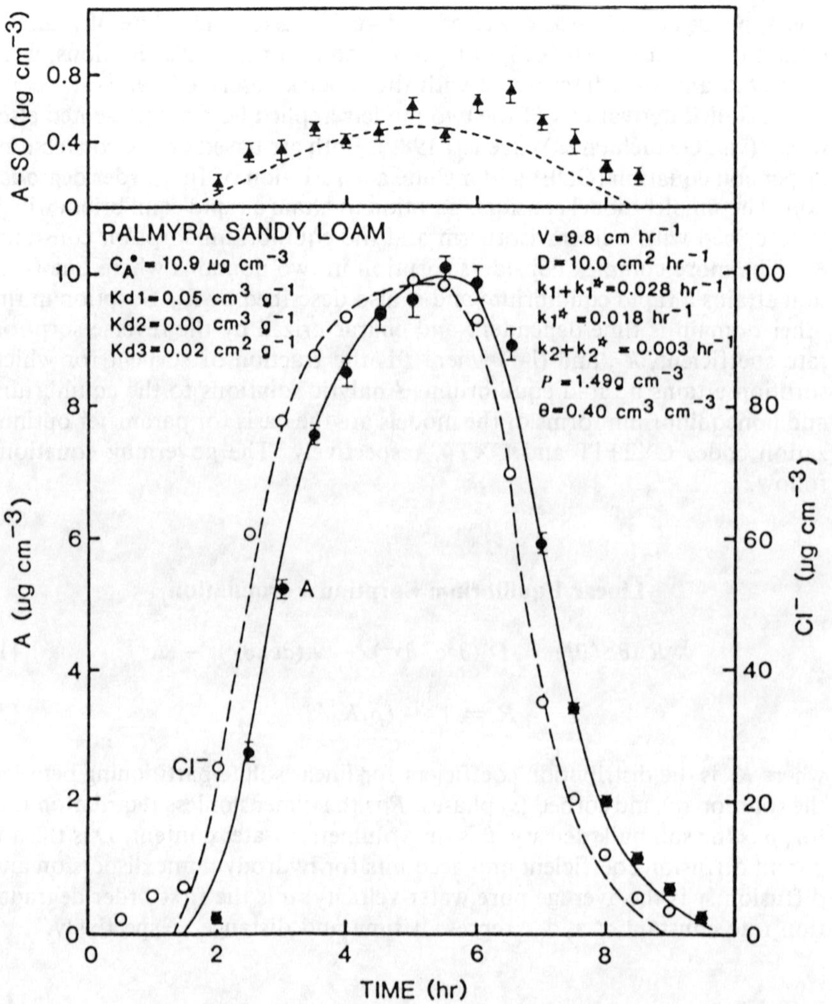

Fig. 7–1. Miscible displacement data and model prediction (equilibrium sorption formulation) for effluent concentrations of chloride (Cl^-), aldicarb (A), and aldicarb sulfoxide (A-SO) (From Zhong et al., 1986).

RESULTS

Breakthrough curves for aldicarb and aldicarb sulfoxide on Palmyra soil (fine-loamy over sandy or sandy skeletal, mixed, mesic Glossoboric Hapludalfs) (Fig. 7–1) were successfully described with the linear equilibrium sorption form of the CDE (CXTFIT). Tracer (Cl^-) data was analyzed to determine the apparent diffusion coefficient, D. Curvefitting was used to determine the retardation factor, R, and the first-order degradation rate constant, μ (Table 7–2). The equilibrium sorption constant, K_d, was then cal-

Table 7-2. Sorption and degradation parameters.

Pesticide	Soil	K_d	k_2†	μ	References
		$cm^3 g^{-1}$	hr^{-1}	hr^{-1}	
Aldicarb	Palmyra	0.06†	--	2.8×10^{-2}	Zhong et al., 1986
Aldicarb	Riverhead 1	0.12†	--	7.0×10^{-2}	Zhong et al., 1986
Methomyl	Riverhead 2	0.30	--	1.7×10^{-2}	Dowling, 1990
Ethoprop	Riverhead 2	1.29	0.048	6.0×10^{-4}	Dowling, 1990
Cyanazine	Valois	1.20	0.054	8.8×10^{-4}	Gamerdinger et al., 1991
Simazine	Valois	1.97	0.019	1.2×10^{-3}	Gamerdinger et al., 1991
Atrazine	Valois	1.98	0.020	9.9×10^{-4}	Gamerdinger et al., 1991
Atrazine	Rhinebeck	3.16	0.005	8.2×10^{-4}	Gamerdinger et al., 1991
Ethoprop	Rhinebeck	2.38	0.026	3.7×10^{-4}	Dowling, 1990

† Determined through curvefitting.

culated from R. Results of a similar experiment on Riverhead soil (coarse-loamy, mixed, mesic Typic Dystrochrepts) are included for comparison.

In applying the linear, equilibrium sorption formulation of the model, the determination of sorption and degradation parameters is essentially independent: sorption is based on retardation or the shift of the BTC to the right, and degradation is calculated based on a reduction in mass or decreased area under the BTC.

The techniques were validated by comparing parameter estimates with those obtained in other studies using other techniques (Zhong et al., 1986). Systematic study of the effects of pH and temperature on the degradation of aldicarb and its degradation products during transport was then accomplished using these methods (Lemley et al., 1988).

Other pesticides were studied as a means of determining the robustness of the miscible displacement and modeling techniques. Like aldicarb, methomyl was not strongly sorbed to soil (Dowling, 1990) and miscible displacement data was modeled with the linear equilibrium sorption form of the CDE. Sorption and degradation parameters for methomyl on Riverhead soil are included in Table 7-2.

When sorption and degradation of the 2-chloro-*s*-triazine herbicides was evaluated (Gamerdinger et al., 1990, 1991), the linear equilibrium sorption form of the CDE did not describe the data (Fig. 7-2). Nonequilibrium sorption is well-documented and several models have been proposed to interpret this behavior (reviewed by Skopp, 1986; van Genuchten & Wagenet, 1989; Brusseau & Rao, 1989a). However, a model that considers nonequilibrium behavior under conditions in which the solute was transformed was not available when experiments with *s*-triazine herbicides were initiated. The experimental work prompted the development of a two-site/two-region model that simultaneously considered nonequilibrium transport and degradation of chemicals (van Genuchten & Wagenet, 1989). A similar model was presented by Lassey (1988).

Briefly, nonequilibrium sorption during transport can result in early breakthrough and extensive tailing of the BTC. Apparent retardation may not equal total retardation, therefore, errors may result if curvefitting is used

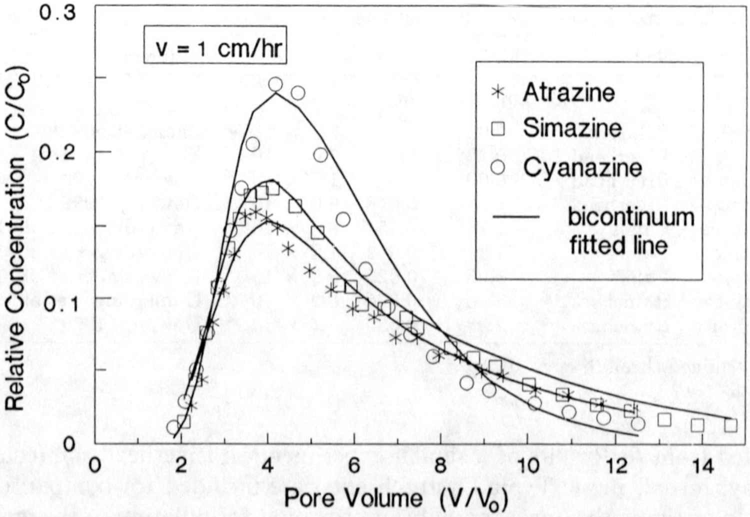

Fig. 7–2. Miscible displacement data and model predictions (nonequilibrium sorption formulation). (From Gamerdinger et al., 1991.)

to determine the retardation factor. In this case, it is necessary to use an independently measured estimate of K_d (e.g., from a sorption isotherm experiment) for calculating the retardation factor; R is then fixed in all model applications. Because degradation shifts the apparent center-of-mass to the left, and changes the area under the BTC, moment analysis (Valocchi, 1985) cannot be used to determine the retardation factor directly from the breakthrough curve.

The increased complexity of a model that simultaneously considers nonequilibrium sorption and degradation is apparent. An increased number of parameters is required; when possible, they should be measured independently. In all cases, R was determined from the sorption isotherm, and tracer data was used to establish the hydrodynamic properties of the soil column.

When parameter optimization code CXT4 was applied for determining nonequilibrium sorption parameters, the degradation term had to be measured directly and fixed. The desorption rate is expected to decrease with increasing time based on the first-order description that is used to characterize nonequilibrium sorption. Thus, under nonequilibrium conditions, low concentrations of pesticide (below the lower limit of detection) continue to elute for several pore volumes after the apparent effluent concentration is "zero". At this point in the experiment, we found that it was necessary to extrude the soil from the column and vigorously extract the soil with a polar solvent to recover parent compound that remained in the column. We considered this noneluted pesticide, in addition to the mass eluted, when determining the first-order degradation rate constant, μ. To summarize, the hydrodynamic, retardation, and degradation parameters were determined independently or directly, and fixed when CXT4 was applied for estimating the reverse sorption rate coefficient, k_2, and the fraction of sorbent for which sorption attained a rapid equilibrium, f.

When nonequilibrium parameters were determined in a separate experiment, we were able to fix k_2 and f, and use curvefitting to estimate the first-order degradation constant. We found good agreement between the model estimate and directly measured values (Gamerdinger et al., 1990). Values of the sorption rate coefficient and the first-order degradation rate constant for atrazine, simazine, cyanazine (Gamerdinger et al., 1991), and ethoprop (Dowling, 1990) on Valois (coarse-loamy, mixed, mesic Typic Dystrochrepts) and Rhinebeck (fine, illitic, mesic Aeric Ochraqualfs) soils are reported in Table 7–2. An example of the fitted lines that were generated with parameter optimization code CXT4 are included in Fig. 7–2.

DISCUSSION

Both of the transport models that were applied in these studies included a first-order description of degradation. This is not always appropriate for environmental systems, particularly when microbial processes are involved (Simkins & Alexander, 1984). All of the compounds selected for study were found to follow pseudo first-order degradation kinetics and were subject to chemical hydrolysis. Aldicarb and aldicarb sulfoxide were also oxidized.

Increased error and variability are associated with parameter estimates from studies where sorption and degradation are considered in the same system. Experiments that consider only sorption under different experimental conditions are often performed on the same soil column. Thus, differences in hydrodynamic properties that arise from variations in column packing and heterogeneities in the porous medium, are eliminated. Destruction of each column upon the completion of each experiment for determining mass balance and recovery is necessary when degradation is considered, thus, we find wider confidence intervals for parameter estimates.

Although the purpose of the studies was not to elucidate rate-limiting mechanisms for sorption and degradation, our results are consistent with those of other studies. The inverse relationship between the reverse sorption rate coefficient and the equilibrium sorption constant (k_2 and K_d) has been oberved by other researchers (Karickhoff & Morris, 1985; Brusseau & Rao, 1989b). Following the technique of Brusseau and Rao (1989b), Fig. 7–3 includes the data presented here and illustrates the dependency of the reverse sorption rate coefficient on the magnitude of the equilibrium sorption constant. To our knowledge, this is the first time that this relationship has been applied in studies where the solute is transformed, and suggests that degradation need not confound the estimation of sorption parameters. We also found that the sorption rate coefficient increased with increasing velocity (Dowling, 1990; Gamerdinger et al., 1991) and do not suggest that the relationship presented here can be used for predictive purposes.

The degradation of the pesticides studied here depends on a variety of environmental conditions including pH, temperature, and soil characteris-

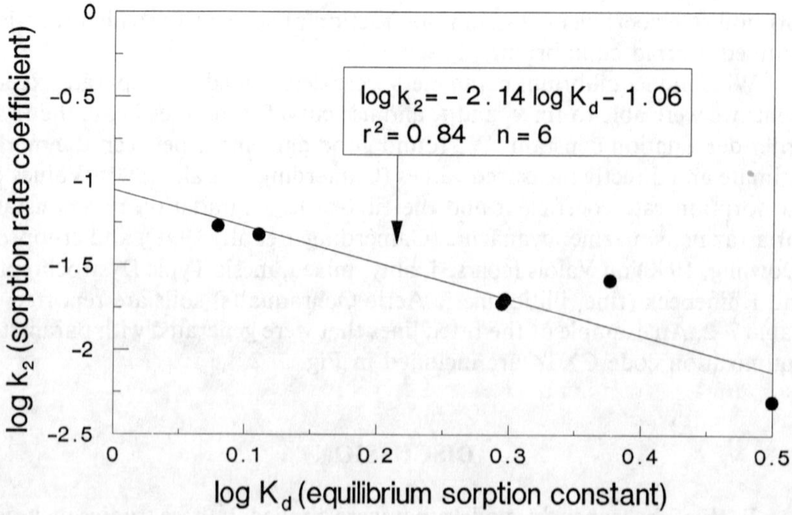

Fig. 7-3. Dependency of the reverse sorption rate coefficient, k_2, on the equilibrium sorption constant, K_d.

tics (Zhong et al., 1986; Lemley et al., 1988; Dowling, 1990) and no clear dependencies on sorption could be established with the limited data presented here.

The experimental and theoretical techniques that are summarized here are complementary to the more familiar batch incubation techniques. Appropriate models of the type used in this study, together with miscible displacement studies, offer a system with which to study complex sorption and degradation processes during transport.

ACKNOWLEDGMENT

This research was performed at Cornell University, Ithaca, NY and was partially funded by The Jesse Smith Noyes Foundation, Cornell University College of Human Ecology, and by the USDA through the Northeast Pesticide Impact Assessment Program, and the Cooperative State Research Service. Contribution no. 2846 of the College of Resource Development, Univ. of Rhode Island, with support from Rhode Island Agric. Exp. Stn.

REFERENCES

Brusseau, M.L., and P.S.C. Rao. 1989a. Sorption nonideality during organic contaminant transport in porous media. Crit. Rev. Environ. Control 19:33–99.

Brusseau, M.L., and P.S.C. Rao. 1989b. The influence of sorbate-organic matter interactions on sorption nonequilibrium. Chemosphere 18:1691–1706.

Dowling, K.C. 1990. The transport and degradation of ethoprop and methomyl in soil-water systems. M.S. thesis. Cornell Univ., Ithaca, NY.

Gamerdinger, A.P., A.T. Lemley, and R.J. Wagenet. 1991. Nonequilibrium sorption, and degradation of three 2-chloro-*s*-triazine herbicides in soil-water systems. J. Environ. Qual. 20:815–822.

Gamerdinger, A.P., R.J. Wagenet, and M.Th. van Genuchten. 1990. Application of two-site/two-region models for studying simultaneous nonequilibrium transport and degradation of pesticides. Soil Sci. Soc. Am. J. 54:957–963.

Karickhoff, S.W., and K.R. Morris. 1985. Sorption dynamics of hydrophobic pollutants in sediment suspensions. Environ. Toxicol. Chem. 4:469–479.

Lassey, K.R. 1988. Unidimensional solute transport incorporating equilibrium and rate-limited isotherms with first-order loss. 1. Model conceptualizations and analytic solutions. Water Resour. Res. 24:343–350.

Lemley, A.T., R.J. Wagenet, and W.Z. Zhong. 1988. Sorption and degradation of aldicarb and its oxidation products in a soil-water flow system as a function of pH and temperature. J. Environ. Qual. 17:408–414.

Rao, P.S.C., and J.M. Davidson. 1979. Adsorption and movement of selected pesticides at high concentrations in soils. Water Res. 13:375–380.

Simkins, S., and M. Alexander. 1984. Models for mineralization kinetics with the variables of substrate concentration and population density. Appl. Environ. Microbiol. 47:1299–1306.

Skopp, J. 1985. Analysis of time-dependent chemical processes in soils. J. Environ. Qual. 15:205–213.

Valocchi, A.L. 1985. Validity of the local equilibrium assumption for modeling sorbing solute transport through homogeneous soils. Water Resour. Res. 21:808–820.

van Genuchten, M.Th., and R.J. Wagenet. 1989. Two-site/two-region models for pesticide transport and degradation: Theoretical development and analytical solutions. Soil Sci. Soc. Am. J. 53:1303–1310.

Zhong, W.Z., A.T. Lemley, and R.J. Wagenet. 1986. Quantifying pesticide adsorption and degradation during transport through soil to groundwater. p. 61–67. *In* W.Y. Garner et al. (ed.) Evaluation of pesticides in groundwater. ACS Symp. Ser. no. 315. Am. Chem. Soc., Washington, DC.

8 Effect of Sorption on the Degradation of Aromatic Acids and Bases

Calvin C. Ainsworth, James K. Fredrickson, and Steven C. Smith

Pacific Northwest Laboratory
Earth and Environmental Sciences Center
Richland, Washington

The biodegradation rate of xenobiotic organic chemicals in soils is often observed to depend on the aqueous substrate concentration, and a substrate's interaction with a solid phase can significantly affect its degradation rate. Several investigators have suggested that desorption from soils and sediments tends to limit the biodegradation rate of pesticides and organic contaminants (Stevens et al., 1990; Heitkamp et al., 1984; Mihelcic & Luthy, 1988; Subba-Rao & Alexander, 1982). Rijnaarts et al. (1990) concluded that desorption is the rate-limiting factor controlling degradation of α-hexachlorocyclohexane, and that intraparticle mass-transfer processes control desorption. The sorption/desorption kinetics of hydrophobic organic compounds (HOCs) from soils, sediments, and surfaces in general are often characterized by a rapid initial uptake (or release) followed by a slow approach toward equilibrium (Wu & Gschwend, 1986; Leenheer & Ahlrichs, 1971; Karickoff, 1981; Brusseau & Rao, 1989). This slow reaction (sorption or desorption) can occur on a time scale of days to months or longer (Coates & Eizerman, 1986). The long equilibration and release times observed in sorption/desorption studies have been characterized as indicating diffusion-controlled processes, such as micropore diffusion in aggregated materials (Wu & Gschwend, 1986; Steinburg et al., 1987) or intraorganic matter diffusion (Brusseau & Rao, 1989). Hence, diffusion or mass-transfer processes are often suggested to limit biodegradation of HOCs.

Unlike HOCs, the reactions that ionizable organic compounds undergo at the solid/water interface are more chemical in nature than the entropy-driven partitioning associated with HOC sorption; these interactions include van der Waals forces, H-bonding, dipole-dipole interactions, ion exchange, covalent bonding, protonation, ligand exchange, cation bridging, and water bridging (Koskinen & Harper, 1990). Depending on the strength of binding, sorbate-solid interactions could cause desorption kinetics to limit biodegradation or, as in the case of diquat, to inhibit degradation totally (Weber & Coble, 1968).

 Sorption and Degradation of Pesticides and Organic Chemicals in Soil. SSSA Special Publication no. 32.

Desorption is a combination of chemical kinetic and mass-transfer processes, and thus, it is difficult to unambiguously distinguish which is the slow step in controlling the overall desorption rate, given a heterogeneous system such as soil. This difficulty primarily arises from the fact that, for a given compound, sorption to soil may be related to a continuum of sorption mechanisms and sorption sites (Koskinen & Harper, 1990). For instance, sorption mechanisms for the triazine herbicides have been suggested to include hydrophobic interactions, cation exchange, van der Waals forces, cation bridging, and charge transfer, and both organic C and clay minerals have been viewed as sorption sites (Hayes, 1970; Koskinen & Harper, 1990). Therefore, to experimentally test hypotheses concerning chemical kinetic effects on biodegradation rates, it is necessary to minimize the mass-transfer processes inherent in multiphase systems. Obtaining information regarding a particular sorption mechanism is best accomplished by minimizing the number of other sorption mechanisms controlling the solid/water partitioning of the target compound. This requirement is best served by investigations conducted in mono-mineralogic systems. However, as one might expect, there is only limited information in the literature regarding sorbate degradation in such systems.

The purpose of this discussion is to illustrate the possible effects that sorption has on the biodegradation of selected ionizable organic compounds sorbed to single, pure mineral phases. In turn, the effects of sorption on degradation kinetics will be related to mechanisms of sorption. This is accomplished using both published data and data from our laboratory. This discussion is separated into four sections: (i) availability of sorbed compounds, (ii) organic cation sorption/degradation, (iii) organic acid sorption/degradation, and (iv) conclusions. Because of the paucity of data on this subject, particularly with regard to pure phase studies, few concrete conclusions may be drawn; hence the discussion is largely illustrative. Reactions of particular interest that are discussed are cation exchange of organic bases to clay minerals and surface complexation of several organic acids to γ-Al_2O_3. In an attempt to isolate the desorption process from the degradation process, fluorescence spectroscopy was used in studies with the organic bases quinoline and 2-hydroxyquinoline sorbed to hectorite and with salicylic acid in the presence of γ-Al_2O_3. The use of fluorescence spectroscopy as a tool to isolate desorption processes from biodegradation processes is discussed, as is the use of this technique in conjunction with $^{14}CO_2$ evolution studies.

AVAILABILITY OF SORBED SUBSTRATES

A common perception is that sorption of a substrate to soil inhibits or at least reduces its availability to microbial attack. Most often, reduced availability of a substrate for microbial utilization is the conclusion drawn from estimates of the aqueous biodegradation rate and observed decreases in that rate in the presence of a solid phase; if a decreased degradation rate is observed, desorption is considered limiting. If one assumes that the sorbed species is unavailable, then a simplified model of the sorption/desorption process is

$$S_1 \underset{k_2}{\overset{k_1}{\rightleftharpoons}} S_2 \overset{k_3}{\rightarrow} \text{Products} \qquad [1]$$

where S_1 and S_2 are substrate in the unavailable (sorbed) and available (aqueous) forms, respectively (Hamaker & Goring, 1976), and k_1, k_2, and k_3 are first-order rate constants for sorption, desorption, and mineralization, respectively. If this model is accurate, then desorption must occur prior to degradation. However, this does not mean that desorption will necessarily control the degradation rate. On the other hand, however, if this model is accurate, (i) sorption could retard biodegradation rates by decreasing the aqueous concentration of the substrate (S_2) to levels below the threshold level necessary for microbial utilization, or (ii) the rate of desorption (k_2) could be rate limiting with respect to the mineralization process (k_3).

Studies in soils and sediments with isopropylphenyl phosphate (Heitkamp et al., 1984), chloroproham and di-*n*-butyl phthalate (Steen et al., 1980), and hexachlorohexane (Rijnaarts et al., 1990) suggest that the sorbed substrate is protected from microbial attack. For hexachlorohexane, Rijnaarts et al. (1990) concluded that intraparticle mass-transfer processes controlled desorption and biodegradation because the degrading microorganism could not penetrate the interior of the soil aggregates. From model evaluation of 2,4-D (2,4-dichlorophenoxy acetic acid) soil desorption/biodegradation rates, Ogram et al. (1985) concluded that sorbed 2,4-D was completely protected from microbial attack. Furthermore, they demonstrated that both sorbed and solution-phase bacteria were able to use aqueous-phase 2,4-D only. In addition, the organisms (whether sorbed or in the aqueous phase) were equally efficient in degrading solution-phase 2,4-D. In general, these studies suggest that sorption protects the substrate from microbial attack and that desorption limits biodegradation. Remberger et al. (1986), however, suggested that sediment-sorbed chloroguaiacols, chlorocatechols, and chloroveratroles may be accessible to biological transformations. Similar results have been observed in pure mineral studies in which the substrate was sorbed to the surface via mechanisms other than hydrophobic partitioning and was apparently accessible to microorganisms. Thus, there are conflicting conclusions with regard to the bioavailability of sorbed organic compounds.

Weber and Coble (1968) found that no degradation of diquat occurred in an aqueous suspension of montmorillonite when diquat was in the sorbed state, indicating that microorganisms could not use the sorbed substrate. In this study, sorption probably decreased the aqueous concentration of diquat to levels below the threshold level necessary for microbial utilization; clearly, sorbed diquat was protected from microbial attack. However, even when the aqueous concentration of diquat was above the threshold level for microbial utilization, degradation was so slow that no conclusions could be drawn regarding the role of desorption from either montmorillonite or kaolinite.

In other pure clay-sorbate systems, dextran (Olness & Clapp, 1972), quinoline (Smith et al., 1992), aspartate, and cysteine (Dashman & Stotzky, 1986), alkylamines (Wszolek & Alexander, 1979), and benzylamine (Miller & Alexander, 1991) sorbed to montmorillonite were apparently at least partially protected from microbial attack. The last two of these investigations, however, demonstrated that the effect of sorption on the biodegradation rate depended on inoculum size. Similarly, degradation of salicylic acid, phthalic acid, and catechol was significantly reduced when they were sorbed to γ-Al_2O_3 (C.C. Ainsworth et al., 1992, unpublished data).

Wszolek and Alexander (1979) observed that at population densities of approximately 2×10^8 cells mL^{-1}, degradation of *n*-decylamine sorbed to montmorillonite was about 32 times slower than for the aqueous phase. However, at greater cell densities (6 and 20×10^9 cells mL^{-1}), degradation was faster in the presence of montmorillonite than in its absence. C.C. Ainsworth et al. (1992, unpublished data) studied the degradation of phthalate by *Pseudomonas putida* NMH102-2 and *P. cepacia* PHK in the presence and absence of γ-Al_2O_3. The rate of phthalate mineralization differed not only between the presence or absence of γ-Al_2O_3, but also between microorganisms. The latter observation was particularly intriguing, given that the numbers of microorganisms were the same [approximately 10^7 colony-forming units (CFU) per milliliter] and that both microorganisms use the same metabolic pathway to mineralize phthalate (i.e., mineralization proceeds through a 4,5-dihydroxyphthalate intermediate). In solution, the *P. cepacia* mineralized the phthalate much faster than did *P. putida*. Phthalate mineralization by *P. putida* (as measured by $^{14}CO_2$ evolution) showed evidence of desorption-limited microbial utilization and hence protection of the sorbed substrate (Fig. 8–1); however, phthalate sorption to γ-Al_2O_3 did not greatly affect the rate of phthalate utilization by *P. cepacia* PHK (Fig. 8–1b). Similarly, Guerin and Boyd (1992) observed that the rate and extent of naphthalene mineralization by *P. putida* (ATCC 17484) and a gram-negative

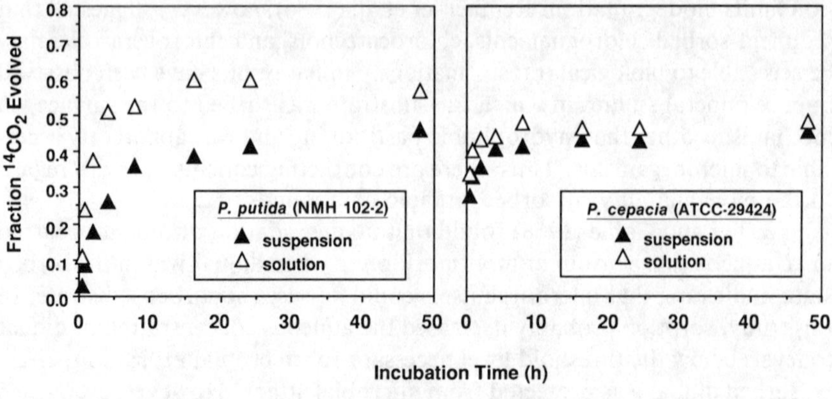

Fig. 8–1. Phthalate mineralization over time in the presence or absence of γ-AL_2O_3 (1.0 g L^{-1}) by (a) *Pseudmonas putida* NMH102-2 and (b) *P. cepacia* PHK at a cell density of 10^7 CFU mL^{-1} and a total concentration of $10^{-6.5}$ *M* phthalate (C.C. Ainsworth et al., 1992, unpublished data).

soil isolate were vastly different. The rate and extent of naphthalene mineralization by *P. putida* were indicative of an enhanced soil desorption rate, but the soil isolate exhibited desorption-limited degradation with regard to both rate and extent of mineralization.

Based on the criteria stated earlier and the data noted above, a substrate in the sorbed state may not be protected from microbial utilization under all conditions. A simplified two-compartment model of this phenomenon is:

$$\begin{array}{ccc} S_1 & \underset{k_2}{\overset{k_1}{\rightleftharpoons}} & S_2 \\ {}_{k_4}\searrow & & \swarrow_{k_3} \\ & \text{products} & \end{array} \qquad [2]$$

where S_1, S_2, k_2, and k_3 have the same definitions as in reaction 1 and k_4 is the first-order rate constant for sorbate utilization directly from the surface (Scow et al., 1986). In fact, this model provided the best estimate of nitrilotriacetic acid and phenol mineralization in soil (Scow et al., 1986). This model has also been used in other investigations (Alexander and Scow, 1989). While the model may provide the best fit of data under certain conditions, no data have been published that substantiate the physical nature of the model. Wszolek and Alexander (1979) speculated that at high population densities of the *n*-decylamine degrader used in their study, extracellular enzymes or microbially mediated pH change facilitated *n*-decylamine desorption from the bentonite. Although the mechanism(s) responsible for the observations in the phthalate study also have not been determined, pH changes mediated by either microorganism in the supernatant were minimal (<0.5 pH units). Also, phthalate mineralization by *P. cepacia* was so rapid that it is unlikely that extracellular enzymes were involved. The phthalate results, however, would suggest that (i) significant differences exist between different microorganisms and their ability to use sorbed substrate, and (ii) while both phthalate degraders use the same pathways for phthalate metabolism, their ability to scavenge substrate differs significantly. Guerin and Boyd (1992) concluded that the observed differential bioavailability of soil-sorbed naphthalene for the two microorganisms is the result of organism-specific properties. From the few examples discussed here and others (Stotzky, 1986), it is evident that little is known mechanistically about coupled processes (chemical and biological) and their interactions regarding sorbed substrate utilization by microorganisms.

SORPTION-DEGRADATION OF ORGANIC BASES

Sorption

Exchange reactions between inorganic cations or anions have been characterized as rate-limited mass-transfer reactions with either diffusion through a water film (film diffusion; FD) or intraparticle diffusion (PD) being the rate-limiting process (Sparks, 1989). The chemical reaction(s) associated with the exchange/desorption process has never been definitively demonstrated in soils or clay systems. Nor have chemical reactions or kinetics been suggested to be rate limiting, as it has always been assumed that exchange reactions are too rapid to affect the overall process rate. Liberti and Passino (1985) reported, however, that the exchange rate is affected by chemical interactions between a sorbate and the exchanger phase in gel-type resins that exhibit high selectivity.

High selectivity for clay surfaces is a characteristic of a variety of organic cations and bases (Zachara et al., 1990; Ainsworth et al., 1987; Hayes et al., 1978; Thompson & Brindley, 1969; Lailach et al., 1968; Weber & Weed, 1968; Weber et al., 1965). Sorption of paraquat (1,1′-dimethyl-4,4′-bipyridinium ion) and diquat (1,1′-ethylene-2,2′-bipyridylium ion) to the surface of clay minerals occurs via cation exchange, and both are sorbed up to the exchange capacity of the clay. These pesticides exhibit a high affinity for the surfaces of montmorillonite, vermiculite, and kaolinite (Hayes et al., 1978; Weed & Weber, 1969; Weber & Weed, 1968; Weber et al., 1965). For fully exchanged montmorillonite and vermiculite, only about 5% of the sorbate is desorbed by 1 *M* $BaCl_2$ (Weber & Weed, 1968). Kaolinite, on the other hand, releases approximately 80% of the sorbate in 1 *M* $BaCl_2$ (Weber & Weed, 1968). The difference in paraquat and diquat desorbability between montmorillonite or vermiculite and kaolinite may be due to adsorption location (basal plant sorption for the 2:1 clays vs. edge sorption with kaolinite) and the sorbate's flat orientation parallel to the basal plane in the 2:1 clay systems. In addition, lamellar collapse of montmorillonite and vermiculite, at high surface loading, onto the flattened organocation further increased their binding intensity. Sorption of a series of *n*-alkylammonium ions (Theng et al., 1967; Vansant & Uytterhoeven, 1972) and α,ω–bistrimethylammoniumalkanes (Maes et al., 1980) to montmorillonite exhibited increased selectivity with increasing molecular weight. Similarly, the selectivity (as denoted by the calculated selectivity coefficient [K_v]) of a series of *N*-heterocyclic compounds with Na-montmorillonite was observed to increase as ring number increased (K_v increases in the order acridine > quinoline > pyridine; Zachara et al., 1990).

Maes et al. (1980) observed a decrease in the enthalpy and free energy of exchange (ΔH^o_{ex} and ΔG^o_{ex}, respectively) with increasing chain length for α,ω–bistrimethylammoniumalkanes exchange to Camp Berteau Na^+- or Ca^{2+}-montmorillonite. For every additional CH_2 group, the ΔH^o_{ex} and ΔG^o_{ex} decreased by approximately -0.6 and -0.88 $kJmol^{-1}$, respectively. Maes et al. (1980) attributed the observed change in ΔH^o_{ex} and ΔG^o_{ex} to delocalization of charge from the N to the alkyl chain. CDNO/2 calculations (Maes et al., 1980) showed that the formal charge on the NH_x head group was increasingly delocalized over the alkyl tail as the number of CH_2 groups increased. Maes et al. (1980) cited charge delocalization as the cause for the observed correlations between enhanced selectivity and (i) molecular weight and (ii) polarizability for exchange of straight-chain ammoniumalkanes on montmorillonite. Likewise, surface stability of pyridine over selected alkylammonium cations on the surface of montmorillonite is hypothesized to be caused by the greater polarizability of pyridine and delocalization of positive charge through the π-ring system (Maes et al., 1978). Similar molecular calculations demonstrate that the formal positive charge is distributed through the ring structure of the *N*-heterocyclic compounds pyridine, quinoline, and acridine and that the increase in selectivity with ring number was closely correlated with formal charge on the ring N and with solvation energy (Zachara et al., 1990). Similar interactions have been suggested to stabilize paraquat and diquat on the surface of montmorillonite (Hayes et al., 1978). The high selectivity associated with these interactions clearly results from forces other than simple coulombic interaction. However, the high binding affinity of organic cations and bases to clays and the effect of binding intensity on sorbate degradation rate is not well understood.

Degradation

Wszolek and Alexander (1979) followed microbial mineralization rates, via evolution of ^{14}C, from a homologous series of *n*-alkylamines in aqueous solution and compared them to the mineralization rates of the same series of *n*-alkylamines sorbed to bentonite. The data suggest that as the molecular weight of the sorbate increased, so did its resistance to microbial attack. The correlation between resistance to microbial attack and molecular weight would appear to be similar to the correlations noted by Maes et al. (1980) and Cowan and White (1958), who demonstrated that ΔG^{o}_{ex} decreased linearly with increasing chain length (molecular weight) for the sorption of a series of *n*-alkylamines to bentonite. Unfortunately, selectivity coefficients were not determined in the Wszolek and Alexander (1979) study, and hence, no definitive relationship can be determined between sorption selectivity of the *n*-alkylamines and their degradation.

The degradation of *n*-decylamine sorbed to bentonite was observed to be about 32 times slower in the presence of bentonite than in the absence of bentonite at high bacterial densities (2×10^8 cells mL^{-1}). Similarly, Miller and Alexander (1991) found that desorption limited degradation of benzylamine that was sorbed to montmorillonite, but only at high population densities ($>10^7$ cells mL^{-1}). The presence of bentonite did not appear to affect the microbial population in a physical sense (e.g., it did not hinder uptake) or cause metabolic problems, given that rates of glucose mineralization by the same microorganism in the presence and absence of the clay were identical.

The results of Wszolek and Alexander (1979) and Miller and Alexander (1991) suggest that desorption probably limited degradation at high isolate densities and that sorbed substrate was protected from microbial attack. However, it is only possible to speculate whether mass transfer or chemical kinetics was the rate-limiting process. The decreasing degradation rate with increasing chain length observed by Wszolek and Alexander (1979) suggests that binding strength is related to the degradation rate and that a chemical kinetic process becomes increasingly important with increasing chain length. However, an interparticle diffusion mass-transfer process analogous to that hypothesized for HOC compounds (Brusseau & Rao, 1989), where the sorbate must go through a series of adsorption/desorption steps prior to arrival in the bulk solution (and subsequent utilization), could explain the observed results just as well as a chemical kinetic process.

Quinoline (Q), an ionogenic *N*-heterocyclic compound ($pk_a = 4.92$), is strongly sorbed to smectite surfaces via cation exchange of the protonated quinolinium ion (Ainsworth et al., 1987; Zachara et al., 1990). Quinoline is further stabilized on the surface as a result of its parallel orientation to the clay surface (Greene-Kelly, 1955) and exhibits increased surface stability because of its π-electron ring structure. The thermodynamic exchange constant (K_{ex}) for Na $\Rightarrow$ Q exchange is 340 (Ainsworth et al., 1987).

Smith et al. (1992) used a Q-degrading bacterium at relatively high population densities ($\approx 10^7$–10^8 CFU mL^{-1}) and five concentrations of Na-montmorillonite (0.0–1.14 cmol$_{(c)}$ L^{-1}) to probe the effects of sorption on Q mineralization. Quinoline is rapidly converted to a 2-hydroxyquinoline intermediate by the bacterium (Brockman et al., 1989). In the absence of clay, both Q and 2-hydroxyquinoline were observed in solution after 30 min. In the presence of Na-montmorillonite, 2-hydroxyquinoline was oberved only within the first 30 min and only in the least dense clay suspensions (Smith et al., 1992). In addition, the overall Q mineralization rate was reduced, with the reduction depending on clay concentration and hence a reflection of the Q solid/aqueous phase partitioning (Fig. 8-2); the initial total Q concentration was constant, but as clay concentration was increased, the aqueous con-

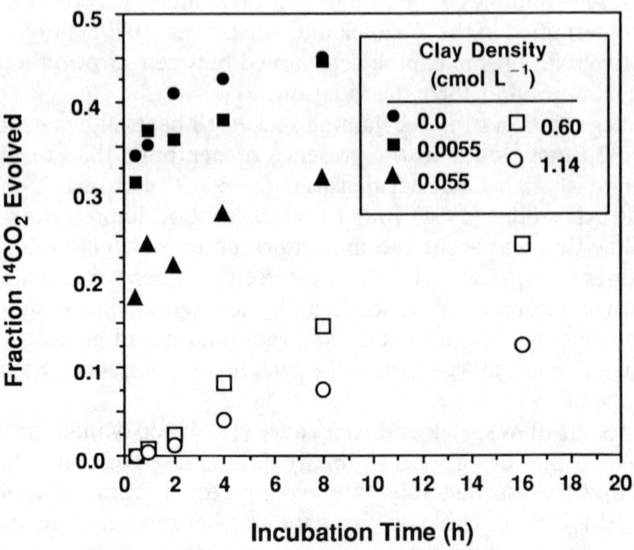

Fig. 8-2. Mineralization over time of 10^{-6} *M* quinoline by a quinoline-degrading bacterium ($\approx 10^7$ CFU mL^{-1}) as a function of Na-montmorillonite suspension density. (After Smith et al., 1992.)

centration decreased from 89 to 6.9% of the total Q and surface coverage decreased from 09.32 to 0.013 cmol(QH^+) kg^{-1}. The evolution of $^{14}CO_2$ apparently proceeded at two distinct rates. The initial rates appeared to be directly related to the aqueous Q concentration (and therefore the clay concentration), and a much slower rate became apparent after aqueous Q was depleted. The latter rate decreased slightly with increasing clay concentration. Modeling the $^{14}CO_2$ data as first-order rates yielded a rate constant (k_{obs}) that was about 20 to 30 times lower in the presence of clay than the rate constant for aqueous mineralization.

The observations of Smith et al. (1992), Miller and Alexander (1991), and Wszolek and Alexander (1979) strongly suggest that desorption from the clay surface limits mineralization of organic bases sorbed through a cation exchange mechanism. However, conclusions about the actual desorption rate and its impact on mineralization cannot be made because such conclusions would be drawn from observations of the end product (i.e., $^{14}CO_2$), which is produced by a series of metabolic reactions that are predominantly internal to the cell.

In an attempt to delineate more clearly the effect of desorption on the overall process of Q degradation, Smith et al. (1982) employed fluorescence spectroscopy to take advantage of the fluorescence properties of Q, the quinolinium ion, and its microbial transformation product, 2-hydroxyquinoline. The quinolinium ion and 2-hydroxyquinoline are both fluorophores whose fluorescence maxima differ by about 40 nm. Quinoline does not fluoresce. Hectorite (a trioctahedral smectite) was used in these studies instead of montmorillonite because hectorite does not quench the quinolinium or 2-hydroxyquinoline fluorescence. In addition, the Q-degrading bacterium does not fluoresce and the fluorescence of both fluorophores is totally quenched when they are taken up by the bacterium. Because of these qualities, the transformation of quinolinium to 2-hydroxyquinoline and its uptake from aqueous solution could be traced directly by simultaneously placing aqueous Q and the bacterium into a cuvette and repeatedly acquiring spectra with time. The same process can be used to trace the changes in concentration of quinolinium ion on the surface of hectorite as the bacterium uses the aqueous-phase Q, causing desorption of the surface-bound substrate.

Using this technique, Smith et al. (1992) observed that the transformation of quinolinium to 2-hydroxyquinoline was extremely rapid, and that loss of 2-hydroxyquinoline from solution was approximately five to eight times slower than the transformation of quinolinium to 2-hydroxyquinoline (Fig. 8–3). In a hectorite suspension (pH 6.0) previously equilibrated with Q (>99% of the Q partitioned to the surface), the disappearance of quinolinium differed from the pattern in the aqueous system in three major aspects (Fig. 8–4): (i) the rate of quinolinium disappearance from the surface of hectorite was about 30 times slower than in solution, (ii) the transient build-up of 2-hydroxyquinoline observed in the aqueous system did not occur in the presence of clay, and (iii) examination of the aqueous phase showed the Q and 2-hydroxyquinoline concentrations were below detection levels ($<10^{-8}$ *M*) over the entire course of the study. Smith et al. (1992) concluded that, in

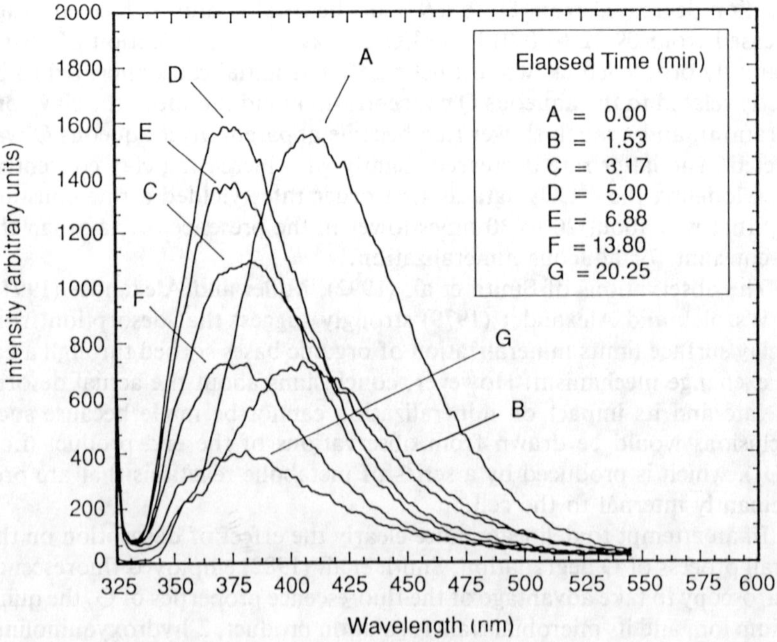

Fig. 8–3. Fluorescence emission spectra with time of aqueous quinolinium transformation to 2-hydroxyquinoline and its subsequent utilization ($10^{-4.30}$ *M* initial quinoline; 10^8 CFU mL^{-1} cell density) (After Smith et al., 1992.)

the absence of clay, the uptake of 2-hydroxyquinoline is the rate-limiting step in Q mineralization, but in the presence of clay, desorption is rate limiting. Put in the context of the Hamaker and Goring (1976) model (Reaction 1), $k_3 >> k_2$ and the back reactions, represented by k_1, does not occur. Interestingly, the difference in the rates of quinolinium disappearance between the aqueous and clay suspensions, as determined by fluorescence, was about the same as that determined using $^{14}CO_2$ mineralization (i.e., about a factor of 30) (Smith et al., 1992). This observation suggests that the biological processes are not affecting the observed rates; that is, microbial processes after uptake are not rate limiting. This is not always true, however, as will be discussed with regard to salicylate biodegradation in the next section.

The disappearance of Q from the surface of hectorite, as a function of surface loading, yielded a family of parallel curves whose slopes appeared to be the same (Fig. 8–5; Smith et al., 1992). Additionally, the rate of quinolinium disappearance from the clay surface was not affected by the saturation cation (Na or Ca) nor by stirring rate. Smith et al. (1992) concluded that desorption was the rate-limiting step in these studies, and that a chemical process rather than a mass-transfer process controlled desorption. Also, data from the fluorescence studies indicate that either (i) Q at surface concentrations ≤ 0.004 cmol(QH^+) kg^{-1} is irreversibly sorbed, or (ii) the aqueous Q concentration supported by such surface concentrations is less than the critical level necessary for microbial utilization. These conclusions reflect

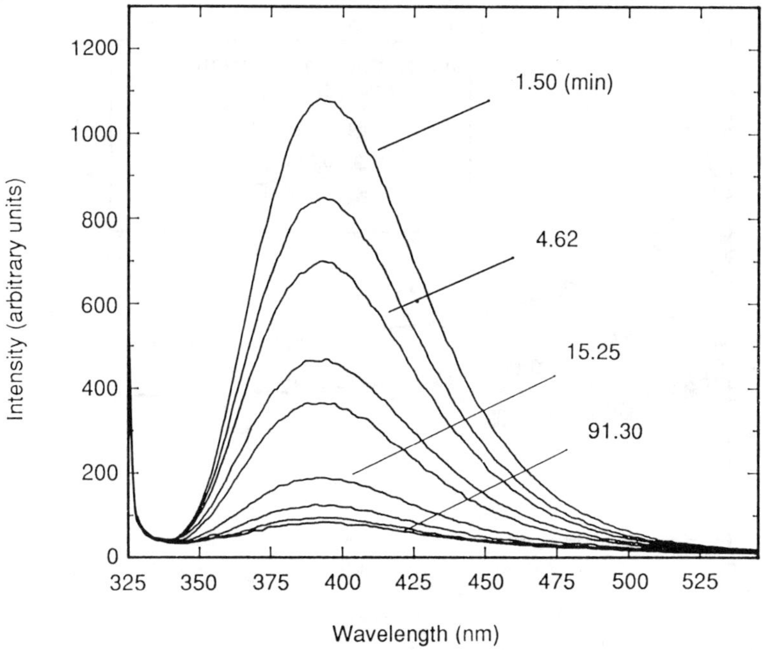

Fig. 8-4. Fluorescence emission spectra with time of quinolinium disappearance from the surface of hectorite as a result of aqueous quinoline biodegradation by a quinoline-degrading bacterium. (After Smith et al., 1992.)

the fact that the quinolinium surface concentration appears to asymptomatically approach 0.004 cmol(QH^+) kg^{-1} at all initial concentrations. Even after 17 h, the quinolinium surface concentration was still slightly above this value.

One of the major determining factors associated with identifying desorption as a rate-limiting step in the biodegradation of an organic cation or base is the interaction between the sorbate and the surface. For instance, the investigations of Wszolek and Alexander (1979) with a homologous series of *n*-alkylamines indicate that there is a relationship between desorption, microbial utilization, and binding intensity. Similarly, there appears to be a considerable difference between the utilization of sorbed Q and that of 2-hydroxyquinoline (C.C. Ainsworth et al., 1991, unpublished data). Sorption of 2-hydroxyquinoline to the hectorite surface occurs via a cation exchange reaction, but the selectivity coefficient is about three to four times less than that of Q. Using fluorescence spectroscopy and the procedure outlined by Smith et al. (1992), the disappearance of quinolinium and 2-hydroxyquinoline in the presence of Na-hectorite and the Q-degrading bacterium was followed with time (Fig. 8-6). 2-Hydroxyquinoline was observed to disappear at a faster rate than Q. These data are in direct contradiction to those for aqueous solution, where Q transformation to 2-hydroxyquinoline was approximately five to eight times faster than the uptake of 2-hydroxyquinoline. Interestingly, in the sorbed state, the quinolinium-surface com-

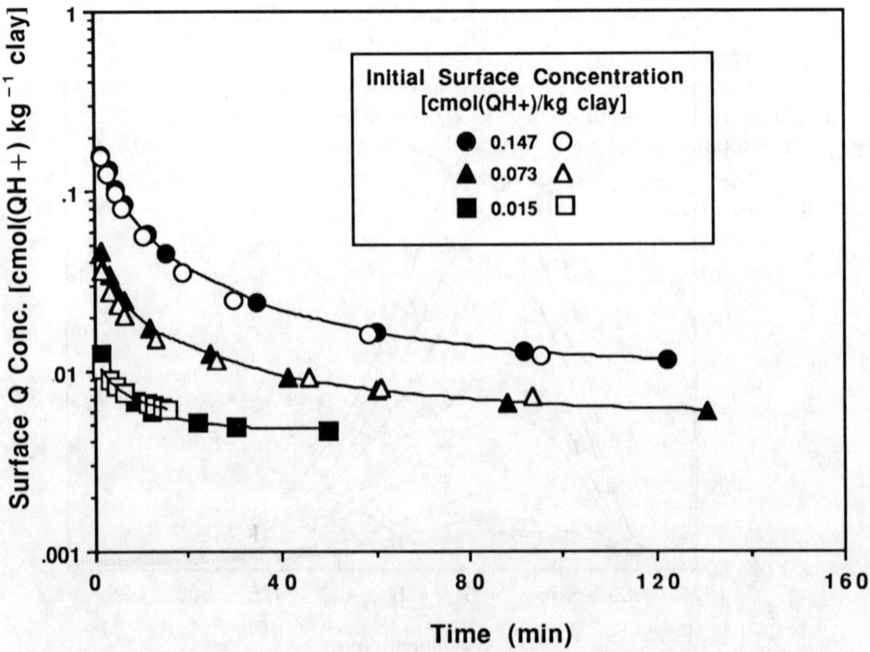

Fig. 8–5. Hectorite-surface concentration of quinolinium over time as a function of initial surface loading. The open and closed symbols represent duplicate runs. Data were collected from fluorescence emission spectra of a suspension over time. (After Smith et al., 1992.)

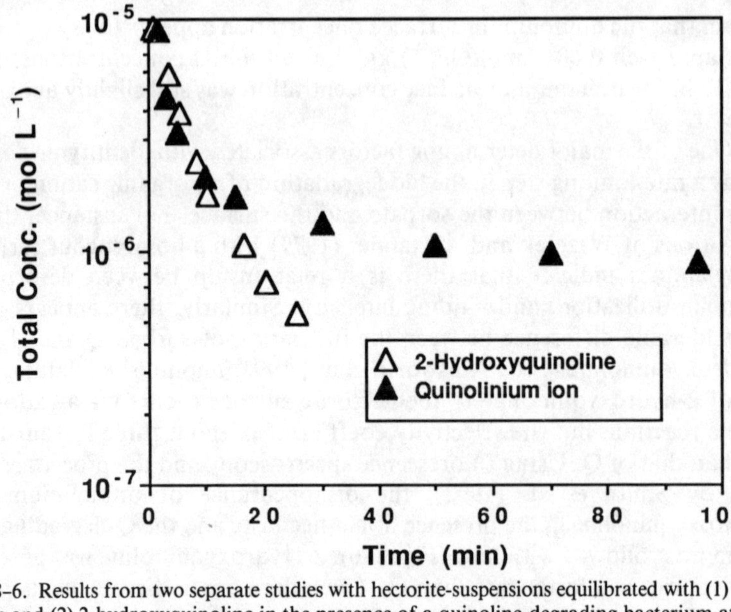

Fig. 8–6. Results from two separate studies with hectorite-suspensions equilibrated with (1) quinoline and (2) 2-hydroxyquinoline in the presence of a quinoline-degrading bacterium at equal cell densities and total concentrations of quinoline and 2-hydroxyquinoline. Data were collected from fluorescence emission spectra of the suspension with time (C.C. Ainsworth et al., 1991, unpublished data).

plex exhibits a 15- to 20-nm blue shift, whereas the 2-hydroxyquinoline-surface complex exhibits no such shift in the maximum wavelength emission. While it is not possible to conclusively demonstrate that this lack of shift is related to binding intensity, Lakowicz (1986) suggests that a blue shift in the emission spectra of a fluorophore bound to a macromolecule can be caused by emission from unrelaxed fluorophores, typically due to a strong interaction between the fluorophore and its surrounding environment.

SORPTION-DEGRADATION OF ORGANIC ACIDS

Salicylate, phthalate, and catechol sorb strongly to γ-Al_2O_3 [salicylic (2-hydroxybenzoic acid) $pK_{a1} = 2.97$, $pK_{a2} = 13.40$; pyrocatechol (1,2-dihydroxybenzene) $pK_{a1} = 9.2$, $pK_{a2} = 13.0$; phthalic (1,2-benzenedicarboxylic acid) $pK_{a1} = 2.8$, $pK_{a2} = 4.9$] (Fig. 8-7). Sorption of cathechol, salicylate, and phthalate (and other organic acids) to Fe- and Al-oxide surfaces occurs via a surface ligand exchange reaction (Kung & McBride, 1989; McBride & Wesselink, 1988; Zeltner et al., 1986; Kummert & Stumm, 1980; Parfitt et al., 1977; Nagarajah et al., 1970). The degree of sorption varies in response to the pH, surface charge, aqueous speciation of the sorbates, and other aqueous- and solid-phase phenomena. Hence, the fractional adsorption of the acids with changes in pH is a reflection of the interactions

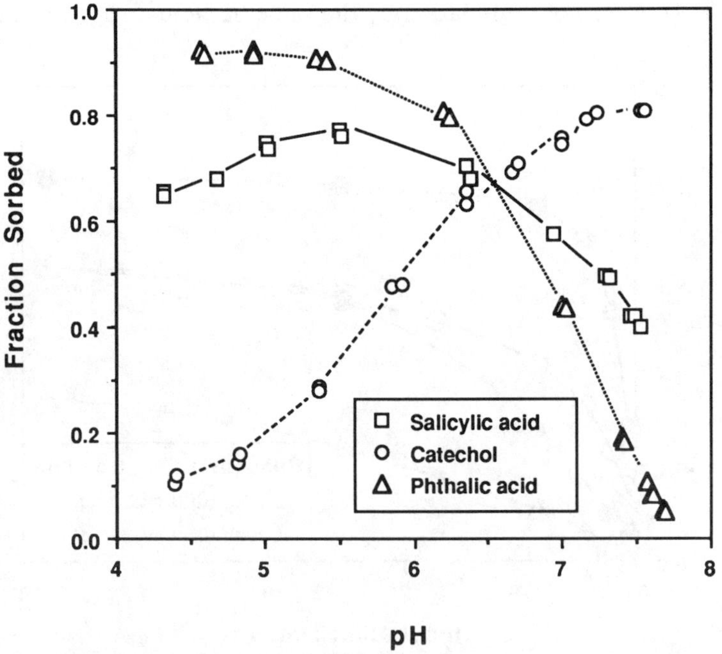

Fig. 8-7. Fractional adsorption of three organic acids ($10^{-6.5}$ *M*) on γ-Al_2O_3 suspension (1.0 g L^{-1}) as a function of pH.

of these phenomena. All experiments reported here were conducted at pH 6.5. Prior to use, the γ-Al_2O_3 (1.0×10^5 m^2 kg^{-1}) was washed in 0.1 *M* NaOH and eluted with distilled water to pH 7.5. Bacteria used in all of the studies were grown in mineral media that contained the specific organic acid substrate as the sole C source. Bacterium were washed three times in a 0.9% saline solution just prior to use, to remove substrate from the cell suspension. For the sorption/degradation experiments, the substrate was equilibrated with γ-Al_2O_3 for 48 hr, during which time the pH was adjusted to value of 6.5. A fresh cell suspension was used for each experiment.

Pseudomonas putida 1901, a methionine auxotroph of *P. putida* P_pG1 (Yen & Gunsalus, 1982) was used in experiments to assess the influence of sorption on the biodegradation of salicylate and catechol. *Pseudomonas putida* 1901 degrades naphthalene through salicylate and catechol intermediates. The mineralization of salicylate by *P. putida* 1901 (10^7 CFU mL^{-1}) may be simulated using a first-order rate whose rate constant does not change over a wide range of substrate concentrations. In the presence of γ-Al_2O_3, where about 80 to 90% of the substrate is sorbed (C.C. Ainsworth et al., 1992, unpublished data), the rate of salicylate mineralization is substantially decreased (Fig. 8–8). In addition, $^{14}CO_2$ evolution studies of salicylate mineralization with and without γ-Al_2O_3 and over a range in *P. putida* 1901 cell densities (10^4–$10^{9.4}$ CFU mL^{-1}) have demonstrated that (i) the aqueous salicylate mineralization rate constant is independent of cell density, and (ii) the rate of salicylate mineralization in the presence of γ-Al_2O_3 is slower than in the absence of γ-Al_2O_3 by about a factor of 5, regardless of cell density (Table 8–1). Catechol degradation by the same bacterium shows a similar

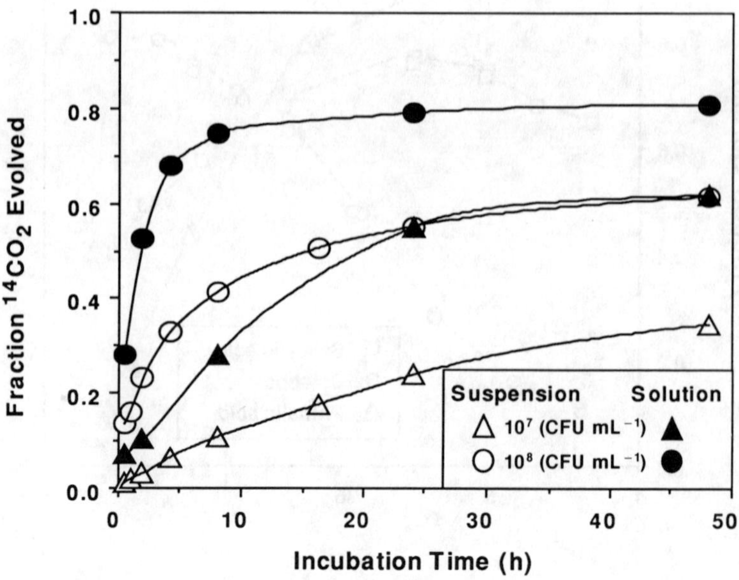

Fig. 8–8. Salicylate mineralization over time in the presence or absence of γ-Al_2O_3 at pH 6.5 and as a function of *Pseydomonas putida* 1901 population density.

Table 8-1. First-order rate constants (h^{-1}) for the mineralization of salicylate from aqueous solution (aqueous rate constant), and in the presence of γ-Al_2O_3 (sorbed rate constant) as a function of *Pseudmonas putida* 1901 cell density (log CFU†mL^{-1}).

Cell density	Aq. rate constant	Sorbed rate constant	Aq./Sorbed
log CFU mL^{-1}	h^{-1}	h^{-1}	
9.36	−0.819	−0.175	5.9
8.52	−0.257	−0.069	3.8
7.49	−0.032	−0.011	2.9
7.08	−0.015	−0.003	5.0

† CFU = colony-forming unit.

response to cell density and the presence of γ-Al_2O_3, but degradation of the sorbed substrate is slightly more rapid than that observed for salicylate.

Direct observation of desorption and microbial uptake of salicylate was obtained using fluorescence spectroscopy as Smith et al. (1992) described. Unlike the $^{14}CO_2$ evolution studies, however, these studies used a salicylate aqueous concentration of 5×10^{-5} *M* of C and 1 g L^{-1} γ-Al_2O_3. At this concentration, only about 25% of the salicylate was partitioned to the solid surface. Interestingly, during the first 200 min of the experiment, the rate of salicylate disappearance due to microbial uptake was the same in either the presence or the absence of γ-Al_2O_3 (Fig. 8-9). However, in the presence of γ-Al_2O_3, the rate of salicylate disappearance decreased by more than an order of magnitude (about a factor of 10.3) in comparison to the aqueous-phase-only salicylate. The 200-min time frame corresponds to the time required for the depletion of aqueous-phase salicylate (Fig. 8-9).

Although it is difficult to precisely determine the actual concentrations in these studies, extrapolation of the data for the later times (>200 min) back to t = 0 yields an intensity (in arbitrary units) that is approximately equal to that associated with the surface concentration at t = 0. These observations suggest that the initial disappearance rate is the result of microbi-

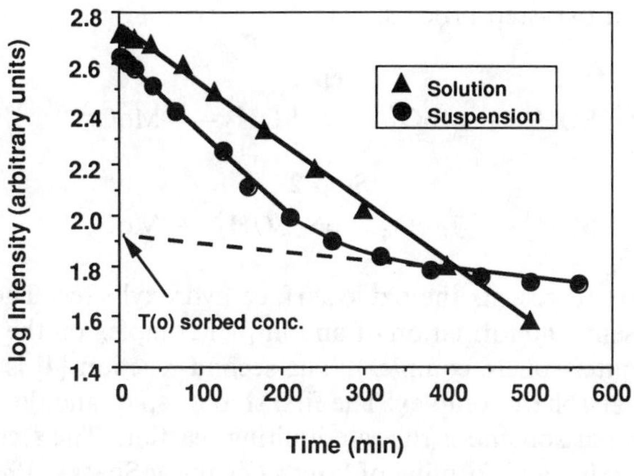

Fig. 8-9. Disappearance of salicylate from aqueous solution and γ-Al_2O_3 suspension (total salicylate = 10^{-5}; pH 6.5; γ-Al_2O_3 1.0 g L^{-1}; *Pseudomonas putida* density = 10^7 CFU mL^{-1}). Data were collected from fluorescence emission spectra taken over time.

al uptake of the aqueous-phase salicylate, and the later rate is the direct result of desorption controlling the rate of microbial uptake of salicylate. The fluorescence and $^{14}CO_2$ evolution data suggest that (i) salicylate is at least partially protected from microbial attack in the sorbed state, (ii) desorption limits degradation in the absence of or in limited amounts of aqueous-phase substrate, and (iii) desorption may be limited by a surface, chemical-controlled desorption reaction.

The difference in the utilization rate of salicylate in the presence and absence of γ-Al_2O_3 as determined by fluorescence spectroscopy (Fig. 8–9) is greater than the corresponding difference determined by complete mineralization (Fig. 8–9). Salicylate utilization as measured by ^{14}C evolution is about one-half that determined by fluorescence, which would suggest that there are rate-limiting steps in the salicylate-to-CO_2 pathway other than salicylate desorption that impact the overall rate as determined by $^{14}CO_2$ evolution. In the Q example, the effect of sorption on uptake was a decrease in rate of about a factor of 30, regardless of whether $^{14}CO_2$ evolution or fluorescence was used in determining the rate, suggesting that desorption is substantially slower than the microbial mineralization process. While this difference is not great, it does suggest that in addition to $^{14}CO_2$ evolution, other techniques must also be used to understand the role of sorption in substrate utilization by microorganisms.

As stated earlier, adsorption of many organic acids to oxide surfaces occurs through the formation of inner-sphere complexes as a result of replacement of surface-coordinated OH groups via ligand exchange. The "ligand-like" adsorption of these acids is typical of many anionic species on oxides, and although the sorption kinetics and mechanisms of these acids have not been investigated, inorganic ligand-like sorbates may serve as a model. Molybdate adsorption to goethite has been described as the formation of an inner-sphere ligand-like surface complex (Zhang & Sparks, 1989). From pressure-jump studies, Zhang and Sparks concluded that the molybdate adsorption mechanism is a two-step process:

Step 1

$$SOH_2^+ + MoO_4^{2-} \rightleftharpoons SOH_2^+\text{——}MoO_4^{2-} \qquad [3]$$

Step 2

$$SOH_2^+\text{——}MoO_4^{2-} \rightleftharpoons SOH_2^+ + MoO_4^{2-} \qquad [4]$$

where the SOH represents the oxide surface hydroxyl sites. The first reaction [3] represents the formation of an ion-pair complex on the protonated surface site (outer-sphere complex). The second reaction [4] is the formation of the inner-sphere complex. The first step is rapid, and the second step is slow by comparison and is the rate-limiting reaction. The second step involves the breaking and forming of bonds (Zhang & Sparks, 1989). It is believed that a similar mechanism is involved in the sorption of salicylate to γ-Al_2O_3 and that the inner-sphere complex may be bidentate or binuclear;

we also believe it is the reversal of the second step that is the rate-limiting step that affects the rate of salicylate utilization in the present study.

Salicylate, phthalate, and catechol have been hypothesized to sorb to the γ-Al_2O_3 surface as a monodentate surface complex at pH 6.5 (Kummert & Stumm, 1980). Catechol, however, has also been suggested to form bidentate or binuclear complexes on Al oxides depending on the oxide surface structure (McBride & Wesselink, 1988). Further, Zeltner et al. (1986) concluded from cylindrical internal reflection-Fourier transform infrared spectroscopy (CIR-FTIR) that salicylate forms a bidentate six-membered ring structure at the surface of goethite. Another type of complex suggested from infrared spectroscopic investigations is a binuclear complex between oxalate and the goethite surface (Parfitt et al., 1977). The formation of these different types of complexes appears to depend on the solid surface structure. We suspect that for a given compound the strength of the surface complexes would vary accordingly. For example, infrared investigations of *p*-hydroxybenzoate sorbed to different Fe oxides suggested that although bidentate binding of the carboxylate formed the dominant surface complex, goethite formed the strongest complex (Kung & McBride, 1989). It is our belief that the magnitude of the effect of desorption on microbial utilization will vary in direct response to the type of surface complex formed (i.e., monodentate, bidentate, or binuclear) and the strength of the bond(s) formed.

CONCLUSION

Sorption of organic compounds to inorganic surfaces and organic matter can protect the compound from microbial attack. This protection may arise from (i) inaccessibility of the micropores to microbes, (ii) surface stabilization against desorption of the compound, or (iii) reduction of aqueous-phase concentrations to levels below that necessary for microbial utilization. Clearly, there are biologically mediated reactions that facilitate sorbed substrate utilization, and these are organism- and cell-density-dependent. It is not known whether these interactions are related to a biologically induced shift in the controlling environmental parameters (e.g., pH), production of extracellular enzymes, or direct microbial utilization of a substrate from the surface. These reactions that facilitate substrate utilization must, however, be investigated more thoroughly and at a mechanistic level. From a scientific viewpoint, these interactions are likely to play a significant role in the behavior of xenobiotic compounds in the natural environment. From a purely applied viewpoint, these interactions may be beneficial in facilitating the remediation of contaminated soils and vadose and saturated zones.

From the few examples discussed here, it is evident that surface-substrate interactions can diminish, enhance, or have no effect on the observed rate of biodegradation of a substrate. In those systems where degradation of ionizable compounds is significantly reduced by sorption (and the sorbate is protected from microbial attack), it appears that the binding intensity and the surface conformation of the sorbate are important, as suggested for a homologous series of *n*-alkylamines (Wszolek & Alexander, 1979) or for Q compared to 2-hydroxyquinoline sorbed to montmorillonite (Smith et al., 1992). For those compounds that interact with oxides through ligand exchange, binding intensity may be related to the surface structure of the sorbent, as Kung and McBride (1989) suggested for *p*-hydroxybenzoate. However, there is a dearth of data that relate binding at the mechanistic level to biodegradation rates. The available data, however, do indicate that the interactions between substrate, surfaces, and microbial utilization are important in our understanding of organic compound fate and transport, and, therefore, more studies linking sorption and degradation at a basic level are needed.

ACKNOWLEDGMENT

This research was supported by the Subsurface Science Program, Office of Health and Environmental Research, U.S. Department of Energy (DOE). The continued support of Dr. F.J. Wobber is greatly appreciated. Pacific Northwest Laboratory is operated for DOE by Battelle Memorial Institute under Contract DE-AC06-76RLO 1830.

REFERENCES

Ainsworth, C.C., J.M. Zachara, and R.L. Schmidt. 1987. Quinoline sorption on Na-montmorillonite: Contributions of the protonated and neutral species. Clays Clay Miner. 35:121–128.

Alexander, M., and K.M. Scow. 1989. Kinetics of biodegradation in soil. p. 243–270. *In* B.L. Sawhney and K. Brown (ed.) Reactions and movement of organic chemical in soils. SSSA Spec. Publ. 22. SSSA and ASA, Madison, WI.

Brockman, F.J., B.A. Denovan, R.J. Hicks, and J.K. Fredrickson. 1989. Isolation and characterization of quinoline-degrading bacteria from subsurface sediments. Appl. Environ. Microbiol. 55:1029–1032.

Brusseau, M.L., and P.S.C. Rao. 1989. Sorption nonideality during organic contaminant transport in porous media. Crit. Rev. Environ. Control 19:33–99.

Coates, J.T., and A.W. Elzerman. 1986. Desorption kinetics for selected PCB congeners from river sediments. J. Contam. Hydrol. 1:191–198.

Cowan, C.W., and D. White. 1958. The mechanism of exchange reactions occurring between sodium montmorillonite and various n-primary aliphatic amine salts. Trans. Faraday Soc. 54:691–697.

Dashman, T., and G. Stotzky. 1986. Microbial utilization of amino acids and a peptide bound on homoionic montmorillonite and kaolinite. Soil Biol. Biochem. 18:5–14.

Greene-Kelly, R. 1955. Sorption of aromatic organic compounds by montmorillonite. Part I. Orientation studies. Trans. Faraday Soc. 51:412–424.

Guerin, W.F., and S.A. Boyd. 1982. Differential bioavailability of soil-sorbed naphthalene to two bacterial species. Appl. Environ. Microbiol. 58:1142–1152.

Hamaker, J.W., and C.A.I. Goring. 1976. Turnover of pesticide residues in soil. p. 219–243. *In* D.D. Kaufman et al. (ed.) Bound and conjugated pesticide residues Am. Chem. Soc., Washington, DC.

Hayes, M.H.B. 1970. Adsorption of triazine herbicides on soil organic matter, including a short review on soil organic matter chemistry. Residue Rev. 31:131–174.

Hayes, M.H.B., M.E. Pick, and B.A. Toms. 1978. The influence of organocation structure on the adsorption of mono- and bipyridinium cations by expanding lattice clay minerals. I. Adsorption by Na^+-montmorillonite. J. Colloid Interface Sci. 65:254–265.

Heitkamp, M.A., J.N. Huckins, J.D. Petty, and J.L. Johnson. 1984. Fate and metabolism of isopropylphenyl diphenyl phosphate in freshwater sediments. Environ. Sci. Technol. 18:434–439.

Karickoff, S.W. 1981. Semi-empirical estimation of sorption of hydrophobic pollutants on natural sediments and soils. Chemosphere 10:883–846.

Koskinen, W.C., and S.S. Harper. 1990. The retention process: Mechanisms. p. 51–78. *In* H.H. Cheng (ed.) Pesticides in the soil environment: Processes, impacts, and modeling. SSSA Book Ser. 2. SSSA, Madison, WI.

Kummert, R., and W. Stumm. 1980. The surface complexation of organic acids on hydrous γ-Al_2O_3. J. Colloid Interface Sci. 75:373–385.

Kung, K-H., and M.B. McBride. 1989. Coordination complexes of *p*-hydroxybenzoate on Fe oxides. Clays Clay Miner. 37:333–340.

Lailach, G.E., T.D. Thompson, and G.W. Brindley. 1968. Adsorption of pyrimidines, purines, and nucleosides by Li-, Na-, Mg-, and Ca-montmorillonite (clay-organic studies XII). Clays Clay Miner. 16:285–293.

Lakowicz, J.R. 1986. Principles of fluorescence spectroscopy. Plenum Press, New York.

Leenheer, J.A., and J.L. Ahlrichs. 1971. A kinetic and equilibrium study of the adsorption of carbaryl and parathion upon soil organic matter surfaces. Soil Sci. Soc. Am. Proc. 35:700–705.

Liberti, L., and R. Passino. 1985. Ion-exchange kinetics in selective systems. p. 175–207. *In* J. Marinsky and Y. Marcus (ed.) Ion exchange and solvent extraction. Marcel Dekker Book Co., New York.

Maes, A., L.V. Leemput, A. Cremers, and J. Uytterhoeven. 1980. Electron density distribution as a parameter in understanding organic cation exchange in montmorillonite. J. Colloid Interface Sci. 77:14–20.

Maes, A., P. Marynen, and A. Cremers. 1978. Thermodynamic stability of the ammonium ions of ethylenediamine and histamine in montmorillonite. Prog. Colloid Polymer Sci. 65:245–250.

McBride, M.B., and L.G. Wesselink. 1988. Chemisorption of catechol on gibbsite, boehmite, and noncrystalline alumina surfaces. Environ. Sci. Technol. 222:703–708.

Mihelcic, J.R., and R.G. Luthy. 1988. Microbial degradation of acenaphthene and naphthalene under denitrification conditions in soil-water systems. Appl. Environ. Microbiol. 54:1188–1198.

Miller, M.E., and M. Alexander. 1991. Kinetics of bacterial degradation of benzylamine in a montmorillonite suspension. Environ. Sci. Technol. 25:240–245.

Nagarajah, S., A.M. Posner, and J.P. Quirk. 1970. Competitive adsorption of phosphate with polygalacturonate, and other organic anions on kaolinite and oxide surfaces. Nature (London) 228:83–84.

Ogram, A.V., R.E. Jessup, L.T. Ou, and P.S.C. Rao. 1985. Effects of sorption on biological degradation rates of (2,4-dichlorophenoxy)acetic acid in soils. Appl. Environ. Microbiol. 49:582–587.

Olness, A., and C.E. Clapp. 1972. Microbial degradation of a montmorillonite-dextran complex. Soil Sci. Soc. Am. Proc. 36:179–181.

Parfitt, R.L., V.C. Farmer, and J.D. Russell. 1977. Adsorption on hydrous oxides. I. Oxalate and benzoate on goethite. J. Soil Sci. 28:29–39.

Remberger, M., A-S. Allard, and A.H. Nielson. 1986. Biotransformations of chloroguaiacols, chlorocatechols, and chloroveratroles in sediments. Appl. Environ. Microbiol. 51:552–558.

Rijnaarts, H.H.M., A. Bachmann, J.C. Jumelet, and A.J.B. Zehnder. 1990. Effect of desorption and intraparticle mass transfer on the aerobic biomineralization of α-hexachlorocyclohexane in a contaminated calcareous soil. Environ. Sci. Technol. 24:1349–1354.

Scow, K.M., S. Simkins, and M. Alexander. 1986. Kinetics of mineralization of organic compounds at low concentrations in soil. Appl. Environ. Microbiol. 51:1028–1035.

Smith, S.C., C.C. Ainsworth, S.J. Traina, and R.J. Hicks. 1992. The effect of sorption on the biodegradation of quinoline. Soil Sci. Soc. Am. J. 56:737–746.

Sparks, D.L. 1989. Kinetics of soil chemical processes. Academic Press, New York.

Steen, W.F., D.F. Paris, and G.L. Baughman. 1980. Effects of sediment sorption on microbial degradation of toxic substances. p. 477–482. *In* R.A. Baker (ed.) Contaminants and sediments. Vol. 1. Fate and transport, case studies, modeling, toxicity. Ann Arbor Science, Ann Arbor, MI.

Steinburg, S.M., J.J. Pignatello, and B.L. Sawhney. 1987. Persistence of 1, 2-dibromoethane in soils: Entrapment in intraparticle micropores. Environ. Sci. Technol. 21:1201–1206.

Stevens, T.O., R.L. Crawford, and D.L. Crawford. 1990. Biodegradation of dinoseb (2-*sec*-butyl-4,6-dinitrophenol) in several Idaho soils with various dinoseb exposure histories. Appl. Environ. Microbiol. 56:133–139.

Stotzky, G. 1986. Influence of soil mineral colloids on metabolic processes, growth, adhesion, and ecology of microbes and viruses. p. 305–428. *In* P.M. Huang and M. Schnitzer (ed.) Interactions of soil minerals with natural organics and microbes. SSSA Spec. Publ. 17; Soil Sci. Soc. Am., Madison, WI.

Subba-Rao, R.V., and M. Alexander. 1982. Effect of sorption on mineralization of low concentrations of aromatic compounds in lake water samples. Appl. Environ. Microbiol. 44:659–668.

Theng, B.K.G., G.J. Greenland, and J.P. Quirk. 1967. Adsorption of alkylammonium cations on montmorillonite. Clay Miner. 7:1–17.

Thompson, T.D., and G.W. Brindley. 1969. Adsorption of pyrimidines, purines, and nucleosides by Na-, Mg-, and Cu(II)-illite (clay-organic studies XVI). Am. Miner. 54:858–868.

Vansant, E.F., and J.B. Uytterhoeven. 1972. Thermodynamics of the exchange of *n*-alkylammonium ions on Na-montmorillonite. Clays Clay Miner. 20:47–54.

Weber, J.B., and H.D. Coble. 1968. Microbial decomposition of diquat adsorbed on montmorillonite and kaolinite clays. J. Agric. Food Chem. 16:475–478.

Weber, J.B., and S.B. Weed. 1968. Adsorption and desorption of diquat, paraquat, and prometone by montmorillonite and kaolinitic clay minerals. Soil Sci. Soc. Am. Proc. 32:485–487.

Weber, J.B., P.W. Perry, and R.P. Upchurch. 1965. The influence of temperature and time on the adsorption of paraquat, diquat, 2,4-D, and prometone by clays, charcoal, and an anion exchange resin. Soil Sci. Soc. Am. Proc. 29:678–688.

Weed, S.B., and J.B. Weber. 1969. The effect of cation exchange capacity on the retention of diquat^{2+} and paraquat $^{2+}$ by three-layer clay minerals. I. Adsorption and release. Soil Sci. Soc. Am. Proc. 33:379–385.

Wszolek, P.C., and M. Alexander. 1979. Effect of desorption rate on the biodegradation of *n*-alkylamines bound to clay. J. Agric. Food Chem. 27:410–414.

Wu, S., and P.M. Gschwend. 1986. Sorption kinetics of hydrophobic compounds to natural sediments and soils. Environ. Sci. Technol. 20:725–730.

Yen, K.-M., and I.C. Gunsalus. 1982. Plasmid gene organization: Naphthalene/salicylate oxidation. Proc. Natl. Acad. Sci. USA 79:874–878.

Zachara, J.M., C.C. Ainsworth, and S.C. Smith. 1990. The sorption of N-heterocyclic compounds on reference and subsurface smectite clay isolates. J. Contam. Hydrol. 6:281–305.

Zeltner, W.A., E.C. Yost, M.L. Machesky, M.I. Tejedor-Tejedor, and M.A. Anderson. 1986. Characterization of anion binding on goethite using titration calorimetry and cylindrical internal reflection-Fourier transform infrared spectroscopy. p. 142–161. *In* J.A. Davis and K.F. Hayes (ed.) Geochemical processes at mineral surfaces. ACS Symp. Ser. 323. Am. Chem. Soc., Washington, DC.

Zhang, P.C., and D.L. Sparks. 1989. Kinetics and mechanisms of molybdate adsorption/desorption at the goethite/water interface using pressure-jump relaxation. Soil Sci. Soc. Am. J. 53:1028–1034.

9 Reductive Dechlorination of Hexachlorobenzene in Wetland Soils[1]

J. H. Pardue, R. D. DeLaune, D. D. Adrian, and W. H. Patrick, Jr.

Louisiana State University
Baton Rouge, Louisiana

Microbially mediated reductive dechlorination processes may be enhanced in wetland environments where anaerobic conditions are common. Reductive dechlorination may serve to detoxify chlorinated organics entering wetland environments although in some instances the process can form metabolites that are more toxic than the parent compound. In general, toxic organics represent reduced forms of C in the environment. Halogenated organic compounds, however, are relatively oxidized and, therefore, their reduction is energetically favorable (Sims et al., 1991) (Table 9–1). Under reducing conditions when oxygen is absent, these compounds can serve as electron acceptors during microbial metabolism. The result of this reduction process is that chlorine is removed and replaced with hydrogen. The end result of these processes is the disappearance of the highly chlorinated parent compound and the appearance of lower chlorinated intermediates.

Hexachlorobenzene (HCB) is a toxic chlorinated aromatic that is a priority pollutant. Reductive dechlorination of HCB has been demonstrated in a sewage sludge under anaerobic conditions (Fathepure et al., 1988), although previous studies showed little degradation of HCB on biofilms (Bouwer & McCarty, 1984) or in anaerobic aquifer material (Kuhn et al., 1985). Indirect evidence of reductive dechlorination of HCB has been reported by examining HCB congener profiles in Great Lakes sediments (Oliver & Nichol, 1982).

The present study was conducted to determine the degradation potential of HCB in soil from two wetland environments; a forested bottomland hardwood and a freshwater marsh. Reductive dechlorination may be an important transformation process in wetlands. In this chapter, evidence is presented on the occurrence of this degradation process in these wetland soils.

[1] This research was supported by a cooperative agreement through the US EPA, Environmental Research Laboratory-Duluth (agreement no. CR816805-01).

Table 9-1. Standard free-energy change for the oxidation of glucose to CO_2 using the following alternate electron acceptors. (Adapted from Sims et al., 1991.)

Oxidant	Reduced product	ΔG, kcal/mol
O_2	Water (H_2O)	−676.10
Hexachlorobenzene	Benzene	−410.16
Sulfate (SO_4^{2-})	Reduced sulfur	−131.78
Carbon dioxide (CO_2)	Methane (CH_4)	−95.63

METHODS AND MATERIALS

Soil Sampling

Wet surface soil (0–10 cm) was sampled from a forested bottomland hardwood wetland in the Spring Bayou Wildlife Management Area (Avoyelles Parish, LA). The sampled soil was a Sharkey Clay (very-fine, montmorillonitic, nonacid, thermic Vertic Haplaquept). The soil was kept refrigerated until use (<1 mo) and was not air dried prior to use. Freshwater marsh soil (0–10 cm) was collected from a *Panicum hemitomon*-dominated marsh near Lake des Allemands, Louisiana. The highly organic marsh soil was homogenized in a blender at low speed before use.

Batch Experiments

Batch incubations were made under anaerobic conditions in small sealed flasks (25-mL Reacti-Flasks,[2] Pierce). Prior to adding soil, HCB was added to the flasks in 1 mL of ether, which was subsequently evaporated off under Ar. A large quantity (350 g) of each wetland soil was made up as a slurry (5:1 water/dry soil ratio for the bottomland hardwood soil, 18:1 water/dry soil ratio for the marsh soil) and incubated under Ar-purged conditions for 14 d before dispensing, without contact with air, into the smaller flasks containing the HCB. Total volume of slurry was 20 mL/flask. Total concentration of HCB was 10 mg/L of suspension. Flasks were incubated in the dark at 25 ± 2 °C.

Microcosms

Additional experiments were conducted with the bottomland hardwood soil in microcosms where the oxidation-reduction potential and pH were continuously monitored. The features of the microcosms have been described previously (Patrick et al., 1973). The microcosms were constructed with 1700 mL of a soil/water suspension at a 5:1 water/soil ratio. Incubations were carried out in the dark at a temperature of 25 ± 2 °C. Two treatments were used for these experiments, aerobic conditions (+500 mV) and anaerobic conditions (−150 to −200 mV). Duplicate soil/water suspensions were in-

[2]Mention of trademark or proprietary products in this chapter does not constitute a guarantee or warranty of the product by the Louisiana State University and does not imply its approval to the exclusion of other products that may also be suitable.

cubated at the desired redox potentials for 7 d. The HCB, unlabeled and ^{14}C-labeled (1 μCi, UL-ring, Sigma Radiochemicals), was dissolved in a small amount of ether (1 mL) and mixed with a small amount of soil (5 g) from the Spring Bayou site. The ether was rapidly evaporated off under a dry stream of Ar. The soil, containing a thin film of hexachlorobenzene on the surface, was added to the microcosms. The small amount of soil did not significantly affect the soil/water ratio. Total concentration of HCB in the microcosms was 10 mg/L of suspension.

After the addition of the HCB to the microcosm, the air stream leaving the flask was connected to a sampling train consisting of (i) a tube containing activated C and (ii) two traps (30 mL each) of 1.0 *N* NaOH to trap evolved $^{14}CO_2$. The CO_2 traps were changed every 2 to 3 d.

Sampling

At selected intervals (1, 2, 5, 7, 11, 13, 15, and 18 wk) replicate flasks from the batch experiments were sacrificed and extracted for 18 h with 20 mL of hexane:acetone (1:1) in Teflon centrifuge tubes. The soil slurry/solvent mixture was subsequently centrifuged to separate the soil from the water and solvents. Finally, the hexane phase was isolated from the water/acetone phase in a separatory funnel. Hexane extracts were cleaned up by passing through a column containing sodium sulfate and were made up to a suitable volume for analysis. Microcosm suspension samples (30 mL) were also removed, extracted, and cleaned up as described. Aliquots of the CO_2 traps were counted on a liquid scintillation counter.

Gas Chromatographic Analyses

Samples were analyzed using high-resolution gas chromatography using an electron capture detector. The chromatographic conditions were as follows: GC-Perkin-Elmer 8410; column-30 m DB-5 capillary column; injection-1 μL splitless at 250 °C; temperature program- 32 °C initially (2 min), ballistic temperature rise to 150 °C, 2 °C/min to 230 °C; total analysis time-48 min; ECD-375 °C. Retention times and response factors for the various chlorinated benzene congeners were determined by analyzing individual standards (1 ng/μL). A single standard containing all of the congeners was made up to calibrate the response factors daily.

RESULTS AND DISCUSSION

Reductive dechlorination of HCB was observed in both the bottomland hardwood soil and the freshwater marsh soil after lag periods of approximately 7 and 5 wk, respectively. Approximately 50 to 60% of HCB was degraded to lower chlorinated intermediates after 18 wk of incubation in both the bottomland hardwood and freshwater marsh soils. High concentrations of methane were measured in the headspace of the batch flasks throughout the experiment. No dechlorination was observed in killed controls treated with $HgCl_2$.

Two groups of lower chlorinated metabolites were observed in the wetland soils. These groups of metabolites correspond to pathways suggested previously for HCB degradation (Fathepure et al., 1988). One possible pathway is degradation of HCB to dichlorobenzenes (1,3;1,4; and 1,2) via pentachlorobenzene, 1,2,4,5 tetrachlorobenzene, and 1,2,4 trichlorobenzene. These metabolites appeared to accumulate preferentially in the bottomland hardwood soil. A second possible pathway is degradation of HCB to 1,3,5 trichlorobenzene via 1,2,3,5 tetrachlorobenzene and pentachlorobenzene. These metabolites appeared to accumulate preferentially in the freshwater marsh soil. This pathway (accumulation of 1,3,5 trichlorobenzene) was also the primary pathway seen in the degradation of HCB in sewage sludge (Fathepure et al., 1988). Example chromatograms of the soil extracts are presented in Fig. 9–1 and 9–2 for the bottomland hardwood and freshwater marsh soil, respectively. These results are further evidence that HCB pathways proposed by Fathepure et al. (1988) are those that operate in the environment.

The presence of different metabolites in the two wetlands suggests that different microbial populations are operating in these soils. The ultimate degradation products of these two pathways have significant implications on the persistence of chlorinated benzenes in wetland soils. In the freshwater marsh soil, accumulation of stable 1,3,5 trichlorobenzene occurs without further removal of the chlorinated benzenes. In the bottomland hardwood soil, however, dechlorination proceeds primarily to the dichlorobenzenes. Aerobic organisms have been isolated that can use these compounds as sole C sources (deBont et al., 1986). Therefore, the potential exists for complete mineralization of chlorinated benzenes via this pathway. In addition, lower chlorinated benzenes have been shown to have lower bioconcentration factors in aquatic organisms in a Louisiana estuary (Pereira et al., 1988).

Further experiments were conducted using the bottomland hardwood soil in controlled redox potential microcosms. Dechlorination was observed in the anaerobic suspension after a lag period of approximately 14 d. Significant concentrations of HCB were dechlorinated under anaerobic conditions (−200 mV) while no lower chlorinated intermediates were detected under anaerobic conditions (+500 mV). No ^{14}C was evolved as $^{14}CO_2$ in either treatment during the course of the first 49 d.

The appearance of dechlorination metabolites is presented in Fig. 9–3. Dechlorination of HCB again appeared to proceed primarily through 1,2,4,5 tetrachlorobenzene to the dichlorobenzenes. The metabolite 1,2,3,5-tetrachlorobenzene was also observed in addition to 1,2,3,4-tetrachlorobenzene suggesting that a third pathway may also be operating. Conversion to pentachlorobenzene is the initial step in HCB reduction. Pentachlorobenzene accumulated until approximately Day 36 when further dechlorination to tetrachlorobenzene became greater than conversion of HCB to pentachlorobenzene. 1,2,4,5-Tetrachlorobenzene began accumulating after several days and was the primary tetrachlorinated congener observed. 1,2,3,5 and 1,2,3,4-Tetrachlorobenzene appeared after 32 and 37 d, respectively. Further dechlo-

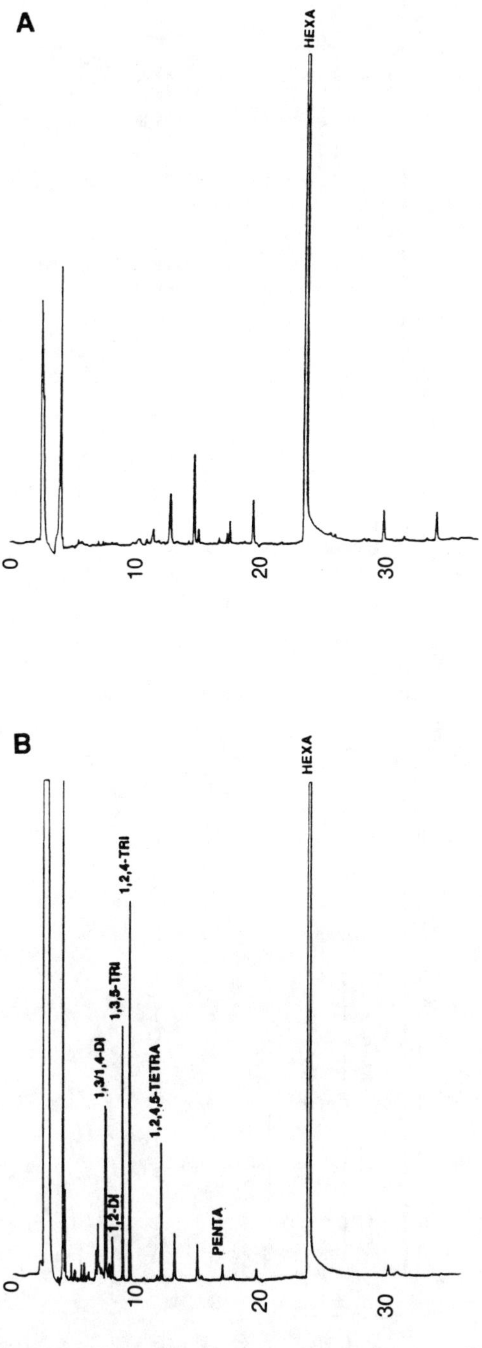

Fig. 9-1. Gas chromatogram of extracted bottomland hardwood forest soil after (A) 1 wk of incubation and (B) 11 wk of incubation under anaerobic conditions. Data show preferential accumulation of dichlorobenzenes through 1,2,4-trichlorobenzene.

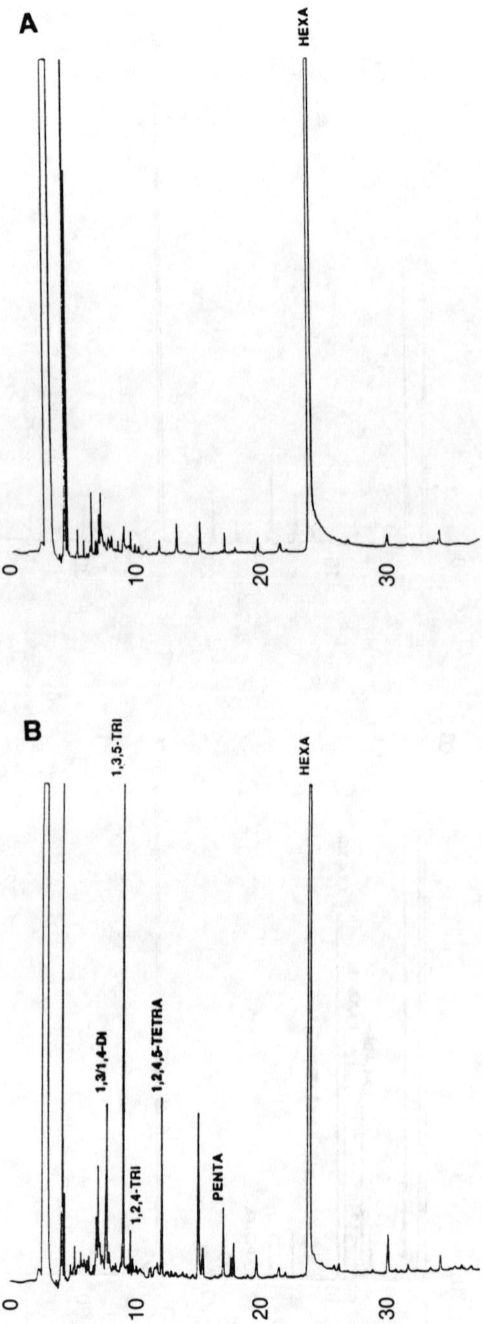

Fig. 9-2. Gas chromatogram of extracted freshwater marsh soil after (A) 1 wk of incubation and (B) 13 wk of incubation under anaerobic conditions. Data show preferential accumulation of 1,3,5-trichlorobenzene.

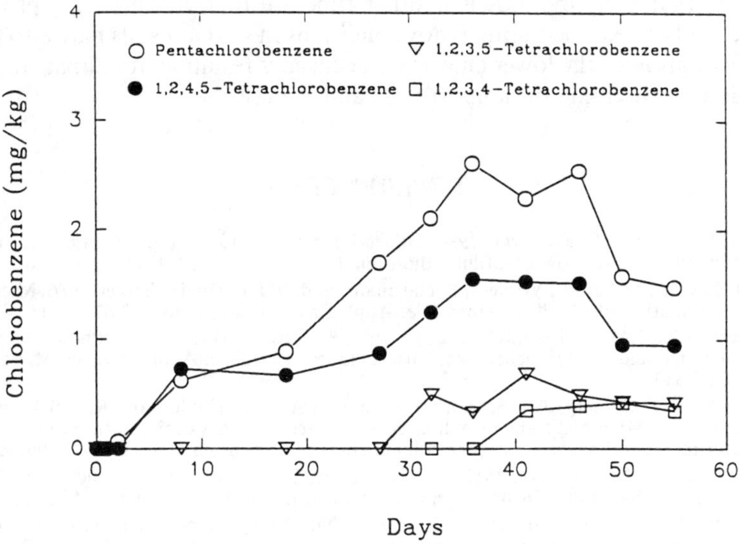

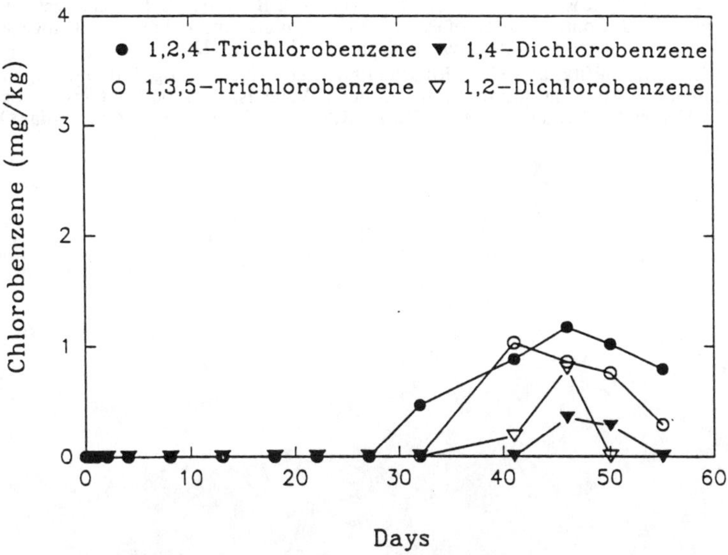

Fig. 9-3. Kinetics of reductive dechlorination of hexachlorobenzene under anaerobic conditions in controlled redox-pH microcosms (−200 mV) using the bottomland hardwood soil.

rination to trichlorobenzenes was detected after 37 d (1,3,5-trichlorobenzene) and 32 d (1,2,4-trichlorobenzenes), respectively. Small, but measurable, concentrations of dichlorinated benzenes began appearing after 42 d of incubation. These results demonstrate the sequential cascading nature of dechlorination reactions.

These results extend the existence of microbially mediated reductive dechlorination processes to the wetland environment. Transformation of chlorinated organic compounds is another function that wetlands can provide on the landscape. Fluctuating redox conditions in wetland soils may also promote oxidation of the lower chlorinated benzenes resulting in ultimate mineralization of the toxic organic to CO_2 and water.

REFERENCES

Bouwer, E.J., and P.L. McCarty. 1984. Utilization rates of trace halogenated organic compounds in acetate-grown biofilms. Biotechnol. Bioeng. 27:1564–1571.

deBont, J.A.M., M.J.A.W. Vorages, S. Hartmans, and W.J.J. van den Tweel. 1986. Microbial degradation of 1,3-dichlorobenzene. Appl. Environ. Microbiol. 52:677–680.

Fathepure, B.Z., J.M. Tiedje, and S.A. Boyd. 1988. Reductive dechlorination of hexachlorobenzene to tri- and dichlorobenzenes in anaerobic sewage sludge. Appl. Environ. Microbiol. 54:327–330.

Kuhn, E.P., P.J. Colberg, J.L. Schnoor, O. Wanner, A.J.B. Zehnder, and R.P. Schwarzenbach. 1985. Microbial transformation of substituted benzenes during infiltration of river water to ground water:laboratory column studies. Environ. Sci. Technol. 19:961–968.

Oliver, B.G., and N.D. Nichol. 1982. Chlorobenzenes in sediments, water, and selected fish from Lakes Superior, Huron, Erie, and Ontario. Environ. Sci. Technol. 16:532–536.

Patrick, W.H., Jr., B.G. Williams, and J.T. Moraghan. 1973. A simple system for controlling redox potential and pH in soil suspension. Soil Sci. Soc. Am. Proc. 37:331–332.

Pereira, W.E., C.E. Rostad, C.T. Chiou, T.I. Brinton, L.B. Barber, D.K. Demcheck, and C.R. Demas. 1988. Contamination of estuarine water, biota, and sediment by halogenated organic compounds: A field study. Environ. Sci. Technol. 22:772–778.

Sims, J.L., J.M. Sulfita, and H.H. Russell. 1991. Reductive dehalogenation of organic contaminants in soils and ground water. Ground Water Issue, Superfund Technology Support Center for Ground Water. USEPA, R.S. Kerr Environ. Res. Lab., Ada, OK.

10 Bioavailability and Bioactivity of Sorbed Organic Chemicals

Jerome B. Weber
North Carolina State University
Raleigh, North Carolina

Jack A. Best
Sandoz AGRO, Inc.
Des Plaines, Illinois

Jane U. Gonese
North Carolina State University
Raleigh, North Carolina

Organic chemicals, including pesticides that reach the soil, react with soil colloids in many different ways. Sorption of pesticides by soil colloids influences both the bioactivity and the mobility of these chemicals. Organic (humus, muck, and peat) and inorganic (clay minerals and metallic hydrous oxides) soil materials are both involved. Soil organic matter content, texture (primarily clay content) and pH are the soil properties most highly associated with pesticide performance and are commonly considered in making pesticide rate recommendations. Sorption of the chemicals reduces the rate of release of pesticides from soil particulate matter and thereby decreases bioactivity and increases persistence. Various edaphic factors, including nutrient and pH status of the soil, temperature, and moisture are also influential, though they differ in importance depending upon the pesticide involved. The properties of the pesticide and soil constituents with which it is in contact are of primary importance in determining the bonding mechanism and hence the bioactivity or mobility of the chemical. The pH status of the soil media affects the ionic status of ionizable compounds, that is, the proportion of the chemical in the ionized or nonionized forms, which in turn affects the type and amount of bonding that occurs, and also may affect the degradation rate of both ionizable and nonionizable forms. Soil moisture and temperature directly affect many biological processes, including plant metabolism and microbial degradation, and thereby influence bioactivity and persistence of the chemicals.

A wide variety of organic chemicals reach the soil in agricultural situations. They represent many different families and they interact with soils in many different ways. Sorption by the soil and bioactivity of each chemical depends upon the mechanism and geometry of binding of the compound. The bioactivity and bioavailability of sorbed organic chemicals will be discussed in the same fashion as in a previous paper (Weber & Weed, 1974), and will be done according to the following classification scheme: quaternary N compounds, weakly basic compounds, weakly acidic compounds, amphoteric compounds, organic As and P pesticides, amide and anilide herbicides, phenylcarbamate and phenylurea pesticides, thiocarbamate herbicides, dinitroaniline herbicides, halogenated hydrocarbons, and miscellaneous nonionic organic chemicals. The chemical names of pesticides and organic chemicals discussed in this chapter are listed in Table 10–1.

Retention of a pesticide by soil is determined by carrying out sorption isotherms using a wide range of aqueous concentrations of the pesticide each equilibrated with a specific amount of soil. After equilibration is attained the solid phase is separated from the solution phase and the amount of pesticide that has disappeared from solution, and assumed sorbed to the soil particles, is determined for each sample. A plot of the amount of pesticide sorbed vs. the amount left in solution is made and the variety of sorption isotherms obtained normally reflects on the sorption mechanism(s) involved. Since soils are mixtures of many different types of materials, both organic and inorganic, the resulting isotherm for a given pesticide is a composite of the retention of the pesticide by the different surfaces present. "Pure" sorption isotherms are obtained when a single soil constituent (e.g., Ca-montmorillonite) is used as the sorbent. The shape of the isotherm for a given pesticide by a given sorbent reflects on the sorption mechanism involved (Giles & MacEwan, 1957). A solvent-affinity or type S isotherm occurs when the sorbent (soil) has lower affinity for the solute (pesticide) than it does for the solvent (water) at low concentrations of the solute. A constant partition or type C isotherm is rare and occurs when the sorbent has an equal affinity for the solute as it does for the solvent at all concentrations employed. A Langmuir or type L isotherm occurs when the sorbent has higher affinity for the solute than it does for the solvent at low concentrations of solute and lower affinity for the sorbent than for the solvent at high concentrations of solute when the sorption sites become limiting. A high affinity or type H isotherm occurs when the sorbent has very high affinity for the solute and removes all of the solute from solution until the surfaces become completely covered by the solute, at which time additional solute remains in solution. The isotherms also reflect on the strength of the bonding mechanism involved and increase in the order type H > type L > type C > type S. The empirical Freundlich equation, $x/m = K_d C_e^{1/n}$, where x/m = nmol/g of pesticide sorbed, K_d = distribution constant, C_e = equilibrium concentration of the pesticide in nmol/mL, and 1/n = slope of the isotherm, is commonly used to describe sorption isotherms of the types L, C, and S. It does not describe type H isotherms. K_d is an index of the sorption capacity of the pesticide for the soil, as it is the ratio of the amount of pesticide retained

Table 10–1. Chemical names of pesticides and organic chemicals.

Common name	Chemical name
Acetochlor	2-Chloro-*N*-(ethoxymethyl)-*N*-(2-ethyl-6-methylphenyl) acetamide
Alachlor	2-Chloro-*N*-(2,6-diethylphenyl)-*N*-(methoxymethyl)acetamide
Aldrin	1,2,3,4,10,10-Hexachloro-1,4,4a,5,8,8a-hexahydro-*exo*-1,4-*endo*-5,8-dimethanonaphthalene
Aminoparathion	*O,O*-diethyl *O*-(4-aminophenyl) phosphorodithioate
Amitrole	1,*H*-1,2,4-triazol-3-amine
Anthracene	Anthracene
Arsenic acid	Orthoarsenic acid
Atrazine	6-Chloro-*N*-ethyl-*N*′-(1-methylethyl)-1,3,5-triazine-2,4-diamine
Azinphosmethyl	*O,O*-dimethyl-*S*-[(4-oxo-1,2,3-benzotriazin-3(4*H*)-yl) methyl] phosphorodithioate
Benefin	*N*-butyl-*N*-ethyl-2,6-dinitro-4-(trifluoromethyl) benzenamine
Bentazon	3-(1-Methylethyl)-(1*H*)-2,1,3-benzothiadiazin-4(3*H*)-one 2,2-dioxide
Benzanthracene	1,2-Benzanthracene
Benzopyrene	3,4-Benzpyrene
Buthidazole	3-[5-(1,1-Dimethylethyl)-1,3,4-thiadiazol-2-yl]-4-hydroxy-1-methyl-2-imidazolidinone
Butachlor	*N*-(butoxymethyl)-2-chloro-*N*-(2,6-diethylphenyl) acetamide
Butralin	4-(1,1-Dimethylethyl)-*N*-(1-methylpropyl)-2-6-dinitrobenzenamine
Cacodylic acid	Dimethylarsinic acid
CDAA	*N,N*-diallyl-2-chloroacetamide
CDEC	2-Chloroallyl-diethyldithiocarbamate
Chloramben	3-Amino-2,5-dichlorobenzoic acid
Chlorazine	6-Chloro-*N,N,N*′,*N*′-tetraethyl-1,3,5-triazine-2,4-diamine
Chlorbromuron	3-(4-Bromo-3-chlorophenyl)-1-methoxy-1-methylurea
Chlordane	1,2,4,5,6,7,8,8-Octachloro-2,3,3a,4,7,7a-hexahydro-4,7-methano-1*H*-indene
Chlornidine	*N, N*-di(2-chloroethyl)-4-methyl-2,6-dinitroaniline
Chlorpropham	1-Methylethyl(3-chlorophenyl)carbamate
Chlorsulfuron	2-Chloro-*N*-[[(4-methoxy-6-methyl-1,3,5-triazin-2-yl)amino] carbonyl]benzenesulfonamide
Crotoxyphos	(E)-1-phenylethyl 3-[(dimethoxyphosphinyl)oxy]-2-butenoate
Cyanazine	2-[[4-Chloro-6-(ethylamino)-1,3,5-triazin-2-yl]amino]-2-methylpropaneniltrile
Cycloate	*S*-ethyl cyclohexylethylcarbamothioate
2,4-D	(2,4-Dichlorophenoxy)acetic acid
DCPA	Dimethyl-2,3,5,6-tetrachloro-1,4-benzenedicarboxylate
DDT	1,1′-(2,2,2-Trichloroethylidene)bis[4-chlorobenzene]
Demeton	*O,O*-diethyl *O*-[2-(ethylthio)ethyl] phosphorothioate mixture with *O,O*-diethyl *S*-[2-(methylthio)ethyl] phosphorothioate
Diallate	*S*-(2,3-dichloro-2-propenyl) bis(1-methylethyl) carbamothioate
Diazinon	*O,O*-diethyl *O*-[6-methyl-2-(1-methylethyl)-4-pyrimidinyl] phosphorothioate
Dibromoethane	1,2-Dibromoethane
Dichlobenil	2,6-Dichlorobenzonitrile
Dichlorobenzene	1,2-Dichlorobenzene
Dichloroethane	1,2-Dichloroethane
Dichloropropane	1,2-Dichloropropane
Dieldrin	*Endo, exo*-1,2,3,4,10,10-hexachloro-6,7-epoxy-1,4,4a,5,6,7,8,8a-octahydro-1,4:5,8-dimethanonaphthalene
Diethylhexylphthalate	Di(2-ethylhexyl) phthalate
Dinitramine	N^3,N^3-diethyl-2,4-dinitro-6-(trifluoromethyl)-1,3-benzenediamine

(continued on next page)

Table 10–1. Continued.

Common name	Chemical name
Dinoseb	2-(1-Methylpropyl)-4,6-dinitrophenol
Diphenamid	*N,N*-dimethyl-α-phenylbenzeneacetamide
Diquat	6,7-Dihydrodipyrido[1,2-a:2′,1′-C]pyrazinediium ion
Disulfoton	*O,O*-diethyl-*S*-[2-(ethylthio)ethyl] phosphorodithioate
Diuron	*N*′-(3,4-dichlorophenyl)-*N,N*-dimethylurea
DNOC	4,6-Dinitro-*o*-cresol
DSMA	Disodium methylarsonate
Endosulfan	6,7,8,9,10,10-Hexachloro-1,5,5a,6,9,9a-hexahydro-6,9-methano-2,4,3-benzodioxathiepin 3-oxide
Endrin	1aα,2β,2aβ,3α,6α,6aβ,7β,7aα-3,4,5,6,9,9-hexachloro-1a,2,2a,3,6,6a,7,7a-octahydro-2,7:3,6-dimethanonaphth[2,3-b]oxirene
EPTC	*S*-ethyl dipropylcarbamothioate
Fenuron	*N,N*-dimethyl-*N*′-phenylurea
Fluchloralin	*N*-(2-chloroethyl)-2,6-dinitro-*N*-propyl-4-(trifluoromethyl) benzenamine
Fluometuron	*N,N*-dimethyl-*N*′-[3-(trifluoromethyl)phenyl]urea
Fluorodifen	*p*-Nitrophenyl α,α,α-trifluoro-2-nitro-*p*-tolyl ether
Fluridone	1-Methyl-3-phenyl-5-[3-(trifluoromethyl)phenyl]-4-(1*H*)-pyridinone
Glyphosate	*N*-(phosphonomethyl)glycine
GS 18183	2-(Tert.-butylamino)-4-chloro-6-(methylamino)-*s*-triazine
Heptachlor	1,4,5,6,7,8,8-Heptachloro-3a,4,7,7a-tetrahydro-4,7-methano-1*H*-indene
Heptachlor epoxide	2,3,4,5,6,7,7-Heptachloro-1a,1b,5,5a,6,6a-hexachloro-2,5-methano-2*H*-indeno[1,2-b]oxirene
Hexadecane	Hexadecane
Imazaquin	(±)-2-[4,5-Dihydro-4-methyl-4-(1-methylethyl)-5-oxo-1*H*-imidazol-2-yl]-3-quinolinecarboxylic acid
Imazethapyr	(±)-2-[4,5-Dihydo-4-methyl-4-(1-methylethyl)-5-oxo-1*H*-imidazol-2-yl]-5-ethyl-3-pyridinecarboxylic acid
Ipazine	6-Chloro-*N,N*-diethyl-*N*′-(1-methylethyl)-1,3,5-triazine-2,4-diamine
Lectophos	*O*-(4-bromo-2,5-dichlorophenyl) *O*-methyl phenylphosphonothioate
Lindane	1,2,3,4,5,6-Hexachlorocyclohexane
Linuron	*N*′-(3,5-dichlorophenyl)-*N*-methoxy-*N*-methylurea
Malathion	*S*-1,2-bis(ethoxycarbonyl)ethyl *O,O* dimethyl phosphorodithioate
MAMA	Monoammonium methanearsonate
MCPA	(4-Chloro-2-methylphenoxy)acetic acid
Mepiquat	1,1-Dimethylpiperidinium chloride
Metamitron	4-Amino-3-methyl-6-phenyl-1,2,4-triazin-5(4*H*)-one
Metazachlor	2-Chloro-*N*-(2,6-dimethylphenyl)-*N*-(1*H*-pyrazol-1-ylmethyl) acetamide
Methoxychlor	1,1′-(2,2,2-Trichloroethylidene)bis[4-methoxybenzene]
Metolachlor	2-Chloro-*N*-(2-ethyl-6-methylphenyl)-*N*-(2-methoxy-1-methylethyl)acetamide
Metribuzin	4-Amino-6-(1,1-dimethylethyl)-3-(methylthio)-1,2,4-triazin-5(4*H*-)-one
Metsulfuron	2-[[[[(4-Methoxy-6-methyl-1,3,5-triazin-2-yl)amino]carbonyl] amino]sulfonyl]benzoate
Mevinphos	Methyl 3-[(dimethoxyphosphinyl)oxy]-2-butenoate
Monuron	*N*′-(4-chlorophenyl)-*N,N*-dimethylurea
MSMA	Monosodium methylarsonate

(continued on next page)

Table 10–1. Continued.

Common name	Chemical name
Naphthalene	Naphthalene
Napropamide	*N,N*-diethyl-2-(1-naphthalenyloxy)propanamide
Neburon	*N*-butyl-*N*-(3,4-dichlorophenyl)-*N*-methylurea
Nitralin	4-(Methylsulfonyl)-2,6-dinitro-*N,N*-dipropylbenzenamine
Norflurazon	4-Chloro-5-(methylamino)-2-[3-(trifluoromethyl) phenyl]-3 (2*H*)-pyridazinone
OCDD	Octachlorodibenzo-*p*-dioxin
Oryzalin	4-(Dipropylamino)-3,5-dinitrobenzenesulfonamide
Oxytetracycline	4-(Dimethylamino)-1,4,4a,5,5a,6,11,12a,octahydro-3,5,6,10,12,12a-hexahydroxy-6-methyl-1,11-dioxo-2-naphthacenecarboxaamide
Paraquat	1,1′-Dimethyl-4,4′-bipyridinium
Parathion	*O,O*-diethyl *O*-(4-nitrophenyl) phosphorothioate
PCB	Polychlorinated biphenyl
Pebulate	*S*-propyl butylethylcarbamothioate
Pendimethalin	*N*-(1-ethylpropyl)-3,4-dimethyl-2,6-dinitrobenzenamine
Phorate	*O,O*-diethyl *S*-[(ethylthio)methyl] phosphorodithioate
Picloram	4-Amino-3,5,6-trichloro-2-pyridinecarboxylic acid
Picolinic acid	2-Pyridinecarboxylic acid
Picric acid	2,4,6-Trinitrophenol
Piperophos	*S*-[2-methyl-1-piperidinyl)-2-oxoethyl]*O,O*-dipropyl phosphorodithioate
PPCl	*N*-(4-pyridyl)-pyridinium chloride
Procyazine	2-[[4-Chloro-6-(cyclopropylamino)-1,3,5-triazin-2-yl]amino]-2-methylpropaneniltrile
Profluralin	*N*-(cyclopropylmethyl)-α,α,α-trifluoro-2,6-dinitro-*N*-propyl-*p*-toluidine
Prometon	6-Methoxy-*N,N′*-bis(1-methylethyl)-1,3,5-triazine-2,4-diamine
Prometryn	*N,N′*-bis-(1-methylethyl)-6-(methylthio)-1,3,5-triazine-2,4-diamine
Pronamide	3,5-Dichloro-*N*-(1,1-dimethyl-2-propynyl)benzamide
Propachlor	2-Chloro-*N*-(1-methylethyl)-*N*-phenylacetamide
Propazine	6-Chloro-*N,N′*-bis-(1-methylethyl)-1,3,5-triazine-2,4-diamine
Propham	1-Methylethyl phenylcarbamate
Prynachlor	2-Chloro-*N*-(1-methyl-2-propynyl)acetanilide
Pyrazon	5-Amino-4-chloro-2-phenyl-3(2*H*)-pyridazinone
Sulfonate	Trimethylsulphoniumcarboxymethyl aminomethyl phosphonate
Schradan	Octamethyldiphosphoramide
Sebuthylazine	2-(*Sec*-butylamino)-4-chloro-6-(ethylamino)-*s*-triazine
Simazine	6-Chloro-*N,N′*-diethyl-1,3,5-triazine-2,4-diamine
Simeton	*N,N′*-diethyl-6-methoxy-1,3,5-triazine-2,4-diamine
2,4,5-T	(2,4,5-Trichlorophenoxy)acetic acid
TCA	Trichloroacetic acid
TCDD	Tetrachlorodibenzo-*p*-dioxin
Terbacil	5-Chloro-3-(1,1-dimethylethyl)-6-methyl-2,4(1*H*,3*H*)-pyrimidinedione
Terbuthylazine	2-(*Tert*-butylamino)-4-chloro-6-(ethylamino)-*s*-triazine
Terbutryn	*N*-(1,1-dimethylethyl)-*N′*-ethyl-6-(methylthio)-1,3,5-triazine-2,4-diamine
Tetracycline	4-(Dimethylamino)-1,4,4a,5,5a,6,11,12a-octahydro-3,6,10,12,12a-pentahydroxy-6-methyl-1,11-dioxo-2-naphthacenecarboxamide
Trietazine	6-Chloro-*N,N,N′*-triethyl-1,3,5-triazine-2,4-diamine
Trifluralin	2,5-Dinitro-*N,N*-dipropyl-4-(trifluoromethyl)benzenamine

by the soil in equilibrium with the amount of pesticide in the solution phase, that is, $K_d = x/m \div C_e^{1/n}$. K_d values of selected pesticides are used to compare the relative sorption of the chemicals in soils. In this chapter, we have used plants to assay the biological concentration of herbicide in the solution phase, and the difference between the amount of herbicide applied and that detected by the plants as a measure of herbicide bound by the soil colloids. From this we have calculated what we refer to as a Bio-K_d value, that is, Bio-K_d = x/m ÷ Bio-C_e. This allowed us to compare the relative bioactivity of selected herbicides bound to specific soil constituents in model soil media.

QUATERNARY NITROGEN COMPOUNDS

Quaternary N compounds, such as the herbicides diquat and paraquat and the growth regulator mepiquat, ionize in aqueous solutions to form cationic species that are biologically active when applied to plant foliage. However, because of their cationic nature these compounds are readily inactivated by soil colloids (Harris & Warren, 1964; Weber et al., 1965; Coats et al., 1966; Weber & Scott, 1966; Knight & Tomlinson, 1967; Scott & Weber, 1967; Tucker et al., 1967). These organic cations are strongly sorbed by clay minerals in exchange for inorganic cations at the mineral surfaces (Weber et al., 1965; Weber & Weed, 1968; Weed & Weber, 1969; Philen et al., 1970, 1971). Sorption isotherms for these compounds are normally of the type H on clay minerals and strongly acid cation exchange resins and of the type L on soil organic matter and weakly acid cation exchange resins (Weber, 1972; Weber & Miller, 1989). X-ray diffraction studies showed diquat and paraquat were sorbed in the interlayer spacings of montmorillonite with the planes of the pyridine rings parallel to the silicate sheets (Weber et al., 1965; Weed & Weber, 1968). Diquat and paraquat were also shown to be readily sorbed to soil organic matter and to cation exchange resins, but not to anion exchange resins and only weakly to activated carbon (Weber et al., 1965, 1968b; Coffey & Warren, 1969; Best et al., 1972). The chemicals could not be desorbed from the negatively charged surfaces with water, but 3 to 11% was displaced from montmorillonite with dilute Al, Ba, Ca, K, Mg, or Na salt solutions. Displacement of 19 to 65% was achieved from vermiculite, 65 to 80% from kaolinite, and 100% from cation exchange resins with dilute salt solutions, whereas Ca was found to be competitive with diquat for sorption to the humic and humin fraction of soil organic matter (Weber & Weed, 1968; Weber et al., 1968b; Weed & Weber, 1969; Best et al., 1972). Additional details concerning competitive sorption between these organic cations and with inorganic cations are discussed elsewhere (Calderbank, 1968; Weber, 1972, 1987; Summers, 1980).

Bioactivity of diquat and paraquat was reported to be reduced in the presence of clay minerals (Coats et al., 1966; Weber & Scott, 1966; Scott & Weber, 1967). The data in Table 10–2 illustrate that 2 to 6 g quantities of kaolinite reduced the bioactivity of paraquat to cucumber (*Cucumis sativus* L.) seedlings in model soil media, and that small quantities (0.1–0.5 g)

Table 10–2. Reduction in herbicidal bioactivity resulting from clay mineral additions to model soil media (160 g of sand, 45 mL of solution, cucumber [*Cucumis sativus* L.] bioassay species). (Calculated from Scott & Weber, 1967.)

Herbicide†	Conc. added	Clay added: Type‡	Clay added: Amount	Plant growth§	Bio-conc., C_e¶	Reduced bioconc, x m^{-1}#	Bio-K_d, (x m^{-1} C_e^{-1})††	Mean Bio-K_d
	nmol mL^{-1}		g pot^{-1}	% of control	nmol mL^{-1}	nmol g^{-1}		
Paraquat	44	Kaol.	2.0	95	0	990	--	
	89			87	35	1 215	35	
	178			80	38	3 150	83	
	355			39	87	6 030	69	
	711			25	151	12 600	83	67
	44		6.0	100	0	330	--	
	89			100	0	667	--	
	178			100	0	1 335	--	
	355			85	30	2 437	81	
	711			59	55	4 920	89	85
							Mean =	76
	44	Mont.	0.1	100	0	19 800	--	
	89			100	0	40 050	--	
	178			95	0	80 100	--	
	355			93	0	159 750	--	
	711			92	0	319 950	--	
	44		0.5	100	0	3 960	--	
	89			100	0	8 010		
	178			100	0	16 020	--	
	355			100	0	31 950	--	
	711			95	0	63 990	--	--
							Mean	--
Prometon	6.7	Kaol.	4.0	100	0	75	--	
	13.3			95	0	150	--	
	26.7			50	14	143	10	
	53.3			15	36	195	5	7
	6.7		8.0	100	0	38	--	
	13.3			100	0	75	--	
	26.7			70	11	88	8	
	53.3			24	25	159	6	7
							Mean =	7
	6.7	Mont.	0.5	100	0	603	--	
	13.3			95	0	1 197	--	
	26.7			92	0	2 403	--	
	53.3			48	15	3 447	230	230
	6.7		1.5	100	0	201	--	
	13.3			100	0	399	--	
	26.7			95	0	801	--	
	53.3			72	11	1 269	115	115
							Mean =	172

† Kaolinite additions of 4.0 and 8.0 g and montmorillonite additions of 0.5 and 1.5 g had no effect on the bioactivity of chlorpropham and 2,4-D.

‡ Kaol. = kaolinite, Mont. = montmorillonite.

§ Percent of control using plant dry weight.

¶ Bioconcentration from standard plant growth curve in Fig. 10–1; values obtained from 10 to 90% of control were used.

Reduced bioconcentration = Concentration added minus bioconcentration available to plants.

†† Bio-K_d = Reduced Bioconc./Bioconc. = Bound herbicide/Bioavailable herbicide = (x m^{-1} C_e^{-1}).

of montmorillonite added to the media made the herbicide completely unavailable (Weber & Scott, 1966; Scott & Weber, 1967). The bioconcentration values for each herbicide were obtained from standard plant growth curves shown in Fig. 10–1. The amount of reduced herbicide bioconcentration (x/m) was calculated by subtracting the bioconcentration (C_e) detected by the plants from the amount added initially. The biological distribution coefficient between that sorbed and that in solution (Bio-K_d) was calculated by dividing the reduced bioconcentration (x/m) by the bioconcentration (C_e). A mean Bio-K_d was calculated for the two levels of additive added to the media. Although the mean Bio-K_d for montmorillonite was large, it only ranged from 35 to 89 (mean = 76) for kaolinite. Additions of organic matter to the model soil media also reduced the bioactivity of paraquat, but by only one-fourth that of kaolinite addition (compare mean Bio-K_d values of 22 and 76, respectively; Tables 10–3 and 10–2). Additions of much smaller quantities of montmorillonite to the growth media resulted in reducing paraquat bioactivity essentially to zero (plant growth equal to 97% of the untreated control) with 0.030 g of the clay mineral (Table 10–4; Weber et al., 1969a). The mean Bio-K_d value calculated for montmorillonite in this study ranged from 10 592 to 23 000 (mean = 17 544), or more than 230 times that for kaolinite. Bio-K_d ranged from 1568 to 3214 (mean = 2613) for vermiculite, which was only one-seventh that for montmorillonite but 34 times that for kaolinite. A similar model soil media study using diquat sorbed on montmorillonite at below the cation exchange capacity showed that the addition to the media of high concentrations (250–1000 nmol/mL) of a nonphytotoxic, monovalent quaternary N compound [PPCl, *N*-(4-pyridyl)-pyridinium chloride] could displace and make bioactive 4 to 15% of the bound herbi-

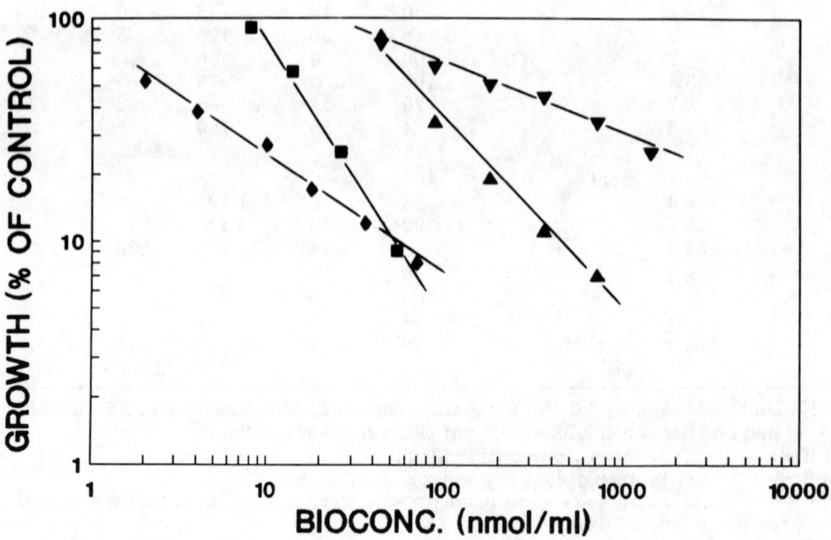

Fig. 10–1. Standard plant growth curves (log scales) for 2,4-D (◆), prometon (■), paraquat (▲), and chlorpropham (▼) added to model soil media (160 g of sand, 40 mL of nutrient solution, cucumber as bioassay species) (Scott & Weber, 1967).

cide (Weber et al., 1969b). It was also shown in a field experiment where high rates of paraquat had been applied to a heavily limed (45 mt/ha) and a lightly limed (3 mt/ha) Portsmouth sandy loam soil (fine-loamy over sandy or sandy skeletal, mixed, thermic Typic Umbraquults). The herbicide provided 25 and 50% weed control at 13 and 22 kg a.i./ha, respectively, on the heavily limed soil but provided 0% weed control at these rates on the lightly limed soil. This suggested that the added Ca reduced paraquat inactivation by the soil (Weber & Best, 1972).

Table 10-3. Reduction in herbicidal bioactivity resulting from soil organic matter (histosol) additions to model soil media (160 g of sand, 45 mL of solution, cucumber [*Cucumis sativus* L.] bioassay species). (Calculated from Scott & Weber, 1967.)

Herbicide	Conc. added	Organic matter added	Plant growth†	Bio-conc., C_e‡	Reduced bioconc., x m^{-1}§	Bio-K_d, (x m^{-1} C_e^{-1})¶	Mean Bio-K_d
	nmol mL^{-1}	g pot^{-1}	% of control	nmol mL^{-1}	nmol g^{-1}		
Chlorpropham	44	2.0	100	0	990	--	
	89		100	0	2 002	--	
	178		100	0	4 005	--	
	355		80	34	7 222	212	
	711		56	112	13 477	120	
	1 422		48	200	27 495	137	156
	44	4.0	100	0	495	--	
	89		100	0	1 001	--	
	178		100	0	2 002	--	
	355		95	20	3 769	188	
	711		75	46	7 481	163	
	1 422		55	125	14 591	117	156
						Mean =	156
2,4-D	2.22	2.0	95	0	50	--	
	4.44		90	0.80	82	102	
	8.88		82	1.05	176	168	
	17.78		57	2.10	353	168	
	35.55		38	4.20	705	168	
	71.11		22	12.00	1 330	111	143
	2.22	4.0	98	0	25	--	
	4.44		95	0	50	--	
	8.88		90	0.80	91	114	
	17.78		85	0.98	189	193	
	35.55		68	1.45	384	264	
	71.11		52	2.50	772	309	220
						Mean =	181
Paraquat	44	2.0	85	36	180	5	
	89		65	49	900	18	
	178		52	62	2 610	42	
	355		19	200	3 487	17	
	711		10	444	6 007	13	19
	44	6.0	92	0	330	--	
	89		90	33	420	13	
	178		65	49	967	20	
	355		50	67	2 160	32	24
	711		28	133	4 335	32	24
						Mean =	22

(continued on next page)

Table 10–3. Continued.

Herbicide	Conc. added	Organic matter added	Plant growth†	Bio-conc., C_e‡	Reduced bioconc., x m^{-1}§	Bio-K_d, (x m^{-1} C_e^{-1})¶	Mean Bio-K_d
	nmol mL^{-1}	g pot^{-1}	% of control	nmol mL^{-1}	nmol g^{-1}		
Prometon	6.7	0.5	100	0	603	--	
	13.3		100	0	1 197	--	
	26.7		95	7	1 773	253	
	53.3		67	13	3 627	279	266
	6.7	2.0	100	0	151	--	
	13.3		100	0	299	--	
	26.7		100	0	601	--	
	53.3		100	0	1 199	--	
						Mean =	266

† Percent of control using plant dry weights.
‡ Bioconcentration from standard plant growth curve in Fig. 10–1; values obtained from 10 to 90% of control were used.
§ Reduced bioconcentration = Concentration added minus bioconcentration available to plants.
¶ Bio-K_d = Reduced Bioconc./Bioconc. = Bound herbicide/Bioavailable herbicide = (x m^{-1} C_e^{-1}).

Table 10–4. Reduction in herbicidal bioactivity of paraquat resulting from additions of clay minerals to model soil media (160 g of sand, 45 mL of solution, cucumber [*Cucumis sativas* L.] bioassay species). (Calculated from Weber et al., 1969a.)

Conc. added	Clay added		Plant growth‡	Bioconc., C_e§	Reduced bioconc, x m^{-1}¶	Bio-K_d, (x m^{-1} C_e^{-1})#	Mean Bio-K_d
	Type†	Amount					
nmol mL^{-1}		g	% of control	nmol mL^{-1}	nmol g^{-1}		
222	Verm.	0.020	0	222	0	--	
		0.025	0	222	0	--	
		0.030	0	222	0	--	
		0.035	10	100	156 857	1 568	
		0.040	15	72	168 750	2 344	
		0.045	20	59	163 000	2 763	
		0.050	25	49	155 700	3 177	
		0.060	30	42	135 000	3 214	2 613
	Mont.	0.010	0	222	0	--	
		0.012	0	222	0	--	
		0.015	25	49	519 000	10 592	
		0.017	50	28	513 529	18 340	
		0.020	60	24	445 500	18 562	
		0.025	75	21	361 800	17 228	
		0.027	90	15	345 000	23 000	
		0.030	97	0	330 000	--	17 544

† Verm. = vermiculite, Mont. = montmorillonite.
‡ Percent of control using plant dry weights.
§ Bioconcentration from standard growth curve similar to that shown in Fig. 10–1.
¶ Reduced bioconcentration = Concentration added minus bioconcentration available to plants.
# Bio-K_d = Reduced Bioconc./Bioconc. = Bound herbicide/Bioavailable herbicide = (x m^{-1} C_e^{-1}).

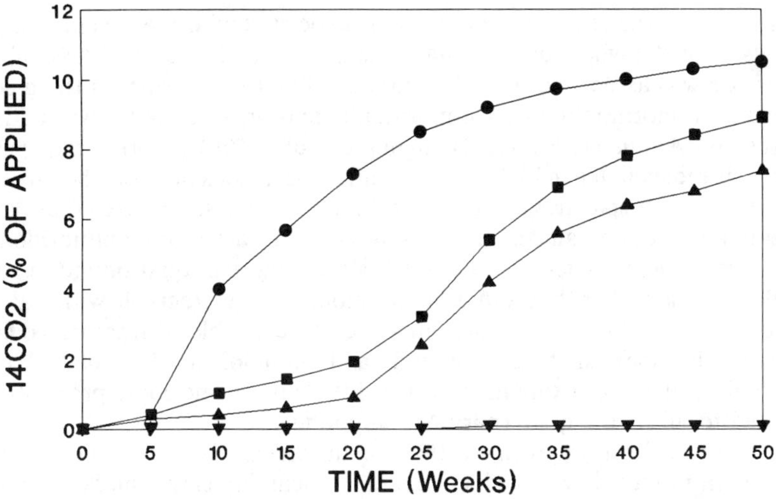

Fig. 10–2. Effects of soil constituents on microbial degradation of diquat in 100 mL of nutrient solution with no added colloids (●), plus 0.5 g of organic matter (■), plus 1.0 g of organic matter (▲), plus 0.05 g of montmorillonite or 0.05 g of montmorillonite and 0.5 g of organic matter (▼). (Weber & Best, 1970, unpublished data).

Microbial degradation studies with ^{14}C-diquat sorbed to kaolinite or montmorillonite in model nutrient media showed that the herbicide was readily degraded when free in solution or sorbed to kaolinite, but was not degraded when sorbed to montmorillonite (Weber & Coble, 1968). In a similar, year-long, microbial degradation study it was found that the incremental addition of small quantities of soil organic matter to the growth media proportionately reduced the rate and quantity of the herbicide degraded. Small amounts of montmorillonite added alone or in combination with small amounts of organic matter completely eliminated microbial degradation of the herbicide (Fig. 10–2; J.B. Weber & J.A. Best, 1970, unpublished data). A mass balance of the dissipation of ^{14}C-diquat over the 1-yr period showed that 90.1 to 101.9% of that applied was recovered in the five treatments (J.B. Weber & J.A. Best, 1970, unpublished data; Table 10–5). Approximately

Table 10–5. Mass balance recovery of ^{14}C-diquat in a microbial degradation study as influenced by additions of organic matter (histosol) or montmorillonite to the media (52 wk) (Weber & Best, 1970, unpublished data).

	Treatment†				
Disposition of ^{14}C-Diquat	A	B	C	D	E
	% ^{14}C-diquat				
Degraded to $^{14}CO_2$	10.5	8.9	7.4	0.1	0.1
Free in solution	72.1	11.5	8.2	0.9	0.9
In the solid pellet	13.1	69.7	80.1	97.6	100.9
Total recovered	95.7	90.1	95.7	98.6	101.9

† Treatments consisted of (A) 100 mL of mineral solution (nutrients, beef extract, and polypeptones) and 4.0 μmol (10 μCi) of ^{14}C-diquat, (B) plus 0.5 g of OM (organic matter), (C) plus 1.0 g of OM, (D) plus 0.05 g of Mont. (montmorillonite), and (E) plus 0.5 g of OM and 0.05 g of Mont.

70 to 80% of the remaining diquat was associated with the organic matter pellets in media where organic matter was added, while 98 to 100% of the herbicide was associated with the montmorillonite clay pellet in the media where montmorillonite or montmorillonite and organic matter was added. Burns and Audus (1970) and Damanakis et al. (1970) reported that paraquat was bioavailable in high organic matter (peat) soils but not when sorbed in the interlayer spacings of montmorillonite. They also reported that paraquat moved from organic matter sites into the clay lattice of montmorillonite in dialysis studies. Tucker et al. (1969) stated that paraquat bound in soils could be classified as "unbound" and "bioactive" (extractable with water), "loosely bound" and "partially bioactive" (extractable with saturated ammonium chloride), and "tightly bound" and "nonbiologically available" (extractable only by refluxing in sulfuric acid). This scheme could probably be applied to all of the quaternary N compounds.

There is general agreement that the quaternary N compounds readily react with negatively charged soil colloids, including clay minerals and organic matter in a manner analogous to that of inorganic cations, such as Ca, K, and Mg. Furthermore, the bioavailability of these compounds may be estimated by use of salt solutions in a manner analogous to that used to estimate the exchangeable cations in soil. It is further suggested that the retention of these compounds in the interlayer spacings of expanding clays and their subsequent lack of bioavailability is not unlike that of "K fixation" by the hydrous micas. Potassium fixation is thought to be the case where exchangeable (and readily bioavailable) K is converted to less available forms by becoming entrapped in the interlayer spacings of the selected clay mineral.

In summary, quaternary N compounds were readily sorbed to clay mineral surfaces in exchange for inorganic cations, sorption isotherms were type H and the chemicals could not be displaced with water but could be displaced from selected clays in varying amounts with salt solutions. The bonding mechanism was ionic supplemented by physical forces. Bio-K_d values ranged from 76 for kaolinite to 2613 for vermiculite and 17 544 for montmorillonite. The compounds were not bioactive when sorbed in the interlayer spacings of montmorillonite. Quaternary N compounds were also readily sorbed by soil organic matter in exchange for inorganic cations and sorption isotherms were type L. Small amounts were displaced with water and large amounts were displaced with salt solutions. The mean Bio-K_d value for quaternary N compounds on organic matter was 22. Clay minerals make up the primary sorption sites for quaternary N compounds in mineral soils but organic matter is the primary inactivator of the compounds in organic soils. Montmorillonite would be extremely effective as a remediation treatment to reduce movement and bioactivity in soils contaminated with quaternary N compounds but activated C would be ineffective.

WEAKLY BASIC COMPOUNDS

Many weakly basic organic compounds have been reported to be sorbed to soil organic matter and clay minerals, including amino acids (Talibudeen,

1955; Dashman & Stotzky, 1986), enzymes (McLaren, 1954; McLaren & Estermann, 1956; Estermann et al., 1959), proteins (Ensminger & Gieseking, 1942; Pinck & Allison, 1951; Pinck et al., 1954; Armstrong & Chesters, 1964), peptides (Dashman & Stotzky, 1986), antibiotics (Siminoff & Gottlieb, 1951; Gottlieb & Siminoff, 1952; Gottlieb et al., 1952; Martin & Gottlieb, 1952; Pinck et al., 1961; Pinck, 1962), triazine herbicides (Frissel & Bolt, 1962; Weber et al., 1965; Weber, 1966, 1970; Cruz et al., 1968; Sullivan & Felback, 1968; Doherty & Warren, 1969; Webner et al., 1969c; Hayes, 1970; Dunigan & McIntosh, 1971; Gilmour & Coleman, 1971; Carringer et al., 1975; Kozak et al., 1983), triazole herbicides (Russell et al., 1968; Nearpass, 1969, 1970; Liu & Weber, 1986), thiadiazoles (Weber, 1980), pyrimidine herbicides (Weber, 1980; Shea & Weber, 1983; Weber et al., 1986), and triazole fungicides (Weber, 1982). Sorption was pH dependent, increasing with decreasing pH, and occurred in maximum amounts in the vicinity of the pK_a of the compound. Sorption isotherms were normally types L or S for molecular species and approached type H for protonated species (Giles & MacEwan, 1957; Weber & Miller, 1989). Bound fractions of atrazine were found in the organic fraction (90%) and in the mineral fraction (10%) of neutral pH soils (Capriel et al., 1985). The bound fractions would likely be different in soils of lower pH, due to protonation and chemical hydrolysis of the compound under acid conditions.

Bioactivity (weed control) of many weakly basic herbicides has been reported to be inversely related to the organic matter content of soils. This included atrazine (Sheets et al., 1962; Parochetti, 1973; Harrison et al., 1976; Rahman, 1976; Blumhorst et al., 1990; Reinhardt et al., 1990), chlorazine, ipazine, propazine, and trietazine (Sheets et al., 1962), cyanazine (Blumhorst et al., 1990), fluridone (Schroeder & Banks, 1986), metribuzin (Schmidt, 1973; Peter & Weber, 1985a; Weber et al., 1987), prometryn (Doherty & Warren, 1969; Weber et al., 1974c), simazine (Burnside & Behrens, 1961; Sheets et al., 1962; Upchurch & Mason, 1962; Harris & Sheets, 1965; Grover, 1966; Harris, C.I., 1966; Upchurch et al., 1966; Day et al., 1968; Doherty & Warren, 1969), and simeton (Slack et al., 1978). Reduction in bioactivity of the weakly basic herbicide cyanazine by increasing organic matter content of soils is shown in Fig. 10–3. Bioactivity of weakly basic herbicides has also been reported to be inversely related to the clay content of soils for the following chemicals: atrazine (Smit & Nel, 1977, 1980), metribuzin (Peter & Weber, 1985a), and simazine (Burnside & Behrens, 1961).

Additions of soil organic matter or montmorillonite to a model soil system consisting of quartz sand and nutrient solution or sandy field soil reduced the bioactivity of the following weakly basic herbicides: atrazine (J.A. Best & J.B. Weber, 1969, unpublished data; Peeper & Weber, 1976), fluridone (Shea & Weber, 1980), procyazine (Peeper & Weber, 1976), prometon (Scott & Weber, 1967), prometryn (Weber et al., 1968a, 1974a, 1974b; J.A. Best & J.B. Weber, 1969, unpublished data; Doherty & Warren, 1969) and simazine (Doherty & Warren, 1969). Reduction in prometon bioactivity by clay minerals and organic matter added to model soil media is illustrated in Tables 10–2 and 10–3, and for atrazine and prometryn in Tables 10–6 and 10–7.

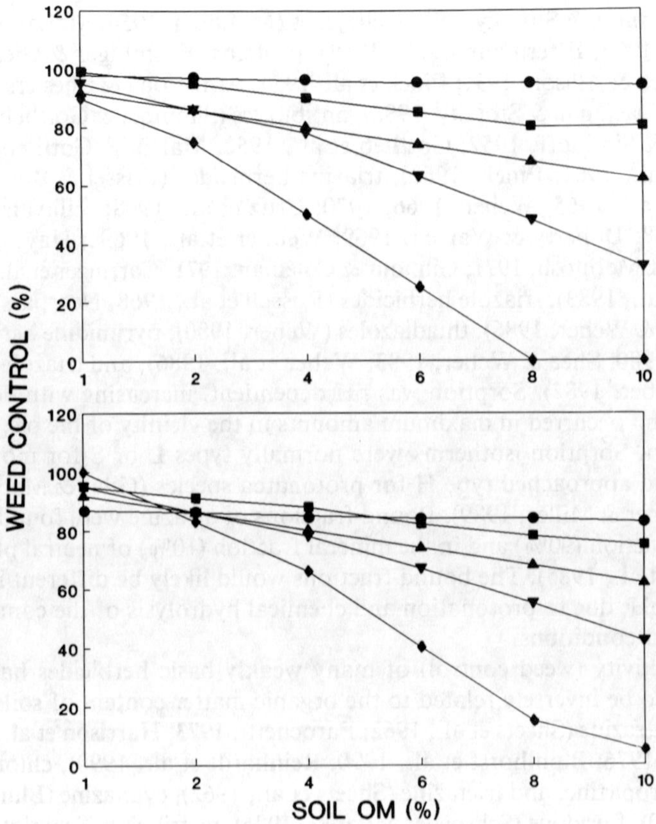

Fig. 10–3. Effect of soil organic matter on bioactivity of herbicides (each herbicide was applied at the same rate to all soils, (upper): chloramben (●), prynachlor (■), alachlor (▲), oryzalin (▼), norflurazon (◆); (lower): propachlor (●), acetochlor (■), alachlor (▲), cyanazine (▼), fluorodifen (◆). (Redrawn from Weber et al., 1974c.)

Organic matter and montmorillonite were effective in reducing bioactivity whereas kaolinite was only slightly effective. Bioactivity of prometryn, atrazine, and procyazine was reduced by organic matter additions to the field soil by about the same degree [compare mean reduction in weed control by −1.1, −1.0, and −1.2%/mt of additive, respectively (Table 10–8)]. Additions of montmorillonite were less effective in reducing bioactivity of the three herbicides, with reductions in weed control of −0.5, −0.1, and −0.5%/mt of additive, respectively. Bioactivity of many basic herbicides has also been shown to be reduced by adding activated C to the soil (Ahrens, 1965; Linscott & Hagin, 1967; Jordan & Smith, 1971; Majek, 1986).

Soil pH was directly related to the bioactivity of the following weakly basic herbicides: atrazine (Sheets et al., 1962; Best & Weber, 1974; Colbert et al., 1975; Smit & Nel, 1977, 1980; Reinhardt et al., 1990), chlorazine, ipazine, propazine and trietazine (Sheets et al., 1962), metribuzin (Ladlie et al., 1976), prometryn (Weber et al., 1968a), simazine (Sheets et al., 1962; Near-

Table 10–6. Reduction in herbicidal bioactivity as influenced by additions of organic matter (histosol) to model soil media (60 g of sand, 20 mL of solution, oat [*Avena sativa* L.] bioassay species). (Best & Weber, 1969, unpublished data.)

Herbicide†	Conc. added	Organic matter added	Plant growth†	Bio-conc., C_e‡	Reduced bioconc, x m^{-1}§	Bio-K_d, (x m^{-1} C_e^{-1})¶	Mean Bio-K_d
	nmol mL^{-1}	g pot^{-1}	% of control	nmol mL^{-1}	nmol g^{-1}		
Atrazine	4.00	0.5	67	1.2	112	93	
		1.0	88	0.9	62	69	
		2.0	100	0	40	--	81
Prometryn	4.00	0.5	89	1.1	116	105	
		1.0	91	1.0	60	60	
		2.0	95	0.5	35	70	78
Dichlobenil	4.00	0.5	83	0.50	140	280	
		1.0	90	0.45	71	158	
		2.0	95	0	40	--	219
Fluometuron	4.00	0.5	76	2.1	156	74	
		1.0	83	1.8	84	47	
		2.0	89	0	60	--	60
Fenuron	8.00	0.5	54	6.0	80	13	
		1.0	60	5.0	60	12	
		2.0	68	4.0	40	10	12
Diphenamid	8.00	0.5	30	5.0	120	24	
		1.0	75	3.0	100	33	
		2.0	95	0	80	--	28
Chlorpropham	1.00	0.5	82	0.39	44	114	
		1.0	93	0.25	25	100	
		2.0	97	0	15	--	107

† Percent of control using plant dry weight.
‡ Bioconcentration from standard growth curves for oat similar to those shown for cucumber in Fig. 10–1.
§ Reduced bioconcentration = Concentration added minus bioconcentration available to plants.
¶ Bio-K_d = Reduced Bioconc./Bioconc. = Bound herbicide/Bioavailable herbicide = (x m^{-1} C_e^{-1}).

pass, 1965), simeton (Slack et al., 1978), and terbutryn (Laerman & Heitefuss, 1974; Colbert et al., 1975).

Lime applied to the soil increased the bioactivity of the following compounds: atrazine (Best et al., 1975; Hiltbold & Buchanan, 1977; Lowder & Weber, 1982), fluridone (Shea & Weber, 1980, 1983), metamitron (Kloskowski & Fuhr, 1987), prometryn (Weber & Best, 1972; Best & Weber, 1974; Best et al., 1975), simazine (Burnside & Behrens, 1961), and terbutryn (Colbert et al., 1975). Decreasing soil pH reduced the bioactivity of the following chemicals: atrazine (Lowder & Weber, 1972), fluridone (Shea & Weber, 1980), prometryn (Weber et al., 1968a), and simazine (Slack et al., 1978). Lowering the pH of model soil media containing organic matter and montmorillonite reduced the bioactivity of prometryn, as illustrated in Table 10–9.

Table 10–7. Reduction in herbicidal bioactivity resulting from additions of clay minerals to model soil media (60 g of sand, 20 mL of solution, oat [*Avena sativa* L.] bioassay species) (Best & Weber, 1969, published data)

Herbicide	Conc. added	Additive Type†	Additive Amount	Plant growth‡	Bio-conc., C_e§	Reduced bioconc., x m^{-1}¶	Bio-K_d, (x m^{-1} C_e^{-1})#	Mean Bio-K_d
	nmol mL^{-1}		g pot^{-1}	% of control	nmol mL^{-1}	nmol g^{-1}		
Atrazine	4.00	Kaol.	2.5	9	4.0	0	--	
			5.0	18	3.0	4.0	1.3	
			10.0	26	2.4	3.2	1.3	1.3
		Mont.	2.5	28	2.2	14.4	6.5	
			5.0	64	1.2	11.2	9.3	
			10.0	95	0	8.0	--	7.9
Prometryn	4.00	Kaol.	2.5	21	3.0	8.0	2.7	
			5.0	51	1.7	9.2	5.4	
			10.0	67	1.4	5.2	3.7	3.9
		Mont.	2.5	90	1.1	23.2	21.1	
			5.0	100	0	16.0	--	
			10.0	100	0	8.0	--	21.1
Dichlobenil	4.00	Kaol.	2.5–10.0	19	4.0	0	0	0
		Mont.	2.5	30	2.0	16.0	8.0	
			5.0	45	1.1	11.6	10.5	
			10.0	75	0.4	7.2	18.0	12.2
Fluometuron	6.00	Kaol.	2.5–10.0	25	6.0	0	0	0
		Mont.	2.5	45	3.4	20.8	6.1	
			5.0	55	2.8	12.8	4.6	
			10.0	78	2.0	8.0	4.0	4.9
Fenuron	8.00	Kaol.	2.5–10.0	40	8.0	0	0	0
		Mont.	2.5	40	8.0	0	--	
			5.0	50	6.0	8.0	1.3	
			10.0	60	4.0	8.0	2.0	1.6
Diphenamid	8.00	Kaol.	2.5	15	8.0	0	--	
			5.0	20	6.0	8.0	1.3	
			10.0	24	5.8	4.4	0.7	1.0
		Mont.	2.5	35	4.6	27.2	5.9	
			5.0	55	3.5	18.0	5.1	
			10.0	88	2.6	10.8	4.1	5.0
Chlorpropham	1.50	Kaol.	2.5–10.0	24	1.5	0	0	
		Mont.	2.5	56	0.85	5.2	6.1	
			5.0	85	0.35	4.6	13.1	
			10.0	93	0	3.0	--	9.6

† Kaol. = kaolinite; Mont. = montmorillonite.
‡ Percent of control using plant dry weight.
§ Bioconcentration from standard plant growth curves for oat similar to those shown for cucumber in Fig. 10–1.
¶ Reduced bioconcentration = Concentration added minus bioconcentration available to plants.
# Bio-K_d = Reduced Bioconc./Bioconc. = Bound herbicide/Bioavailable herbicide = (x m^{-1} C_e^{-1}).

Organic matter was the more effective additive in reducing bioactivity, but lowering the soil pH had more pronounced effectg on the media where the clay mineral had been added.

Table 10–8. Herbicidal bioactivity as influenced by addition of organic matter (histosol) and montmorillonite to a sandy loam soil in the field. (From Weber et al., 1974b; Peeper & Weber, 1976.)

Herbicide	Rate applied	Additive Type†	Additive Amount	Weed control‡	Change in weed control by additive		Mean change in weed control
	kg ha^{-1}		mt ha^{-1}	%	%	% mt^{-1}	% mt^{-1}
Flumeturon	1.0	None	0	77	--	--	
	2.0	None	0	94	--	--	
	1.0	OM	27	42	−35	−1.3	
			54	9	−68	−1.3	
	2.0	OM	27	65	−29	−1.1	
			54	45	−49	−0.9	−1.1
	1.0	Mont.	45	73	−4	−0.1	
			90	69	−8	−0.1	
	2.0	Mont.	45	93	−1	--	
			90	92	−2	--	−0.1
Prometryn	2.6	None	0	55	--	--	
	5.2	None	0	73	--	--	
	2.6	OM	27	14	−41	−1.5	
			54	4	−51	−0.9	
	5.2	OM	27	41	−32	−1.2	
			54	17	−56	−1.0	−1.1
	2.6	Mont.	45	42	−13	−0.3	
			90	22	−33	−0.4	
	5.2	Mont.	45	42	−31	−0.7	
			90	25	−48	−0.5	−0.5
Trifluralin	0.5	None	0	90	--	--	
	1.0	None	0	93	--	--	
	0.5	OM	27	61	−29	−1.1	
			54	50	−40	−0.7	
	1.0	OM	27	81	−12	−0.4	
			54	54	−39	−0.7	−0.7
	0.5	Mont.	45	95	+5	+0.1	
			90	97	+7	+0.1	
	1.0	Mont.	45	99	+6	+0.1	
			90	95	+2	0	+0.1
Atrazine	2.24	None	0	100	--	--	
		OM	27	75	−25	−0.9	
			54	33	−67	−1.2	−1.0
		Mont.	45	94	−6	−0.1	
			90	89	−11	−0.1	−0.1
Procyazine	3.36	None	0	99	--	--	
		OM	27	70	−29	−1.1	
			54	25	−74	−1.4	−1.2
		Mont.	45	78	−21	−0.5	
			90	63	−36	−0.4	−0.5
Buthidazole	0.56	None	0	98	--	--	
		OM	27	95	−3	−0.1	
			54	55	−43	−0.8	−0.5
		Mont.	45	91	−7	−0.2	
			90	45	−53	−0.6	−0.4

† OM = organic matter (histosol), Mont. = montmorillonite; additives were incorporated to 15 cm with a power rotary hoe.

‡ Mean weed control ratings for 3 yr for fluometuron, prometryn, and trifluralin (Weber et al., 1974b) and 1-yr weed control for atrazine, procyazine, and buthidazole (Peeper & Weber, 1976).

Table 10–9. Reduction in herbicidal bioactivity of prometryn resulting from organic matter (histosol) and montmorillonite additions and lowering soil pH in model soil media (160 g of sand, 45 mL of solution, wheat [*Triticum aestivum* L.] bioassay species). (Calculated from Weber et al., 1968a.)

Herb. conc. added	Additive Type†	Additive Amount	pH	Plant growth†	Bio-conc., C_e§	Reduced bioconc., x m^{-1}¶	Bio-K_d, (x m^{-1} C_e^{-1})	Mean Bio-K_d
nmol mL^{-1}		g		% of control	nmol mL^{-1}	nmol g^{-1}		
0.49	OM	0.5	6.5	100	0	44	--	
1.00				100	0	90	--	
2.00				72	0.7	117	167	
2.67				54	1.3	123	95	
4.00				50	1.5	225	150	137
0.49	OM	0.5	4.1	100	0	44		
1.00				100	0	90	--	
2.00				93	0	180	--	
2.67				66	0.8	168	210	
4.00				61	1.0	270	270	240
0.49	Mont.	1.9	6.8	100	0	12	--	
1.00				84	0.5	12	24	
2.00				55	1.3	16	13	
2.67				48	1.7	23	14	
4.00				45	2.0	47	24	19
0.49	Mont.	1.9	4.3	100	0	12	--	
1.00				99	0	24	--	
2.00				97	0	47	--	
2.67				81	0.5	51	103	
4.00				78	0.6	80	134	118

† OM = organic matter, Mont. = montmorillonite.
‡ Percent of control using plant dry weight.
§ Bioconcentration from standard plant growth curves for wheat similar to that shown for cucumber in Fig. 10–1.
¶ Reduced bioconcentration = Concentration added minus bioconcentration available to plants.
# Bio-K_d = Reduced Bioconc./Bioconc. = Bound herbicide/Bioavailable herbicide = (x m^{-1} C_e^{-1}).

Bio-K_d values for prometon resulting from kaolinite and montmorillonite additions to model soil media ranged from 5 to 10 (mean = 7) and from 115 to 230 (mean = 172), respectively (Table 10–2). Bio-K_d values for prometon resulting from organic matter additions ranged from 253 to 279 (mean = 266) (Table 10–3). Bio-K_d values for atrazine resulting from kaolinite, montmorillonite, and organic matter additions to model soil media were 1.3, 6.5 to 9.3 (mean = 7.9), and 69 to 93 (mean = 81), respectively (Tables 10–7 and 10–6, respectively). Bio-K_d values for prometryn resulting from the soil additives were 2.7 to 5.4 (mean = 3.9), 21.1, and 60 to 105 (mean = 78), respectively (Tables 10–7 and 10–6, respectively). Prometryn Bio-K_d values were pH dependent and increased with decreasing pH. Bio-K_d values for prometryn where montmorillonite was added to model soil media increased more than six times, from 13 to 24 (mean = 19) at pH 6.8 to 103 to 134 (mean = 118) at pH 4.3 (Table 10–9). Bio-K_d values for the herbicide in organic matter amended media increased as pH was lowered, ranging from 95 to 167 (mean = 137) at pH 6.4 to 210 to 270 (mean = 240) at pH 4.1. Higher Bio-K_d values at lower pH levels is a reflection of the greater binding of cationic species at lower pH, as compared with physical binding of molecular species at higher pH levels.

Chemical degradation of atrazine was reported to be inversely related to soil pH (Armstrong & Chesters, 1967; Harris, 1967; Skipper et al., 1967; Skipper & Volk, 1972; Best et al., 1975; Kells et al., 1980), whereas biological degradation of cyanazine (Blumhorst & Weber, 1988) and prometryn (Best & Weber, 1974) was directly related to pH, that is, increasing pH resulted in increasing biological degradation.

Smit et al. (1980, 1981) reported increased bioactivity of atrazine when KH_2PO_4 fertilizer was applied to soil and Hermosin et al. (1982) reported that prometryn, sebuthylazine, terbuthylazine, and GS 18183, were released from montmorillonite and made bioactive by displacement with NH_3. Apparently, mass action replacement of the triazine herbicides from soil colloids by nutrients occurs and plants and other organisms respond to the released materials.

The amount of amitrole extracted from soil with water or dilute salt solutions was reported to be proportional to that bioavailable to plants (Sund, 1956) and the amount of bioactive atrazine and prometryn in soils was proportional to that extracted with methanol (Best & Weber, 1974).

The amount of triazine herbicides (atrazine and prometryn) extractable from soil with methanol decreased with time, suggesting that "fixation" of the chemicals occurred (Best & Weber, 1974; Khan & Hamilton, 1980; Khan & Ivarson, 1981). Bioactivity of the chemicals was related to their extractability.

In summary, weakly basic compounds were sorbed to soil organic matter and clay minerals and sorption was pH dependent, increasing with decreasing pH. Sorption isotherms were type L at neutral pH where molecular species were retained through physical forces and became more like the type H in acidic systems where protonated species were retained by ionic forces. Water was effective in displacing molecular bound species but not ionically bound ones. Salt solutions were effective in displacing ionically bound species. Bio-K_d values for weakly basic compounds ranged from 7 for kaolinite, to 19 for montmorillonite and 137 for soil organic matter at neutral pH levels. They increased to 118 for montmorillonite and 240 for organic matter under acidic conditions. Organic matter and montmorillonite contribute to the binding and reduction in bioactivity of weakly basic compounds in soils, with organic matter contributing the greatest under neutral or alkaline pH conditions and montmorillonite contributing greatly under acidic conditions. Activated C additions would be effective as a remediation treatment for soils contaminated with weakly basic compounds, and would be most effective under neutral or alkaline conditions. Montmorillonite could be used as a remediation treatment but would be effective only under acidic conditions.

WEAKLY ACIDIC COMPOUNDS

Because of their anionic charge in the ionized form, many weakly acidic compounds, including 2,4-D, 2,4,5-T, MCPA, DNOC, dinoseb, and picric and picolinic acids, have been reported to be repelled from negatively charged clay minerals (Bolt & Warkentin, 1958; Frissell, 1961). Many compounds, including bentazon; chlorosulfuron; phenoxy-, benzoic-, phenylacetic- and aliphatic acids; and phenols were sorbed in low to moderate amounts by soil organic matter (Talbert et al., 1970; Adams, 1972; Weber, 1972; Mercie & Foy, 1986), metallic hydrous oxides (Weber, 1972), and, in high amounts, by strongly basic anion exchange resins and activated C (Weber et al., 1965, 1968b; Ward & Getzen, 1970; Abernathy & Wax, 1973). Sorption was pH dependent and was greater at low pH where the less water soluble, uncharged molecular species predominated. Sorption isotherms were of the type L on organic matter and activated C and of the type S on clay minerals (Weber & Miller, 1989).

Bioactivity of weakly acidic herbicides has been reported to be inversely related to soil organic matter content. Chemicals involved included chloramben (Schliebe et al., 1965; Weber et al., 1974c; Harrison et al., 1976); 2,4-D (Jorgensen & Hamner, 1948); picloram (Herr et al., 1966; Grover, 1968); TCA (Blouch & Fults, 1953; Rai & Hamner, 1953); dinoseb (Dowler et al., 1958); terbacil (Kratky & Warren, 1973); chlorsulfuron (Anderson, 1985; Anderson & Barrett, 1985); metsulfuron (Anderson, 1985); and oryzalin (Weber et al., 1974c). Clay minerals were not related to herbicide inactivation. A survey of weed performance for a variety of herbicides, including the weakly acidic herbicides chloramben and oryzalin, when applied at the same rate on a variety of soils, showed that soil organic matter was the soil constituent most highly (and inversely) related to bioactivity (Weber et al., 1974c; Fig. 10–3). Inactivation by soil organic matter was generally inversely related to the water solubility of the chemicals.

Bioactivity of 2,4-D was reduced greatly by additions of strongly basic anion exchange resins to growth media (Weaver, 1947; Scott & Weber, 1967) and was reduced moderately by the addition of soil organic matter (Scott & Weber, 1967), but additions of strongly acidic cation exchange resins (Weaver, 1947) or clay minerals (Scott & Weber, 1967) had no effect. Table 10–3 shows the moderate Bio-K_d values (102–309) (mean = 181) for 2,4-D calculated from the additions of soil organic matter to model soils and Table 10–10 shows the large Bio-K_d values (1110–2127) (mean = 1601) calculated from the addition of strongly basic anion exchange resins. High Bio-K_d values probably reflect sorption of high amounts of the herbicide by ionic bonding mechanisms and moderate Bio-K_d probably reflect sorption of the herbicide by physical bonding mechanisms.

Bioactivity of several weakly acidic herbicides, including chloramben (Coffey & Warren, 1969; Ogg, 1978); 2,4-D (Lucas & Hamner, 1947; Arle et al., 1948; Weaver, 1948; Coffey & Warren, 1969); DCPA (Coffey & Warren, 1969); and terbacil (Elmstrom, 1972; Ogg, 1982) was reduced or eliminated by additions of activated C to the growth media.

Table 10-10. Reductions in herbicidal bioactivity resulting from additions of a strongly basic anion exchange resin (Cl-form) to model soil media (160 g of sand, 45 mL of solution, cucumber [*Cucumis sativas* L.] bioassay species.) (Calculated from Scott & Weber, 1967.)

Herbicide†	Conc. added	Anion exch. resin	Plant growth‡	Bio.- conc., x m^{-1}§	Reduced bioconc. x m^{-1}¶	Bio-K_d (x m^{-1} C_e#)	Mean Bio-K_d
	nmol mL^{-1}	g pot^{-1}	% of control	nmol mL^{-1}	nmol g^{-1}		
Chlorpropham	44	0.5	85	28	1 440	51	
	89		82	31	5 220	168	
	178		58	90	7 920	88	
	355		52	150	18 450	123	
	711		44	260	40 590	156	
	1 422		42	310	100 080	323	151
	44	1.5	97	0	1 320	--	
	89		92	0	2 670	--	
	178		90	23	4 650	202	
	355		82	31	9 720	313	
	711		63	80	18 930	237	
	1 422		52	150	38 160	254	251
						Mean =	201
2,4-D	2.2	0.2	95	0	495	--	
	4.4		91	0	990	--	
	8.9		67	1.5	1 665	1 110	
	17.8		47	2.9	3 352	1 156	
	35.5		38	4.4	6 997	1 590	
	71.1		30	6.8	14 467	2 127	1 496
	2.2	0.5	96	0	198	--	
	4.4		92	0	396	--	
	8.9		91	0	801	--	
	17.8		76	1.1	1 503	1 366	
	35.5		65	1.6	3 051	1 907	
	71.1		44	3.3	6 102	1 849	1 707
						Mean =	1 601

† Anion exchange resin additions of 0.5 and 1.5 g had no effect on the bioactivity of paraquat or prometon.
‡ Percent of control using plant dry weights.
§ Bioconcentration from standard plant growth curves similar to those shown in Fig. 10-1.
¶ Reduced bioconcentration = Concentration added minus bioconcentration available to plants.
# Bio-K_d = Reduced Bioconc./Bioconc. = Bound herbicide/Bioavailable herbicide = (x m^{-1} C_e^{-1}).

Reducing soil pH from 6.5 to 5.0 reduced the bioactivity of picloram due to increasing binding of the molecular species by soil organic matter (Grover, 1968). Increasing soil pH from 6.5 to 7.0 reduced bioactivity of the herbicide due to lower uptake of the anionic species by plant roots. Analogous results were reported for phenols (Shea et al., 1983). Liming the soil has been reported to reduce the volatilization and vapor injury to crop plants by dinoseb (Davis, 1956). This was probably because the lime increased soil pH and the equilibrium of dinoseb was shifted from the volatile molecular species to the less volatile anionic phenate species.

In summary, weakly acidic compounds were sorbed primarily to soil organic matter under a wide range of pH conditions and isotherms were type L. The compounds were repelled by clays under neutral conditions but sorbed through physical bonding mechanisms under acidic conditions when the compounds were in the molecular form. Bio-K_d values for weakly acidic compounds ranged from 181 for soil organic matter where binding was primarily through physical forces to 1601 for strongly basic anion exchange resins where binding was through ionic bonds. Soil organic matter was the primary contributor to retaining and bioinactivating weakly acidic compounds in soils. Activated C additions would be the most effective remediation treatment for soils contaminated with weakly acidic compounds and the process would be most effective under acidic conditions where the chemicals are in the molecular form.

AMPHOTERIC COMPOUNDS

Amphoteric compounds contain functional groups that protonate at low pH to form cationic species, remain uncharged at neutral pH and ionize to form anionic species at high pH. Sorption of these compounds, including buthidazole (Weber, 1980), imazaquin (Liu & Weber, 1985; Renner et al., 1988; Stougaard et al., 1990), and imazethapyr (Renner et al., 1988; Stougaard et al., 1990) was inversely related to soil pH. High amounts of imazaquin were sorbed by strongly basic anion exchange resins, suggesting anionic properties, and moderate amounts were sorbed by strongly acid cation exchange resins in the H form but not in the Na form, suggesting very weakly basic properties (Liu & Weber, 1985). Very low amounts were sorbed by montmorillonite and low to moderate amounts were sorbed by soil organic matter, suggesting that physical forces were involved. Sorption isotherms were type L (Weber & Miller, 1989).

Bioactivity of imazaquin and imazethapyr was inversely related to soil organic matter content (Loux et al., 1989; Stougaard et al., 1990). Additions of soil organic matter to model soils reduced the bioactivity of imazaquin in greenhouse studies (McKinnon, 1989) and buthidazole in the field (Peeper & Weber, 1976). Bioactivity of buthidazole was reduced by additions of organic matter and montmorillonite to about the same degree (compare reductions in weed control of −0.5 and −0.4%/mt of additive, respectively (Table 10–8). Additions of montmorillonite or hydrous mica (illite) to model soil media reduced the bioactivity of the amphoteric antibiotics oxytetracycline and tetracycline (Martin & Gottlieb, 1952).

Bioactivity of imazaquin has been reported to be both directly related to soil pH (Loux et al., 1989; Stougaard et al., 1990) and inconsistently related to soil pH (Renner et al., 1988). It would be expected that the chemicals would be less bioactive in acid soils where they would be bound as molecular or protonated species, than in neutral or alkaline soils where they would be present in the anionic form. It is also likely that the molecular forms of the chemicals are more readily taken up by plants than either the protonated or anionic forms, thus the reported differences in pH effects.

Degradation of imazaquin and imazethapyr has been reported to be inversely related to the organic matter and clay contents of soils (Basham et al., 1987; Cantwell et al., 1989). Activated C added to soil media also reduced degradation of imazethapyr (Cantwell et al., 1989).

In summary, amphoteric compounds were sorbed primarily by soil organic matter and isotherms were type L. Montmorillonite also sorbed amphoteric compounds with weakly basic pK_A values of 2 or higher and sorption was greatest under acidic conditions. Organic matter was the primary contributor to retaining and inactivating amphoteric compounds in soils. Activated C additions would be the most effective remediation treatment for soils contaminated with amphoteric compounds.

ORGANIC ARSENIC AND PHOSPHORUS PESTICIDES

Organic herbicides with ionizable As or P acid moieties, such as arsenic acid compounds and glyphosate, are readily sorbed by clay minerals and hydrous oxides (Woolson et al., 1971a, b; Woolson, 1975; McConnell & Hossner, 1985). Extraction of As acid herbicides from soils, including DSMA, MAMA, and MSMA, showed them to be associated with the clay mineral fraction (Dickens & Holtbold, 1967; Johnson & Hiltbold, 1969). Sorption isotherms were types H and L (Weber & Miller, 1989).

Because of the high sorption by soil minerals, bioactivity of arsenic acid, cacodylic acid, and MSMA occurred only when high rates (10 times normal) were applied (Woolson & Isensee, 1981). Bioactivity of arsenic acid, DSMA, and glyphosate was increased when P fertilizer was added to soil) Schweizer, 1967; Woolson et al., 1973; Grossbard & Atkinson, 1985) and glyphosate bioactivity was reduced when clay minerals or Fe or Al salts were added to soils (Grossbard & Atkinson, 1985). Phosphorus fertilizer appears to compete with the As or P herbicides for sites on the colloidal soil surfaces making the herbicides more bioactive due to less binding. Iron and Al salts, however, react with the As and P herbicides to form water-insoluble salts that have low bioactivity.

Ease of extraction of As herbicides from soils with dilute acid or base salt solution has been correlated with the bioactivity of the chemicals (Woolson et al., 1971a, b).

Residual bioactivity of glyphosate and sulfosate has been reported to occur in sandy soils (Sprankle et al., 1975; Torstensson & Aamisepp, 1977; Rodrigues et al., 1982; Salazar & Appleby, 1982; Devlin et al., 1986) and was likely due to inadequate amounts of clay minerals present to inactivate the chemicals.

Organophosphorous pesticides were reported to be sorbed by organic matter and clays in moderate to high amounts. This included azinphosmethyl (Martin & Camazane, 1984), leptophos (Carringer et al., 1975), parathion (Saltzman et al., 1972; Yaron & Saltzmann, 1978), and piperophos (Hata

& Akashi, 1980). Sorption isotherms were type L (Weber & Miller, 1989). Aminoparathion, a metabolite of parathion, was reported to be very strongly bound to soil colloids (Katan & Lichtenstein, 1977).

Bioactivity of many organophosphorous pesticides has been reported to be inversely related to the organic matter and clay contents of soils. Included were crotoxyphos (Konrad & Chesters, 1969), demeton, mevinphos, and schraden (Getzin & Chapman, 1960), diazinon (Harris, 1966), disulfoton (Graham-Bryce, 1967), malathion (Konrad et al., 1969; Bowman et al., 1970), parathion (Weidhaas et al., 1961; Harris, C.R., 1966), and phorate (Getzin & Chapman, 1960; Kirk & Wilson, 1960). Bioactivity of the organophosphorous pesticides was reported to increase with soil moisture levels (Burt et al., 1975). All of the compounds were effective in moist soils but only the more volatile compounds were bioactive under very dry conditions. Addition of manure to soil-bound residues of ^{14}C-parathion increased soil microbial populations and degradation of the bound materials (Racke & Lichtenstein, 1985).

In summary, ionizable organic As and P pesticides were bound primarily by clay minerals and hydrous metallic oxides in soils or formed insoluble salts, and isotherms were types H and L. Phosphorous fertilizer additions released the chemicals from the sorption sites and enhanced their bioactivity in soils. Nonionizable P pesticides were sorbed by and bioinactivated by soil organic matter and montmorillonite in soils, but acidic metabolites were most strongly associated with the clay mineral. Activated C additions would be the most effective remediation treatment for soils contaminated with nonionizable P pesticides and montmorillonite would be the most effective remediation treatment for ionizable As and P pesticides and for acidic metabolites of nonionizable P pesticides.

AMIDE AND ANILIDE HERBICIDES

Many amide and anilide herbicides have been reported to be sorbed by soil organic matter in moderate amounts and by montmorillonitic clays in low to moderate amounts (Weber, 1972). Chemicals involved included: acetochlor (Weber & Peter, 1982), alachlor (Strek & Weber, 1981, 1982a; Weber et al., 1981; Weber & Peter, 1982), butachlor (Sato et al., 1987), diphenamid (Coffey & Warren, 1969), metolachlor (Strek & Weber, 1981, 1982a; Kozak et al., 1983), and napropamide (Wu et al., 1975). Sorption isotherms were type L on soil organic matter and soils and type S on montmorillonite (Weber & Miller, 1989).

Bioactivity of amide and anilide herbicides has been reported to be inversely related to organic and humic matter contents of soils; this included acetochlor (Weber et al., 1974c), alachlor (Parochetti, 1973; Rahman, 1976; Weber et al., 1974c, 1981, 1987; Blumhorst et al., 1990), CDAA (Upchurch et al., 1966; Parochetti, 1973), diphenamid (Duby et al., 1966), metolachlor (Weber et al., 1981, 1987; Blumhorst et al., 1990), pronamide (Carlson et al., 1975; Dutt & Harvey, 1980), propachlor (Parochetti, 1973; Weber et al., 1974c; Harrison et al., 1976), and prynachlor (Parochetti, 1973; Weber et al., 1974c). The relationship of soil organic matter content and bioactivity (weed control) of acetochlor, alachlor, propachlor, and prynachlor is shown in Fig. 10-3.

Organic matter and clay content have both been reported to be inversely related to bioactivity of acetochlor (Weber & Peter, 1982; Reinhardt & Nel, 1990), alachlor (Ballard & Santelmann, 1973; Weber & Peter, 1982; Peter & Weber, 1985b; Reinhardt & Nel, 1988, 1989), metazachlor (Reinhardt & Nel, 1990), and metolachlor (Weber & Peter, 1982; Peter & Weber, 1985b; Reinhardt & Nel, 1989).

Other soil parameters that have been reported to be inversely related to the bioactivity of amide and anilide herbicides include specific surface area (Weber & Peter, 1982; Peter & Weber, 1985b), cation exchange capacity (Ballard & Santelmann, 1973; Dutt & Harvey, 1980; Reinhardt & Nel, 1989, 1990), soil moisture-holding capacity (Harrison et al., 1976; Dutt & Harvey, 1980), and pH (Reinhardt & Nel, 1988, 1989). These parameters are normally highly correlated with organic matter or clay contents of soils, however.

Additions of soil organic matter to model soil media reduced the bioactivity of diphenamid moderately (Bio-K_d = 24 to 33 [mean = 28]; Table 10-6), whereas addition of montmorillonite reduced bioactivity slightly (Bio-K_d = 4.1 to 5.9 [mean = 5.0]; Table 10-7), and addition of kaolinite had almost no effect (Bio-K_d = 0.7 to 1.3 [mean = 1.0]; Table 10-7) (J.A. Best and J.B. Weber, 1969, unpublished data). Addition of activated C to soil greatly reduced the bioactivity of diphenamid (Coffey & Warren, 1969).

In summary, amide and anilide herbicides were sorbed by soil organic matter and isotherms were type L while sorption of the compounds by montmorillonite resulted in isotherms of type S. Sorption of the compounds by soils resulted in composite isotherms of types L and C. Both soil constituents reduced bioactivity of the compounds and Bio-K_d values ranged from 5 for montmorillonite to 28 for soil organic matter. Activated C addition would be an effective remediation treatment for amide- and anilide-contaminated soils. Montmorillonite would be much less effective.

PHENYLCARBAMATE AND PHENYLUREA PESTICIDES

Phenylcarbamate herbicides were reported to be sorbed by organic matter in moderate to high amounts (Harris & Warren, 1964; Weber, 1972), by montmorillonite and exchange resins in low to moderate amounts (Harris & Warren, 1964; Coffey & Warren, 1969; Weber, 1972) and by activated C in very high amounts (Schwartz, 1967; Coffey & Warren, 1969). Phenylurea herbicides were also reported to be adsorbed by organic matter in moderate to high amounts (Hance, 1969; Weber, 1972; Carringer et al., 1975), montmorillonite and exchange resins in low amounts (Frissel & Bolt, 1962; Hance, 1969) and activated C in high amounts (Leopold et al., 1960). Sorption isotherms were type L (Weber & Miller, 1989).

Bioactivity of the phenylcarbamate herbicides chlorpropham and propham was reported to be inversely related to the organic matter content of soils (Blouch & Fults, 1953; Robocker & Canode, 1964; Harris & Sheets, 1965; Roberts & Wilson, 1965; Upchurch et al., 1966) and bioactivity of phenylurea herbicides was likewise reported to be inversely related to soil matter content. The latter included diuron (Upchurch, 1958; Upchurch & Mason, 1962; Harris & Sheets, 1965; Upchurch et al., 1966; Darding & Freeman, 1968; Nash, 1968), fluometuron (Darding & Freeman, 1968; Harrison et al., 1976), linuron (Dubey et al., 1966; Doherty & Warren, 1969; Rahman, 1976; Mapplebeck & Waywell, 1983), monuron (Harris & Sheets, 1965; Harris, C.I., 1966), and neburon (Obien et al., 1966).

Organic matter added to model soils greatly reduced the bioactivity of phenylcarbamate herbicide chlorpropham (Scott & Weber, 1967; Table 10–3; J.A. Best & J.B. Weber, 1969, unpublished data; Table 10–6); as did activated C (Coffey & Warren, 1969), or basic anion exchange resin (Scott & Weber, 1967; Table 10–10). Addition of montmorillonite reduced bioactivity of the herbicide slightly and kaolinite additions had no effect (J.A. Best & J.B. Weber, 1969, unpublished data; Table 10–7). Reduction in chlorpropham bioactivity by soil organic matter additions, as indicated by Bio-K_d values, is shown in Table 10–3 where Bio-K_d ranged from 117 to 212 (mean = 156) and in Table 10–6 where Bio-K_d ranged from 100 to 114 (mean = 107) (overall mean = 131). Differences in Bio-K_d for the two studies was because of the different plant species used and probably other factors such as temperature and watering regime. Reduction in bioactivity of chlorpropham by additions of montmorillonite is shown in Table 10–7 where Bio-K_d ranged from 6.1 to 13.1 (mean = 9.6), and reductions in bioactivity by addition of anion exchange resins is shown in Table 10–10, where Bio-K_d ranged from 51 to 323 (mean = 201).

Organic matter additions to model soil systems reduced the bioactivity of the phenylureas fenuron (J.A. Best & J.B. Weber, 1969, unpublished data), fluometuron (J.A. Best & J.B. Weber, 1969, unpublished data), and linuron (Doherty & Warren, 1969). Organic matter additions to a sandy soil in the field reduced bioactivity of fluometuron (Weber et al., 1974b), but also extended its persistence (Weber et al., 1974a). Additions of activated C to model soils also reduced bioactivity of chlorbromuron (Ogg, 1978), diuron (Jordan & Smith, 1971; Burr et al., 1972; Chandler et al., 1978), and linuron (Ogg, 1978). Montmorillonite additions to model soils significantly reduced bioactivity of diuron, monuron, and neburon (Coggins & Crafts, 1959), fenuron (Coggins & Crafts, 1959; J.A. Best & J.B. Weber, 1969, unpublished data), and fluometuron (J.A. Best & J.B. Weber, 1969, unpublished data), and additions of the clay to a sandy soil in the field reduced fluometuron bioactivity slightly (Weber et al., 1974b), but extended its persistence (Weber et al., 1974a). Additions of kaolinite to model soil media had no effect on bioactivity of fenuron or fluometuron (J.A. Best & J.B. Weber, 1969, unpublished data).

Reduction in bioactivity by organic matter additions to model soil media resulted in Bio-K_d values ranging from 47 to 74 (mean = 60) for fluometuron to 10 to 13 (mean = 12) for fenuron (Table 10–6). Reduction in bioactivity of fluometuron, as indicated by reduced weed control, for organic matter additions to a field soil ranged from −29 to −68% or from −0.9 to −1.3 (mean = −1.1) %/mt of organic matter added (Table 10–8). Reduction in bioactivity by montmorillonite added to model soil media resulted in Bio-K_d values ranging from 4.0 to 6.1 (mean = 4.9) for fluometuron to 1.3 to 2.0 (mean = 1.6) for fenuron (Table 10–7). Reduction of fluormeturon bioactivity, as indicated by reduced weed control, for montmorillonite additions to a field soil ranged from −1 to −8% or about −0.1%/mt of montmorillonite added (Table 10–8). Additions of kaolinite to model soil media did not affect bioactivity of fluometuron or fenuron (Table 10–7). Organic matter was 10 times more effective than montmorillonite in reducing bioactivity of fluometuron.

In summary, phenylcarbamate and phenylurea pesticides were sorbed by soil organic matter and, to a much lesser extent, by montmorillonite and isotherms were type L. Sorption and bioinactivation of the compounds was inversely related to water solubility and Bio-K_d values ranged from 1.6 to 13.1 for montmorillonite to 12 to 131 for soil organic matter. Activated C additions would be effective as a remediation treatment for phenylcarbamate and phenylurea pesticide contaminated soils.

THIOCARBAMATE HERBICIDES

Sorption of thiocarbamate herbicides, including cycloate, diallate, and pebulate (Koren et al., 1969), and EPTC (Ashton & Sheets, 1959; Koren et al., 1969) was related to the organic matter and clay contents of soils. Sorption of EPTC by dry montmorillonite was reported to occur, but the chemical was readily released with water, and it was bioactive in the sorbed state (Mortland & Meggitt, 1966). Sorption of EPTC by soil was also reported to be reduced by increasing moisture content of soils and vapor losses were greater from moist soils than dry ones (Fang et al., 1961; Gray & Weierich, 1965). Loss of CDEC from soil has also been reported to increase with moisture content and temperature (Menges & Hubbard, 1966; Taylorson, 1966). This is a characteristic of all of the volatile thiocarbamate compounds, and necessitates that the chemicals be incorporated into the soil to be effective.

Bioactivity of CDEC (Sheets, 1959), cycloate (Koren et al., 1968; Stanger & Appleby, 1971), EPTC (Ashton & Sheets, 1959; Sheets, 1959; Jordon & Day, 1962; Koren et al., 1968), and pebulate (Koren et al., 1968) was reported to be inversely related to the organic matter and clay contents of soils. Danielson et al. (1961) also reported that EPTC persistence increased with increased organic matter content of soils due to the greater retention of the herbicide by organic colloids than by other soil colloids. Soil organic matter binds thiocarbamate herbicides more tenaciously than other soil colloids and reduces bioactivity and vapor losses of the chemicals, and thereby extends their longevity in soils. Vapor losses of thiocarbamates decrease with increased clay content of soils because of slower diffusion through the soil, but the chemicals are retained by clays so weakly that they are readily bioavailable when wetted. Activated C additions would be effective as a remediation treatment for soils contaminated with thiocarbamate pesticides.

DINITROANILINE HERBICIDES

Sorption of dinitroaniline herbicides, including benefin (Jacques & Harvey, 1979), butralin (Carringer et al., 1975; Peter & Weber, 1985c), dinitramine (Jacques & Harvey, 1979; Okafor et al., 1983), fluchloralin, oryzalin, and profluralin (Jacques & Harvey, 1979), nitralin (Lambert, 1968), and trifluralin (Hollist & Foy, 1971; Carringer et al., 1975; Jacques & Harvey, 1979; Peter & Weber, 1985c) was reported to occur in high amounts by soil organic matter. Butralin (Carringer et al., 1975) and trifluralin (Hollist & Foy, 1971; Grover, 1974; Carringer et al., 1975) were reported to be sorbed to montmorillonite in low amounts. Sorption isotherms for dinitroaniline herbicides by soils were type L (Weber & Miller, 1989).

Bioactivity of the dinitroaniline herbicides, including butralin (Peter & Weber, 1985c; Weber et al., 1987), dinitramine (Okafor et al., 1983), oryzalin (Weber et al., 1974c), pendimethalin (Blumhorst et al., 1990), and trifluralin (Menges & Hubbard, 1970; Harrison et al., 1976; Rahman, 1976; Peter & Weber, 1985c; Weber et al., 1987) was inversely related to the organic matter content of soils. The effect of soil organic matter on reducing weed control by oryzalin is shown in Fig. 10–3. Bioactivity of nitralin (Ogg, 1978) and trifluralin (Coffey & Warren, 1969) was greatly reduced by the additions of activated C to soil, and the addition of organic matter to a sandy soil in the field reduced bioactivity and increased persistence of trifluralin; whereas montmorillonite had no effect (Weber et al., 1974a, b). Helling and Krivonak (1978) reported that only small amounts of butralin, chlornidine, dinitramine, fluchloralin, profluralin, and trifluralin bound in soil was biologically available.

Reduction in bioactivity of trifluralin by organic matter additions, as indicated by reduced weed control, ranged from −12 to −40% or from −0.4 to −1.1 (mean = −0.7) %/mt of organic matter added (Table 10–8). Additions of montmorillonite did not reduce trifluralin bioactivity. Additional details on the fate and behavior of dinitroaniline herbicides in soils is published elsewhere (Weber, 1990).

In summary, dinitroaniline herbicides were strongly sorbed and bioinactivated by soil organic matter due to their low water solubility and high propensity to H bond to proteinaceous substances. The compounds owe their effectiveness to the low concentrations in soil solution required to inhibit seed germination. Clay minerals had no effect on the bioactivity of the compounds. Activated C addition would be an effective remediation treatment for dinitroaniline herbicide-contaminated soils.

HALOGENATED HYDROCARBONS

Sorption of halogenated hydrocarbons has been reported to occur on both organic and inorganic colloidal surfaces in soils and soil materials. The following compounds have been reported to be sorbed in high amounts by organic matter, humic matter, peat, muck, aquatic sediments, and soils containing moderate amounts of organic matter: DDT (Barlow & Hadaway, 1956, 1958; Carter & Suffet, 1982), dieldrin (Barlow & Hadaway, 1956, 1958), dibromoethane, dichlorobenzene, dichloroethane and dichloropropane (Chiou et al., 1979), and lindane (Barlow & Hadaway, 1956, 1958; Kay & Elrick, 1967; Lotse et al., 1968; Mills & Biggar, 1969; Miller & Weber, 1986). Sorption in moderate to low amounts by clays has been reported for the following: aldrin (Yaron et al., 1967), DDT, dieldrin and lindane (Barlow & Hadaway, 1956, 1958), and methoxychlor (Karickhoff et al., 1979). Sorption of all but the aliphatic halogenated compounds by organic colloids was type L, whereas sorption by clays was type S (Weber & Miller, 1989). Sorption of the aliphatic halogenated compounds by soils was type C but the concentration range was narrow (Chiou et al., 1979).

Bioactivity of many chlorinated hydrocarbons was inversely related to the organic matter content of soils. This included aldrin (Edwards et al., 1957; Lichtenstein et al., 1960; Bowman et al., 1965; Bruce & Decker, 1966; Harris & Sans, 1967), chlordane (Fleming & Maines, 1954), DDT (Fleming & Maines, 1953; Barlow & Hadaway, 1958; Lichtenstein et al., 1960; Bowman et al., 1965; Edwards, 1966; Beall & Nash, 1969; Peterson et al., 1971), dieldrin (Barlow & Hadaway, 1958; Bowman et al., 1965; Bruce & Decker, 1966; Hermanson & Forbes, 1966; Harris & Sans, 1967; Beall & Nash, 1969), endosulfan (Bowman et al., 1965), endrin and heptachlor (Bowman et al., 1965; Beall & Nash, 1969), heptachlor epoxide (Bowman et al., 1965; Bruce & Decker, 1966), lindane (Edwards et al., 1957; Barlow & Hadaway, 1958; Lichtenstein et al., 1960; Bowman et al., 1965), and PCB (Strek & Weber, 1982b). Bioactivity was inversely related to clay content of dry soils but not moist soils for DDT, dieldrin, and lindane (Barlow & Hadaway, 1958). Increasing relative humidity or moisture content was reported to increase bioactivity of aldrin and heptachlor in soils (Burkhardt & Fairchild, 1967).

Additions of activated C to soils reduced the bioactivity of aldrin, chlordane, dieldrin, heptachlor, heptachlor epoxide (Lichtenstein et al., 1968, 1971), DDT (Thurston, 1953), and PCB (Weber & Mrozek, 1979; Strek et al., 1981; Strek & Weber, 1982b). Additions of montmorillonite to model soil media reduced bioactivity of PCB slightly (Strek & Weber, 1982b).

In summary, halogenated hydrocarbons are low in water solubility and are readily sorbed and bioinactivated by soil organic matter. Clay minerals bind the chemicals in low amounts but the chemicals are released by water. Activated C addition would be an effective remediation treatment for soils contaminated with halogenated hydrocarbons.

MISCELLANEOUS NONIONIC ORGANIC CHEMICALS

Bioactivity of several miscellaneous nonionic organic herbicides was reported to be inversely related to soil organic matter content. This included dichlobenil (Miller et al., 1966; Briggs & Dawson, 1970), pyrazon (Doherty

& Warren, 1969), norflurazon (Weber et al., 1974c; Schroeder & Banks, 1986), and fluorodifen (Weber et al., 1974c). The relationship between herbicide bioactivity and soil organic matter content for norflurazon and flurodifen is shown in Fig. 10–3. Organic matter additions to model soils greatly reduced the bioactivity of dichlobenil (J.A. Best & J.B. Weber, 1969, unpublished data) and pyrazon (Doherty & Warren, 1969), while additions of montmorillonite reduced dichlobenil bioactivity slightly and kaolinite had no effect (J.A. Best & J.B. Weber, 1969, unpublished data). Reduction in bioactivity of dichlobenil by organic matter and clay additions is shown in tables 10–6 and 10–7, respectively. Bio-K_d values for dichlobenil ranged from 158 to 280 (mean = 219) for organic matter additions (Table 10–6) and from 8.0 to 18.0 (mean = 12.2) for montmorillonite additions (Table 10–7).

Bioactivity of many other nonionic, hydrophobic organic compounds has been reported to be inversely related to humic matter levels in aquatic systems. The subject was reviewed by McCarthy (1989) and included anthracene, benzanthracene, benzopyrene, diethylhexyl phthalate, hexadecane, naphthalene, OCDD, and TCDD. A review of the postulated mechanisms of uptake by plants of nonionic organic compounds has also been published (Patterson et al., 1990).

In summary, miscellaneous nonionic organic chemicals were sorbed and bioinactivated primarily by soil organic matter in soils. Activated C addition would be an effective remediation treatment for soils contaminated by these chemicals.

RELATIONSHIP BETWEEN BIO-K_{oc} and SOIL K_{oc}

Table 10–11 contains the mean Bio-K_d values, computed Bio-K_{oc} values (from Bio-K_{oc} = Bound herbicide/Bioavailable herbicide) for herbicides used in model soil systems and the soil K_{oc} values reported in the literature (Weber, 1992). Mean Bio-K_d values for herbicides where kaolinite was added to the media were as follows: Bio-K_d = 0 (chlorpropham, 2,4-D, dichlobenil, fenuron, fluometuron), 1.0 (diphenamid), 1.3 (atrazine), 3.9 (prometryn), 7.0 (prometon), and 76.0 (paraquat).

The values were smallest for the nonionic compounds chlorpropham, dichlobenil, fenuron, and fluometuron, and the weak acid 2,4-D, and largest for the strongly basic quaternary N compound paraquat. The weakly basic herbicides atrazine, prometryn, and prometon had intermediate Bio-K_d values. Retention of the herbicides was probably through physical forces for nonionic compounds and through ionic or physical forces for compounds with basic properties, depending on the pH.

Bio-K_d values for montmorillonite additions to the model soil media were much larger than for kaolinite and were as follows (Table 10–11): Bio-K_d = 0 (2,4-D), 1.6 (fenuron), 4.9 (fluometuron), 5.0 (diphenamid), 7.9 (atrazine, pH 7), 9.6 (chlorpropham), 12.2 (dichlobenil), 19.0 to 118 (prometryn, pH 7–4, respectively), 172 (prometon, pH 6), and 17 544 (paraquat).

Table 10-11. Mean Bio-K_d values and calculated Bio-K_{oc} values for herbicides from model soil media studies, and soil K_{oc} values from the literature.

Herbicide	Soil additive†		Mean Bio-K_d‡	Computed Bio-K_{oc}§	Soil K_{oc}¶	K_{sp}¶	pK_a¶
						mg L^{-1}	
Fenuron	Kaol		0				
	Mont.		1.6				
	OM		12	21	10	3 850	
Diphenamid	Kaol.		1.0				
	Mont.		5.0				
	OM		28	48	210	260	
Fluometuron	Kaol.		0				
	Mont.		4.9				
	OM		60	103	100	110	
Atrazine	Kaol.		1.3				
	Mont.		7.9				
	OM		81	140	150	33	1.68
Chlorpropham	Kaol.		0				
	Mont.		9.6				
	OM		131	226	350	89	
	AER		201				
Prometryn	Kaol.		3.9				
	Mont.	(pH 7)	19.0				
		(pH 6)	21				
		(pH 4)	118				
	OM	(pH 7)	78	134	300	48	4.05
		(pH 6)	137	236	350		
		(pH 4)	240	414	600		
2,4-D	Kaol.		0				
	Mont.		0				
	OM		181	312	100	890	3.0
	AER		1 601				
Dichlobenil	Kaol.		0				
	Mont.		12.2				
	OM		219	377	400	25	
Prometon	Kaol.		7.0				
	Mont.		172				
	OM		266	458	400	750	4.28
	AER		0				
Paraquat	Kaol.		76				
	Verm.		2 613				
	Mont.		17 544				
	OM		22	36	10^6	7×10^4	11
	AER		0				

† Kaol. = kaolinite, Mont. = montmorillonite, OM = organic matter (histosol), AER = strongly basic anion exchange resin, Verm. = vermiculite.
‡ Bio-K_d = Herbicide bound/Bioavailable herbicide.
§ From Bio-K_{oc} = Bio-K_d/fraction organic C (assuming organic matter was 58% C).
¶ From Weber (1992).

Inactivation generally increased with decreased water solubility (K_{sp}) and increased basicity (pK_a) of the chemicals and was pH dependent for basic compounds.

The values increased for the herbicides in a pattern similar to that for kaolinite additions, but at much higher values due to the higher cation exchange capacity and surface area of montmorillonite. The mechanism(s) of retention of the herbicides was probably similar to that for kaolinite additions. This is reinforced by the effect of decreasing pH on the increased retention of the compounds with weakly basic properties. However, the geometry of binding of the compounds by the two clays was different, that is, the herbicides were sorbed on the exterior surfaces of kaolinite and in the interlayer spacings of montmorillonite.

Vermiculite also had a high Bio-K_d for paraquat at 2613; that is, 34 times that for kaolinite, but only one-seventh that for montmorillonite (Table 10–11). Paraquat was sorbed on the interlayer spacings of vermiculite and montmorillonite and on the external surfaces of kaolinite.

Anion exchange resins added to the model soil media did not reduce bioactivity of prometon or paraquat (Bio-K_d = 0), but did reduce bioactivity of chlorpropham to a moderate degree (Bio-K_d = 201) and were very effective in reducing the bioactivity of 2,4-D (Bio-K_d = 1601) (Table 10–11). Chlorpropham was probably retained through H bonding or other physical forces and 2,4-D anions were probably retained through anion exchange reactions.

Soil organic matter additions to the model soil media reduced the bioactivity of all of the herbicides and with the exception of paraquat was the most effective soil additive (Table 10–11). Bio-K_d values for selected herbicides by organic matter additions were as follows: Bio-K_d = 12 (fenuron), 22 (paraquat), 28 (diphenamid), 60 (fluometuron), 81 (atrazine), 131 (chlorpropham), 78 to 240 (prometryn, pH 7–4, respectively), 181 (2,4-D), 219 (dichlobenil), and 266 (prometon).

Bio-K_{oc} values for added organic matter increased with decreasing water solubility (K_{sp}) of nonionic herbicides (fenuron, diphenamid, fluometuron, chlorpropham, dichlobenil; Fig. 10–4) and with increasing basicity (pK_a) and decreasing K_{sp} of the weakly basic compounds (atrazine, prometryn, and prometon; Table 10–11). Bioactivity of paraquat was reduced by organic matter additions, but not nearly as much as by additions of clay minerals.

Bio-K_{oc} values calculated for all of the herbicides were moderately well correlated with the reported soil K_{oc} values for all except paraquat (r = 0.70; Fig. 10–5). The data suggest that the concentration of bioavailable herbicide is inversely related to that retained by organic soil colloids. The relationship of Bio-K_{oc} and soil K_{oc} is improved substantially to $r = 0.86$ when nonionic compounds only are considered. This suggests that the soil K_{oc} values should probably be restricted to nonionic herbicides, or to pH conditions where ionizable compounds would be present only in the molecular form.

SUMMARY

Organic chemicals react with organic and inorganic soil colloids in many different ways, depending upon the nature and properties of the chemicals,

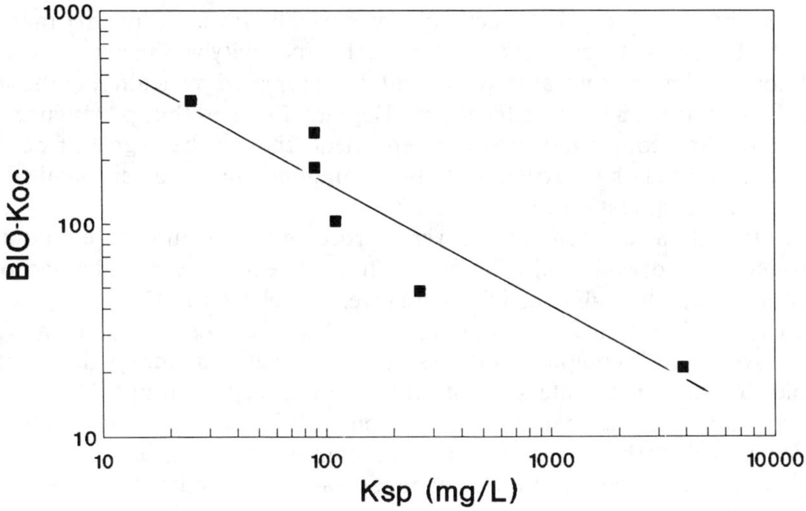

Fig. 10–4. Relationship between Bio-K_{oc} and K_{sp} for nonionic herbicides (r = −0.56).

the colloids, and the media. Soil sorption affects not only the mobility of organic compounds, but their bioactivity as well. Quaternary N compounds were ionically bound to clay minerals and soil organic matter in very high amounts. They were bioactive in the sorbed state, but to a lesser degree than when free in solution. However, where quaternary N compounds were retained on the internal surfaces of montmorillonitic clay minerals, they were neither bioactive nor bioavailable.

Weakly basic compounds were ionically and physically bound to organic and inorganic soil colloids in moderate to high amounts, and sorption was

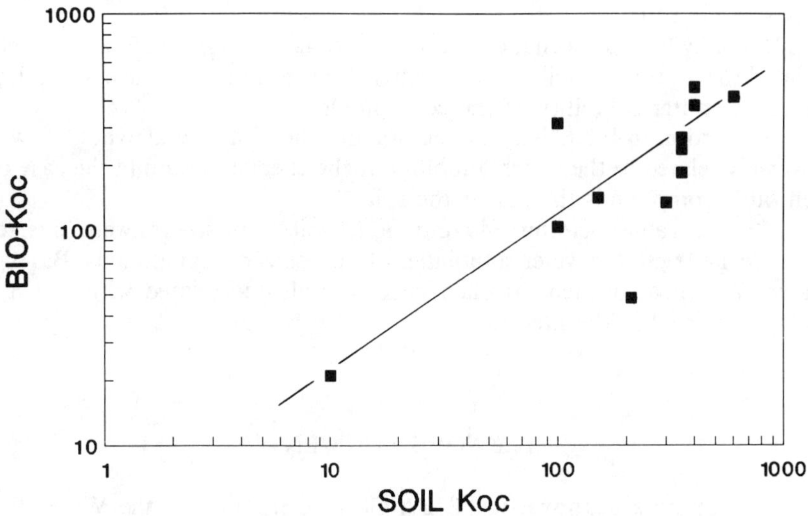

Fig. 10–5. Relationship between Bio-K_{oc} and soil K_{oc} for all herbicides except paraquat (r = 0.70).

inversely related to pH. Bioactivity was normally lower at low pH than at high pH due to higher retention at low pH. Bioactivity was increased by additions of lime or neutral fertilizer and was decreased by acidifying the soil or by adding acid-forming fertilizers. Degradation, and thus persistence, of weakly basic compounds was pH dependent, though the degree of persistence depended on the structure of the compound, and whether chemical or biological degradation predominated.

Weakly acidic compounds were sorbed in low to moderate amounts primarily by organic soil colloids and to some extent by metallic hydrous oxides. Sorption and bioactivity were inversely related to pH, due to greater retention of molecular species than anionic species by soil organic matter.

Amphoteric compounds were sorbed by organic and inorganic soil colloids in low to moderate amounts and retention depended upon the pH of the system and the charge of the predominating species present. Cationic species were sorbed in greater amounts than molecular or anionic species, respectively. Bioactivity of the compounds was inversely related to the amount of sorption by the soil.

Organic As and P pesticides were retained in moderate to high amounts primarily by soil clays, and bioactivity was inversely related to sorption. Bioactivity was increased by addition of P fertilizer to soils.

Amide and anilide herbicides and phenylcarbamate and phenylurea pesticides were sorbed by organic matter in moderate to high amounts and by expanding-type clay minerals in low to moderate amounts. Bioactivity was inversely related to sorption.

Bioactivity of thiocarbamate pesticides, dinitroaniline herbicides, and halogenated hydrocarbons was most highly and inversely associated with soil organic matter content. Persistence was related to inactivation by the soil, that is, longevity of the chemicals increased with increased organic matter levels in the soil.

Bioactivity of miscellaneous nonionic organic chemicals was inversely related to sorption by soil organic matter. Soil retention was inversely related to the water solubility of the compounds.

Reduction in bioactivity of the organic chemicals by activated C was inversely related to the water solubility of the chemicals, and in the case of ionizable compounds the pH of the soil.

Bio-K_{oc} values determined from model soil/plant studies were inversely correlated with the water solubilities of nonionizable compounds. Bio-K_{oc} values for an assortment of chemicals were also correlated with soil K_{oc} values reported in the literature.

ACKNOWLEDGMENT

The authors acknowledge CIBA-Geigy Corporation, the Water Resources Research Institute, and the U.S. Geological Survey for financial support of research on the behavior and fate of toxic organic in the environment,

Director and Associate Dean J.C. Wynne for project support and Marie K. Hall for typing this manuscript.

REFERENCES

Abernathy, J.R., and L.M. Wax. 1973. Bentazon mobility and adsorption in twelve Illinois soils. Weed Sci. 21:224–227.

Adams, R.S., Jr. 1972. Effect of soil organic matter on the movement and activity of pesticides in the environment. p. 81–93. *In* D.D. Hemphill (ed.) Trace substance in environmental health-V. A symposium. Univ. of Missouri, Columbia, Feb. 7–9.

Aherns, J.F. 1965. Detoxification of simazine- and atrazine-treated soil with activated carbon. Proc. Northeast Weed Contr. Conf. 19:364–365.

Anderson, R.L. 1985. Environmental effects of metsulfuron and chlorsulfuron bioactivity in soil. J. Environ. Qual. 14:517–521.

Anderson, R.L., and M.R. Barrett. 1985. Residual phytotoxicity of chlorsulfuron in two soils. J. Environ. Qual. 14:111–114.

Arle, H.F., O.A. Leonard, and V.C. Harris. 1948. Inactivation of 2,4-D on sweet potato slips with activated carbon. Science 107:247–248.

Armstrong, D.E., and G. Chesters. 1964. Properties of protein-bentonite complexes as influenced by equilibration conditions. Soil Sci. 98:39–51.

Armstrong, D.E., and G. Chesters. 1967. Atrazine hydrolysis in soil. Soil Sci. Soc. Am. Proc. 31:61–66.

Ashton, F.M., and T.J. Sheets. 1959. The relationship of soil adsorption of EPTC to oats injury in various soil types. Weeds 7:88–90.

Ballard, J.L., and P.W. Santelmann. 1973. Influence of selected soil properties on alachlor activity. Proc. South. Weed Sci. Soc. 26:385–388.

Barlow, F., and A.B. Hadaway. 1956. Effect of changes in humidity on the toxicity and distribution of insecticides sorbed by some dried soils. Nature (London) 178:1299–1300.

Barlow, F., and A.B. Hadaway. 1958. Studies on the aqueous suspensions of insecticides. Part VI. Further notes on the sorption of insecticides by soils. Bull. Entomol. Res. 49:315–338.

Basham, G.W., T.L. Lavy, L.R. Oliver, and H.D. Scott. 1987. Imazaquin persistence and mobility in three Arkansas soils. Weed Sci. 35:576–582.

Beall, M.L., Jr., and R.G. Nash. 1969. Crop seedling uptake of DDT, dieldrin, endrin, and heptachlor from soils. Agron. J. 61:571–575.

Best, J.A., and J.B. Weber. 1974. Disappearance of *s*-trazines as affected by soil pH using a balance-sheet approach. Weed Sci. 22:364–373.

Best, J.A., J.B. Weber, and T.J. Monaco. 1975. Influence of soil pH on *s*-triazine availability to plants. Weed Sci. 23:378–382.

Best, J.A., J.B. Weber, and S.B. Weed. 1972. Competitive adsorption of diquat^{2+}, and Ca^{2+} on organic matter and exchange resins. Soil Sci. 114:444–450.

Blouch, R., and J. Fults. 1953. The influence of soil type on the selective action of chloro-IPC and sodium TCA. Weeds 2:119–124.

Blumhorst, M.R., and J.B. Weber. 1988. The longevity of cyanazine and atrazine as affected by soil type and pH. Proc. South. Weed Sci. Soc. 41:339–341.

Blumhorst, M.R., J.B. Weber, and L.R. Swain. 1990. Efficacy of selected herbicides as influenced by soil properties. Weed Technol. 4:279–283.

Bolt, G.H., and B.P. Warkentin. 1958. The negative adsorption of anions by clay suspensions. Kolloid Z. 156:41–46.

Bowman, B.T., R.S. Adams, Jr., and S.W. Fenton. 1970. Effect of water upon malathion adsorption onto five montmorillonite systems. J. Agric. Food Chem. 18:723–727.

Bowman, M.C., M.S. Schechter, and R.L. Carter. 1965. Behavior of chlorinated insecticides in a broad spectrum of soil types. J. Agric. Food Chem. 13:360–365.

Briggs, G.G., and J.E. Dawson. 1970. Hydrolysis of 2,6-dichlorobenzonitrile in soils. J. Agric. Food Chem. 18:97–99.

Bruce, W.N., and G.C. Decker. 1966. Insecticide residues in soybeans grown in soil containing various concentrations of aldrin, dieldrin, and heptachlor epoxide. J. Agric. Food Chem. 14:395–398.

Burkhardt, C.C., and M.L. Fairchild. 1967. Toxicity of insecticides to house crickets and bioassay of treated soils in the laboratory. J. Econ. Entomol. 60:1496–1503.

Burns, R.G., and L.J. Audus. 1970. Distribution and breakdown of paraquat in soil. Weed Res. 10:49–58.

Burnside, O.C., and R. Behrens. 1961. Phytotoxicity of simazine. Weeds 9:145–157.

Burr, R.J., W.O. Lee, and A.P. Appleby. 1972. Factors affecting use of activated carbon to improve herbicide selectivity. Weed Sci. 20:180–183.

Burt, P.E., R. Bardner, and P. Etheridge. 1975. The influence of volatility and water solubility of systemic insecticides on their movement through soil and absorption by plant roots. Ann. Appl. Biol. 56:411–418.

Calderbank, A. 1968. The bipyridylium herbicides. Adv. Pest. Control Res. 8:127–229.

Cantwell, J.R., R.A. Liebl, and F.W. Slife. 1989. Biodegradation characteristics of imazaquin and imazethapyr. Weed Sci. 37:815–819.

Capriel, P., A. Haisch, and S.U. Khan. 1985. Distribution and nature of bound (nonextractable) residues of atrazine in a mineral soil nine years after the herbicide application. J. Agric. Food Chem. 33:567–569.

Carlson, W.C., E.M. Lignowski, and H.J. Hopen. 1975. Uptake, translocation, and adsorption of pronamide. Weed Sci. 23:148–154.

Carringer, R.D., J.B. Weber, and T.J. Monaco. 1975. Adsorption-desorption of selected pesticides by organic matter and montmorillonite. J. Agric. Food Chem. 23:568–572.

Carter, C.W., and J.H. Suffet. 1982. Binding of DDT to dissolved humic materials. Environ. Sci. Technol. 16:735–740.

Chandler, J.M., O.B. Wooten, and F.E. Fulgham. 1978. Influence of placement of charcoal on protection of cotton (*Gossypium hirsutum*) from diuron. Weed Sci. 26:239–244.

Chiou, C.T., L.J. Peters, and V.H. Freed. 1979. A physical concept of soil-water equilibria for nonionic organic compounds. Science 206:831–832.

Coats, G.E., H.H. Funderburk, Jr., J.M. Lawrence, and D.E. Davis. 1966. Factors affecting persistence and inactivation of diquat and paraquat. Weed Res. 6:58–66.

Coffey, D.L., and G.F. Warren. 1969. Inactivation of herbicides by activated carbon and other adsorbents. Weed Sci. 17:16–19.

Coggins, C.W., Jr., and A.C. Crafts. 1959. Substituted urea herbicides: Their electrophoretic behavior and the influence of clay colloid in nutrient solution on their phytotoxicity. Weeds 7:349–358.

Colbert, F.O., V.V. Volk, and A.P. Appleby. 1975. Sorption of atrazine, terbutryn, and GS-14254 on natural and lime-amended soils. Weed Sci. 23:390–394.

Cruz, M., J.L. White, and J.D. Russell. 1968. Montmorillonite-*s*-triazine interactions. Israel J. Chem. 6:315–320.

Damanakis, M., D.S.H. Drennan, J.D. Fryer, and K. Holly. 1970. The adsorption and mobility of paraquat on different soils and soil constituents. Weed Res. 19:264–277.

Danielson, L.L., W.A. Gentner, and L.L. Jansen. 1961. Persistence of soil-incorporated EPTC and other carbamates. Weeds 9:463–476.

Darding, R.L., and J.F. Freeman. 1968. Residual phytotoxicity of fluometuron in soils. Weed Sci. 16:226–229.

Dashman, T., and G. Stotzky. 1986. Microbial utilization of amino acids and a peptide bound on homoionic montmorillonite and kaolinite. Soil Biol. Biochem. 18:5–14.

Davis, D.E. 1956. Some factors that affect the phytotoxicity of water soluble DNBP. Weeds 4:227–234.

Day, B.E., L.S. Jordan, and V.A. Jolliffe. 1968. The influence of soil characteristics on the adsorption and phytotoxicity of simazine. Weed Sci. 16:209–213.

Devlin, R.M., S.J. Karczmarczyk, I.I. Zbiec, and Z.K. Koszanski. 1986. Initial and residual activity of glyphosate and SC-0224 in a sandy soil. Crop Prot. 5:293–296.

Dickens, R., and A.E. Hiltbold. 1967. Movement and persistence of methanearsonates in soil. Weeds 15:299–304.

Doherty, P.J., and G.F. Warren. 1969. The adsorption of four herbicides by different types of organic matter and a bentonite clay. Weed Res. 9:20–26.

Dowler, C., N.M. Baughman, and C. Veatch. 1958. The effect of soil type, soil pH, and simulated rainfall on the distribution of DNBP in the soil. Weeds 6:281–288.

Dubey, H.D., R.E. Sigafus, and J.F. Freeman. 1966. Effect of soil properties on the persistence of linuron and diphenamid in soils. Agron. J. 58:228–231.

Dunigan, E.P., and T.H. McIntosh. 1971. Atrazine-soil organic matter interactions. Weed Sci. 19:279–282.

Dutt, T.E., and R.G. Harvey. 1980. Pronamide phytotoxicity in ten Wisconsin soils. Weed Sci. 28:429–432.

Edwards, C.A. 1966. Insecticide residues in soils. Res. Rev. 13:83–132.

Edwards, C.A., S.D. Beck, and E.P. Lichtenstein. 1957. Bioassay of aldrin and lindane in soil. J. Econ. Entomol. 50:622–626.

Elmstrom, G.W. 1972. Protection of watermelon from terbacil injury with activated charcoal. Proc. South. Weed Sci. Soc. 25:204–208.

Ensminger, L.E., and J.E. Gieseking. 1942. Resistance of clay adsorbed proteins to proteolytic hydrolysis. Soil Sci. 53:205–209.

Estermann, E.F., G.H. Peterson, and A.D. McLaren. 1959. Digestion of clay-protein and silica-protein complexes by enzymes and bacteria. Soil Sci. Soc. Am. Proc. 23:31–36.

Fang, S.G., P. Theisen, and V.H. Freed. 1961. Effects of water evaporation, temperature, and rates of application on the retention of ethyl-*N,N*-di-*n*-propylthiocarbamate in various soils. Weeds 9:569–574.

Fleming, W.E., and W.W. Maines. 1953. Persistence of DDT in soils infested by the Japanese beetle. J. Econ. Entomol. 46:445–449.

Fleming, W.E., and W.W. Maines. 1954. Persistence of chlordane in soils of the area infested by the Japanese beetle. J. Econ. Entomol. 47:165–169.

Frissel, M.J. 1961. The adsorption of some organic compounds, especially herbicides, on clay minerals. Versl. Landbouwkd. Onderz., n.r. 67.3.

Frissel, M.J., and G.H. Bolt. 1962. Interaction between certain ionizable organic compounds (herbicides) and clay minerals. Soil Sci. 94:284–291.

Getzin, L.W., and R.K. Chapman. 1960. The fate of phorate in soils. J. Econ. Entomol. 53:47–51.

Giles, G.H., and T.H. MacEwan. 1957. Classification of isotherm types for adsorption from solution. Proc. Int. Congr. Surf. Act., 2nd, 3:457–461.

Gilmour, J.T., and N.T. Coleman. 1971. *s*-Triazine adsorption studies: Ca-H-humic acid. Soil Sci. Soc. Am. Proc. 35:256–259.

Gottlieb, D., and P. Siminoff. 1952. The production and role of antibiotics in the soil. II. Chloromycetin. Phytopathology 42:91–97.

Gottlieb, D., P. Siminoff, and M.M. Martin. 1952. The production and role of antibiotics in soil. IV. Actidione and clavacin. Phytopathology 42:493–496.

Graham-Bryce, I.J. 1967. Adsorption of disulfoton by soil. J. Sci. Food Agric. 18:72–77.

Gray, R.A., and A.J. Weierich. 1965. Factors affecting the vapor loss of EPTC from soils. Weeds 13:141–147.

Grossbard, E., and D. Atkinson. 1985. The herbicide glyphosate. Butterworths, London.

Grover, R. 1966. Influence of organic matter, texture, and available water on the toxicity of simazine in soil. Weeds 14:148–151.

Grover, R. 1968. Influence of soil properties on phytotoxicity of 4-amino-3,5,6-trichloropicolinic acid (picloram). Weed Res. 8:226–232.

Grover, R. 1974. Adsorption and desorption of trifluralin, triallate, and diallate by various adsorbents. Weed Sci. 22:405–408.

Hance, R.J. 1969. Influence of pH, exchangeable cation, and the presence of organic matter on the adsorption of some herbicides by montmorillonite. Can. J. Soil Sci. 49:357–364.

Harris, C.I. 1966. Adsorption, movement, and phytotoxicity of monuron and *s*-triazine herbicides in soil. Weeds 14:6–10.

Harris, C.I. 1967. Fate of 2-chloro-*s*-triazines in soil. J. Agric. Food Chem. 15:157–162.

Harris, C.I., and T.J. Sheets. 1965. Influence of soil properties on adsorption and phytotoxicity of CIPC, diuron, and simazine. Weeds 13:215–219.

Harris, C.I., and G.F. Warren. 1964. Adsorption and desorption of herbicides by soil. Weeds 12:120–126.

Harris, C.R. 1966. Influence of soil type on the activity of insecticides in soil. J. Econ. Entomol. 59:1221–1225.

Harris, C.R., and W.W. Sans. 1967. Absorption of organochlorine insecticide residues from agricultural soils by root crop. J. Agric. Food Chem. 15:861–863.

Harrison, G.W., J.B. Weber, and J.V. Baird. 1976. Herbicide phytotoxicity as affected by selected properties of North Carolina soils. Weed Sci. 24:120–126.

Hata, Y., and K. Akashi. 1980. Effects of properties of soil colloids on adsorption of piperophos by soils. J. Pestic. Sci. 5:473–479.

Hayes, M.H.B. 1970. Adsorption of triazine herbicides on soil organic matter. Res. Rev. 32:131–174.

Helling, C.S., and A.E. Krivonak. 1978. Biological characteristics of bound dinitroaniline herbicides in soils. J. Agric. Food Chem. 26:1164–1172.

Hermanson, H.P., and C. Forbes. 1966. Soil properties affecting dieldrin toxicity of *Drosophila melanoqastis*. Soil Sci. Soc. Am. Proc. 30:748–752.

Hermosin, M.C., J. Cornejo, J.L. White, and F.D. Hess. 1982. Bioavailability of *s*-triazines adsorbed on montmorillonite. J. Agric. Food Chem. 30:728–733.

Herr, D.E., E.W. Stroube, and D.A. Ray. 1966. The movement and persistence of picloram in soil. Weeds 14:248–250.

Hiltbold, A.E., and G.A. Buchanan. 1977. Influence of pH on persistence of atrazine in the field. Weed Sci. 25:515–520.

Hollist, R.L., and C.L. Foy. 1971. Trifluralin interactions with soil constituents. Weed Sci. 19:11–16.

Jacques, G.L., and R.G. Harvey. 1979. Adsorption and diffusion of dinitroaniline herbicides in soils. Weed Sci. 27:450–455.

Johnson, L.R., and A.E. Hiltbold. 1969. Arsenic content of soil and crops following methanearsonate herbicides. Soil Sci. Soc. Am. Proc. 33:279–282.

Jordan, L.S., and B.E. Day. 1962. Effects of soil properties on EPTC phytotoxicity. Weeds 10:212–215.

Jordan, P.D., and L.W. Smith. 1971. Adsorption and deactivation of atrazine and diuron by charcoals. Weed Sci. 19:541–544.

Jorgansen, C.J.C., and C.L. Hamner. 1948. Weed control in soils with 2,4-dichlorophenoxyacetic acid and related compounds and their residual effects under varying environmental conditions. Bot. Gaz. 109:324–333.

Karickhoff, S.W., D.S. Brown, and T.A. Scott. 1979. Sorption of hydrophobic pollutants on natural sediments. Water Res. 13:241–248.

Katan, J., and E.P. Lichtenstein. 1977. Mechanisms of production of soil-bound residues of [^{14}C] parathion by microorganisms. J. Agric. Food Chem. 25:1404–1407.

Kay, B.D., and D.E. Elrick. 1967. Adsorption and movement of lindane in soils. Soil Sci. 104:314–322.

Kells, J.J., C.E. Rieck, R.L. Blevins, and M.M. Muir. 1980. Atrazine dissipation as affected by surface pH and tillage. Weed Sci. 28:101–104.

Khan, S.U., and H.A. Hamilton. 1980. Extractable and bound (nonextractable) residues of prometryn and its metabolites in an organic soil. J. Agric. Food Chem. 28:126–132.

Khan, S.U., and K.C. Ivarson. 1981. Microbiological release of unextracted (bound) residues from an organic soil treated with prometryn. J. Agric. Food Chem. 29:1301–1303.

Kirk, R.E., and M.C. Wilson. 1960. The effect of soil type and moisture on germination and growth from wheat seed treated with phorate. J. Econ. Entomol. 53:771–774.

Kloskowski, R., and F. Fuhr. 1987. Aged and bound herbicide residues in soil and their bioavailability: Part 1. Uptake of aged and non-extractable (bound) ^{14}C-metamitron residues by sugar beets. J. Environ. Sci. Health B22(5):509–535.

Knight, B.A.G., and T.E. Tomlinson. 1967. The interaction of paraquat (1:1′-dimethyl 4:4′-dipyridylium dichloride) with mineral soils. J. Soil Sci. 18:233–243.

Konrad, J.G., and G. Chesters. 1969. Degradation in soils of ciodrin, an organophosphorous insecticide. J. Agric. Food Chem. 17:226–230.

Konrad, J.G., G. Chesters, and D.E. Armstrong. 1969. Soil degradation of malathion, a phosphorodithioate insecticide. Soil Sci. Soc. Am. Proc. 33:259–262.

Koren, E., C.L. Foy, and F.M. Ashton. 1968. Phytotoxicity and persistence of four thiocarbamates in five soil types. Weed Sci. 16:172–175.

Koren, E., C.L. Foy, and F.M. Ashton. 1969. Adsorption, volatility, and migration of thiocarbamate herbicides in soil. Weed Sci. 17:148–153.

Kozak, J., J.B. Weber, and T.J. Sheets. 1983. Adsorption of prometryn and metolachlor by selected soil organic matter fractions. Soil Sci. 136:94–101.

Kratky, B.A., and G.F. Warren. 1973. Water-soil-plant interactions with terbacil. Weed Sci. 21:451–454.

Ladlie, J.S., W.F. Meggitt, and D. Penner. 1976. Effect of pH on metribuzin activity in the soil. Weed Sci. 24:505–507.

Laermann, H.T., and R. Heitefuss. 1974. The effect of terbutryn and methabenzthiazuron in relation to different soil properties. Z. Pflanzenkr. Pflanzenschutz. 81:177–195.

Lambert, S.M. 1968. Omega, a useful index of soil sorption equilibria. J. Agric. Food Chem. 16:340–343.

Leopold, A.C., P. Van Schaik, and M. Neal. 1960. Molecular structure and herbicide adsorption. Weeds 8:48–54.

Lichtenstein, E.P., L.J. DePew, E.L. Eshbaugh, and J.P. Sleesman. 1960. Persistence of DDT, aldrin, and lindane in some midwestern soils. J. Econ. Entomol. 53:136–142.

Lichtenstein, E.P., T.W. Fuhremann, and K.R. Schultz. 1968. Use of carbon to reduce the uptake of insecticidal soil residues by crop plants. J. Agric. Food Chem. 16:348–355.

Lichtenstein, E.P., K.R. Schultz, and T.W. Fuhremann. 1971. Long-term effects of carbon in reducing uptake of insecticidal soil residues by crops. J. Econ. Entomol. 64:585–588.

Linscott, D.L., and R.D. Hagin. 1967. Protecting alfalfa seedlings from a triazine with activated charcoal. Weeds 15:304–306.

Liu, S.L., and J.B. Weber. 1985. Retention and mobility of AC 252,214 (imazaquin), chlorsulfuron, prometryn, and SD 95481 (cinmethylin) in soils. Proc. South. Weed Sci. Soc. 38:465–474.

Liu, S.L., and J.B. Weber. 1986. Adsorption/desorption, and mobility of propiconazole and prometryn in soils. Proc. South. Weed Sci. Soc. 39:460–471.

Lotse, E.G., D.A. Graetz, C. Chesters, G.B. Lee, and L.W. Newland. 1968. Lindane adsorption by lake sediments. Environ. Sci. Technol. 5:353–357.

Loux, M.M., R.A. Liebl, and F.W. Slife. 1989. Availability and persistence of imazaquin, imazethapyr, and clomazone in soil. Weed Sci. 37:259–267.

Lowder, S.W., and J.B. Weber. 1982. Atrazine efficacy and longevity as affected by tillage, liming, and fertilizer type. Weed Sci. 30:273–280.

Lucas, E.H., and C.L. Hamner. 1947. Inactivation of 2,4-D by adsorption on charcoal. Science 105:340.

Majek, B.A. 1986. Activated carbon for safening peppers (*Capsicum annuum* L.) in soils treated with metribuzin. Weed Sci. 34:467–470.

Mapplebeck, L., and C. Waywell. 1983. Detection and degradation of linuron in organic soils. Weed Sci. 31:8–13.

Martin, M.J.S., and M.S. Camazane. 1984. Aspects of adsorption of azinphosmethyl by smectites. J. Agric. Food Chem. 32:720–725.

Martin, N., and D. Gottlieb. 1952. The production and role of antibiotics in the soil. III. Terramycin and aureomycin. Phytopathology 42:294–296.

McCarthy, J.F. 1989. Bioavailability and toxicity of metals and hydrophobic organic contaminants. Adv. Chem. Ser. 219:263–277.

McConnell, J.S., and L.R. Hossner. 1985. pH-dependent adsorption isotherms of glyphosate. J. Agric. Food Chem. 33:1075–1078.

McKinnon, E.J. 1989. The behavior and bioactivity of imazaquin in soils. Ph.D. diss. North Carolina State Univ., Raleigh (Diss. Abstr. AAC 8918115).

McLaren, A.D. 1954. The adsorption and reactions of enzymes and proteins on kaolinite. J. Phys. Chem. 58:129–137.

McLaren, A.D., and E.F. Estermann. 1956. The adsorption and reactions of enzymes and proteins on kaolinite. III. The isolation of enzyme-substrate complexes. Arch. Biochem. Biophys. 61:158–173.

Menges, R.M., and J.L. Hubbard. 1966. Herbicidal performances of CDEC and EPTC incorporated to various depths in furrow-irrigated soils. Weeds 14:215–219.

Menges, R.M., and J.L. Hubbard. 1970. Phytotoxicity of bensulide and trifluralin in several soils. Weed Sci. 18:244–247.

Mercie, W., and C.L. Foy. 1986. Adsorption, desorption, and mobility of chlorsulfuron in soils. J. Agric. Food Chem. 34:89–92.

Miller, C.T., and W.J. Weber, Jr. 1986. Sorption of hydrophobic organic pollutants in saturated soil systems. J. Contam. Hydrol. 1:243–261.

Miller, C.W., I.E. Demoranville, and A.J. Charig. 1966. Persistence of dichlobenil in cranberry bogs. Weeds 14:296–298.

Mills, A.C., and J.W. Biggar. 1969. Solubility-temperature effect on the adsorption of gamma- and beta-BHC from aqueous and hexane solutions by soil materials. Soil Sci. Soc. Am. Proc. 33:210–216.

Mortland, M.M., and W.F. Meggitt. 1966. Interaction of ethyl *N,N*-di-*n*-propylthiocarbamate (EPTC) with montmorillonite. J. Agric. Food Chem. 14:126–129.

Nash, R.G. 1968. Plant uptake of ^{14}C-diuron in modified soil. Agron. J. 60:177–179.

Nearpass, D.C. 1965. Effects of soil acidity on the adsorption, penetration, and persistence of simazine. Weeds 13:341–346.

Nearpass, D.C. 1969. Exchange adsorption of 3-amino-1,2,4-triazole by an organic soil. Soil Sci. Soc. Am. Proc. 33:524–528.

Nearpass, D.C. 1970. Exchange adsorption of 3-amino-1,2,4-triazole by montmorillonite. Soil Sci. 109:77–84.

Obien, S.R., R.H. Suehisa, and O.R. Younge. 1966. The effects of soil factors on the phytotoxicity of neburon to oats. Weeds 14:105–109.

Ogg, A.G., Jr. 1978. Herbicides and activated carbon for weed control in direct-seeded asparagus. Weed Sci. 26:284–286.

Ogg, A.G., Jr. 1982. Effect of activated carbon on phytotoxicity of terbacil to several crops. Weed Sci. 30:683–687.

Okafor, L.I., G.A. Sagar, and V.M. Shorrocks. 1983. Biological activity of dinitramine in soils. II. Soil organic matter and soil moisture content. Weed Res. 23:199–206.

Parochetti, J.V. 1973. Soil organic matter effect on activity of acetamides, CDAA, and atrazine. Weed Sci. 21:157–160.

Patterson, S., D. MacKay, D. Tam, and W.Y. Shiu. 1990. Uptake of organic chemicals by plants: A review of processes, correlations and models. Chemosphere 21:297–331.

Peeper, T.F., and J.B. Weber. 1976. Activity and persistence of atrazine, procyazine, and VEL-5026 (buthidazole) influenced by soil organic matter and clay. Proc. South. Weed Sci. Soc. 29:387–398.

Peter, C.J., and J.B. Weber. 1985a. Adsorption, mobility, and efficacy of metribuzin as influenced by soil properties. Weed Sci. 33:868–873.

Peter, C.J., and J.B. Weber. 1985b. Adsorption, mobility, and efficacy of alachlor and metolachlor as influenced by soil properties. Weed Sci. 33:874–881.

Peter, C.J., and J.B. Weber. 1985c. Adsorption and efficacy of trifluralin and butralin as influenced by soil properties. Weed Sci. 33:861–867.

Peterson, J.R., R.S. Adams, Jr., and L.K. Cutkomp. 1971. Soil properties influencing DDT bioactivity. Soil Sci. Soc. Am. Proc. 35:72–78.

Philen, O.D., Jr., S.B. Weed, and J.B. Weber. 1970. Estimation of surface charge density of mica and vermiculite by competitive adsorption of diquat^{2+} vs. paraquat^{2+}. Soil Sci. Soc. Am. Proc. 34:527–531.

Philen, O.D., Jr., S.B. Weed, and J.B. Weber. 1971. Surface charge characterization of layer silicates by competitive adsorption of two organic divalent cations. Clays Clay Miner. 19:295–302.

Pinck, L.A. 1962. Adsorption of proteins, enzymes, and antibiotics by montmorillonite. Clays Clay Miner. 9:520–529.

Pinck, L.A., and F.E. Allison. 1951. Resistance of a protein-montmorillonite complex to decomposition by soil microorganisms. Science 114:130–131.

Pinck, L.A., R.S. Dyal, and F.E. Allison. 1954. Protein-montmorillonite complexes and their decomposition. Soil Sci. 78:109–118.

Pinck, L.A., D.A. Soulides, and F.E. Allison. 1961. Antibiotics in soils. II. Extent and mechanism of release. Soil Sci. 91:94–98.

Racke, K.D., and E.P. Lichtenstein. 1985. Effects of soil microorganisms on the release of bound ^{14}C residues from soils previously treated with [^{14}C] parathion. J. Agric. Food Chem. 33:938–943.

Rahman, A. 1976. Effect of soil organic matter on the phytotoxicity of soil-applied herbicides-glasshouse studies. N.Z. J. Exp. Agric. 4:85–88.

Rai, G.S., and C.L. Hamner. 1953. Persistence of sodium trichloracetate in different soil types. Weeds 2:271–279.

Reinhardt, C.F., J.G. Ehlers, and P.C. Nel. 1990. Persistence of atrazine as affected by selected soil properties. S. Afr. Tydskr. Plant Grond 7:182–187.

Reinhardt, C.F., and P.C. Nel. 1988. Effect of alachlor on *Sorghum bicolor* L. Moench and some factors influencing bioactivity in soil. p. 251–256. *In* Proc. Eur. Weed Res. Counc. Symp., Jan. 6–12, Begbroke Hill, Yarnton, Oxford; Page Bros., Ltd., Norwich, England.

Reinhardt, C.F., and P.C. Nel. 1989. Importance of selected soil properties on the bioactivity of alachlor and metolachlor. S. Afr. Tydskr. Plant Grond 6:120–123.

Reinhardt, C.F., and P.C. Nel. 1990. Importance of selected soil properties on the bioactivity of acetochlor and metazachlor. S. Afr. J. Plant Soil 7:101–104.

Renner, K.A., W.F. Meggitt, and D. Penner. 1988. Effect of soil pH on imazaquin and imazethapyr adsorption to soil and phytotoxicity to corn (*Zea mays*). Weed Sci. 36:78–83.

Roberts, H.A., and B.J. Wilson. 1965. Adsorption of chlorpropham by different soils. Weed Res. 5:348–350.

Robocker, W.C., and C.L. Canode. 1964. Effects of litter, incorporated organic matter, soil type, and selective herbicides on seedlings of downy brome. Weeds 12:8–20.

Rodrigues, J.J., A.D. Worsham, and F.T. Corbin. 1982. Exudation of glyphosate from wheat (*Triticum aestivum*) plants and its effects on interplanted corn (*Zea mays*) and soybean (*Glycine max*). Weed Sci. 30:316–320.

Russell, J.D., M.I. Cruz, and J.L. White. 1968. The adsorption of 3-aminotriazole by montmorillonite. J. Agric. Food Chem. 16:21–23.

Salazar, L.C., and A.P. Appleby. 1982. Herbicidal activity of glyphosate in soil. Weed Sci. 30:463–466.

Saltzman, S., L. Kliger, and B. Yaron. 1972. Adsorption-desorption of parathion as affected by soil organic matter. J. Agric. Food Chem. 20:1224–1226.

Sato, T., S. Kohnosu, and J.F. Hartwig. 1987. Adsorption of butachlor to soils. J. Agric. Food Chem. 35:397–402.

Schliebe, K.A., O.C. Burnside, and T.L. Lavy. 1965. Dissipation of amiben. Weeds 13:321–325.

Schmidt, R.R. 1973. Calculations on the herbicidal activity of metribuzin in relation to various soil properties. p. 24–30. *In* Proc. Eur. Weed Res. Council Symp. Herbicides-Soil, Jan. 8–14, Stuttgart, Germany, Page Bros., Ltd., Norwich, England.

Schroeder, J., and P.A. Banks. 1986. Persistence and activity of norflurazon and fluridone in five Georgia soils under controlled conditions. Weed Sci. 34:599–606.

Schwartz, H.G. 1967. Adsorption of selected pesticides on activated carbon and mineral surfaces. Environ. Sci. Technol. 1:332–337.

Schweizer, E.E. 1967. Toxicity of DSMA soil residues to cotton and rotational crops. Weeds 15:72–76.

Scott, D.C., and J.B. Weber. 1967. Herbicide phytotoxicity as influenced by adsorption. Soil Sci. 104:151–158.

Shea, P.J., and J.B. Weber. 1980. Effect of pH and soil constituents on the persistence and availability of fluridone. Proc. South. Weed Sci. Soc. 33:240–246.

Shea, P.J., and J.B. Weber. 1983. Fluridone adsorption on mineral clays, organic matter, and modified Norfolk soil. Weed Sci. 31:528–532.

Shea, P.J., J.B. Weber, and M.R. Overcash. 1983. Biological activities of 2,4-dinitrophenol in plant-soil systems. Res. Rev. 87:1–41.

Sheets, T.J. 1959. Effects of soil type and time on the herbicidal activity of CDAA, CDEC, and EPTC. Weeds 7:442–448.

Sheets, T.J., A.S. Crafts, and H.R. Drever. 1962. Influence of soil properties on the phytotoxicities of the *s*-triazine herbicides. J. Agric. Food Chem. 10:458–462.

Siminoff, P., and D. Gottlieb. 1951. The production and role of antibiotics in the soil. I. The fate of streptomycin. Phytopathology 41:420–430.

Skipper, H.D., C.M. Gilmour, and W.R. Furtick. 1967. Microbial versus chemical degradation of atrazine in soils. Soil Sci. Soc. Am. Proc. 31:653–656.

Skipper, H.D., and V.V. Volk. 1972. Biological and chemical degradation of atrazine in three Oregon soils. Weed Sci. 20:344–347.

Slack, C.H., R.L. Blevins, and C.E. Rieck. 1978. Effect of soil pH and tillage on persistence of simazine. Weed Sci. 26:145–148.

Smit, N.S.H., and P.C. Nel. 1977. The activity of atrazine on two South African soils. Crop Prod. 6:67–71.

Smit, N.S.H., and P.C. Nel. 1980. Factors influencing availability of atrazine in South African soils. Proc. Natl. Weeds Conf. S.A. 3:151–158.

Smit, N.S.H., P.C. Nel, and W.J. Folscher. 1980. Faktore wat sorpsie en ontbinding van atrasien in grond beinvloed. Gewasproduksie 9:135–139.

Smit, N.S.H., P.C. Nel, and W.J. Folsche. 1981. Fosforvalegging as moontlike kriterium vir biologiese aktiwiteit van atrasien in grond. Gewasproduksie 10:209–213.

Sprankle, P., W.F. Meggitt, and D. Penner. 1975. Rapid inactivation of glyphosate in the soil. Weed Sci. 23:224–228.

Stanger, C.E., Jr., and A.P. Appleby. 1971. Effect of soil type and irrigation method on lateral movement of cycloate. Weed Sci. 19:709–711.

Stougaard, R.N., P.J. Shea, and A.R. Martin. 1990. Effect of soil type and pH on adsorption, mobility, and efficacy of imazaquin and imazethapyr. Weed Sci. 38:67–73.

Strek, J.H., and J.B. Weber. 1981. Alachlor (Lasso) and metolachlor (Dual) comparisons in conventional and reduced tillage systems. Proc. South. Weed Sci. Soc. 34:33–40.

Strek, H.J., and J.B. Weber. 1982a. Adsorption, mobility, and activity comparisons between alachlor (Lasso) and metolachlor (Dual). Proc. South. Weed Sci. Soc. 35:332–338.

Strek, H.J., and J.B. Weber. 1982b. Adsorption and reduction in bioactivity of polychlorinated biphenyl (Aroclor 1254) to redroot pigweed by soil organic matter and montmorillonite clay. Soil Sci. Soc. Am. J. 46:318–322.

Strek, H.J., J.B. Weber, P.J. Shea, E. Mrozek, Jr., and M.R. Overcash. 1981. Reduction of polychlorinated biphenyl toxicity and uptake of carbon-14 activity by plants through the use of activated carbon. J. Agric. Food Chem. 29:288–293.

Sullivan, J.D., Jr., and G.T. Felbeck, Jr. 1968. A study of the interaction of *s*-triazine herbicides with humic acids from three different soils. Soil Sci. 106:42–52.

Summers, L.A. 1980. The bipyridylium herbicides. Academic Press, New York.

Sund, K.A. 1956. Residual activity of 3-amino-1,2,4-triazole in soils. J. Agric. Food Chem. 4:57–60.

Talbert, R.E., R.L. Runyan, and H.R. Baker. 1970. Behavior of amiben and dinoben derivatives in Arkansas soils. Weed Sci. 18:10–15.

Talibudeen, O. 1955. Complex formation between montmorillonoid clays and amino acids and proteins. Trans. Faraday Soc. 51:582–587.

Taylorson, R.B. 1966. Influence of temperature and other factors on loss of CDEC from soil surfaces. Weeds 14:294–296.

Thurston, R. 1953. The effects of some soil characteristics on DDT phytotoxicity. J. Econ. Entomol. 46:545–550.

Torstensson, N.T.L., and A. Aamisepp. 1977. Detoxification of glyphosate in soil. Weed Res. 17:209–212.

Tucker, B.V., D.E. Pack, and J.N. Ospenson. 1967. Adsorption of bipyridylium herbicides in soil. J. Agric. Food Chem. 15:1005–1008.

Tucker, B.V., D.E. Pack, J.N. Ospenson, A. Omid, and W.D. Thomas, Jr. 1969. Paraquat soil binding and plant response. Weed Sci. 17:448–451.

Upchurch, R.P. 1958. The influence of soil factors on the phytotoxicity and plant selectivity of diuron. Weeds 6:161–171.

Upchurch, R.P., and D.D. Mason. 1962. The influence of soil organic matter on the phytotoxicity of herbicides. Weeds 10:9–14.

Upchurch, R.P., F.L. Selman, D.D. Mason, and E.J. Kamprath. 1966. Correlations of herbicidal activity with soil and climatic factors. Weeds 14:42–49.

Ward, T.M., and F.W. Getzen. 1970. Influence of pH on the adsorption of aromatic acids on activated carbon. Environ. Sci. Technol. 4:64–67.

Weaver, R.J. 1947. Reaction of certain plant growth-regulators with ion exchangers. Bot. Gaz. 109:72–84.

Weaver, R.J. 1948. Some uses of activated carbon in contratoxification of plant growth-regulators. Bot. Gaz. 110:300–312.

Weber, J.B. 1966. Molecular structure and pH effects on the adsorption of 13 *s*-triazine compounds on montmorillonite clay. Am. Mineral. 51:1657–1670.

Weber, J.B. 1970. Mechanisms of adsorption of *s*-triazines by clay colloids and factors affecting plant availability. Res. Rev. 32:93–130.

Weber, J.B. 1972. Interactions of organic pesticides with particulate matter in aquatic and soil systems. *In* R.F. Gould (ed.) Fate of organic pesticides in the aquatic environment. Adv. Chem. Ser. 111:55–120.

Weber, J.B. 1980. Adsorption of buthidazole. VEL-3510, tebuthiuron, and fluridone by organic matter, montmorilloinite clay, exchange resins, and a sandy loam soil. Weed Sci. 28:478–483.

Weber, J.B. 1982. Ionization and adsorption-desorption of tricyclazole by soil organic matter, montmorillonite clay, and Cape Fear sandy loam soil. J. Agric. Food Chem. 30:584–588.

Weber, J.B. 1987. Physical/chemical interactions of herbicides with soil. Proc. Calif. Weed Conf. 39:96–109.

Weber, J.B. 1990. Behavior of dinitroaniline herbicides in soils. Weed Technol. 4:394–406.

Weber, J.B. 1993. Properties and behavior of pesticides in soil. *In* R. Honeycutt and D. Schabaker (ed.) Advances in understanding the mechanisms of movement of pesticides into groundwater. Lewis Publ., Chelsea, MI.

Weber, J.B., and J.A. Best. 1972. Activity and movement of 13 soil-applied herbicides as influenced by soil reactions. Proc. South. Weed Sci. Soc. 25:403–413.

Weber, J.B., J.A. Best, and W.W. Witt. 1974a. Herbicide residues and weed species shifts on modified-soil field plots. Weed Sci. 22:427–433.

Weber, J.B., and H.D. Coble. 1968. Microbial decomposition of diquat adsorbed on montmorillonite and kaolinite clays. J. Agric. Food Chem. 16:475–478.

Weber, J.B., H.D. Coble, T.J. Monaco, A.D. Worsham, A. Mehlich, A. Hatfield, D.W. Eaddy, and C.J. Peter. 1981. Soil tests and herbicide rate recommendations for metolachlor and alachlor in agronomic and horticultural crops. Proc. South. Weed Sci. Soc. 34:265–273.

Weber, J.B., R.C. Meek, and S.B. Weed. 1969a. The effect of cation-exchange capacity on the retention of diquat^{2+} and paraquat^{2+} by three-layer type clay minerals. II. Plant availability of paraquat. Soil Sci. Soc. Am. Proc. 33:382–385.

Weber, J.B., and C.T. Miller. 1989. Organic chemical movement over and through soil. p. 305–334. *In* B.L. Sawhney and K. Brown (ed.) Reactions and movement of organic chemicals in soils. SSSA Spec. Publ. 22. SSSA and ASA, Madison, WI.

Weber, J.B., T.J. Monaco, R.D. Carringer, and D.W. Eaddy. 1974c. Effect of organic matter and volume weight of North Carolina soils on herbicide performance. Proc. South. Weed Sci. Soc. 27:371–377.

Weber, J.B., and E. Mrozek, Jr. 1979. Polychlorinated biphenyls: Phytotoxicity, absorption and translocation by plants, and inactivation by activated carbon. Bull. Environ. Contam. Toxicol. 23:412–417.

Weber, J.B., P.W. Perry, and K. Ibaraki. 1968a. Effect of pH on the phytoxicity of prometryne applied to synthetic soil media. Weed Sci. 16:134–136.

Weber, J.B., P.W. Perry, and R.P. Upchurch. 1965. The influence of temperature and time on the adsorption of paraquat, diquat, 2,4-D, and prometone by clays, charcoal, and an anion-exchange resin. Soil Sci. Soc. Am. Proc. 29:678–688.

Weber, J.B., and C.J. Peter. 1982. Adsorption, bioactivity, and evaluation of soil tests for alachlor, acetochlor, and metolachlor. Weed Sci. 30:14–20.

Weber, J.B., and D.C. Scott. 1966. Availability of a cationic herbicide adsorbed on clay minerals to cucumber seedlings. Science 152:1400–1402.

Weber, J.B., P.J. Shea, and S.B. Weed. 1986. Fluridone retention and release in soils. Soil Sci. Soc. Am. J. 50:582–588.

Weber, J.B., M.R. Tucker, and R.A. Isaac. 1987. Making herbicide recommendations based on soil tests. Weed Technol. 1:41–45.

Weber, J.B., T.M. Ward, and S.W. Weed. 1968b. Adsorption and desorption of diquat, paraquat, prometone, and 2,4-D by charcoal and exchange resins. Soil Sci. Soc. Am. Proc. 32:197–200.

Weber, J.B., and S.B. Weed. 1968. Adsorption and desorption of diquat, paraquat, and prometone by montmorillonitic and kaolinitic clay minerals. Soil Sci. Soc. Am. Proc. 32:485–487.

Weber, J.B., and S.B. Weed. 1974. Effects of soil on the biological activity of pesticides. p. 223–256. *In* W.D. Guenzi (ed.) Pesticides in soil and water. SSSA, Madison, WI.

Weber, J.B., S.B. Weed, and J.A. Best. 1969b. Displacement of diquat from clay and its phytotoxicity. J. Agric. Food Chem. 17:1075–1076.

Weber, J.B., S.B. Weed, and T.W. Waldrep. 1974b. Effect of soil constituents on herbicide activity in modified-soil field plots. Weed Sci. 22:454–459.

Weber, J.B., S.B. Weed, and T.M. Ward. 1969c. Adsorption of *s*-triazines by soil organic matter. Weed Sci. 17:417–421.

Weed, S.B., and J.B. Weber. 1968. The effect of adsorbent charge on the competitive adsorption of divalent organic cations by layer-silicate minerals. Am. Mineral. 53:478–490.

Weed, S.B., and J.B. Weber. 1969. The effect of cation exchange capacity on the retention of diquat^{2+} and paraquat^{2+} by three-layer clay minerals. I. Adsorption and release. Soil Sci. Soc. Am. Proc. 33:379–382.

Weidhaus, D.E., M.C. Bowman, and C.H. Schmidt. 1961. Loss of parathion and DDT to soil from aqueous dispersions and vermiculite granules. J. Econ. Entomol. 54:175–177.

Woolson, E.A. 1975. Arsenical pesticides. ACS Symp. Ser. no. 7. Am. Chem. Soc., Washington, DC.

Woolson, E.A., J.H. Axley, and P.C. Kearney. 1971a. Correlation between available soil arsenic estimated by six methods, and response of corn (*Zea mays* L.). Soil Sci. Soc. Am. Proc. 35:101–105.

Woolson, E.A., J.H. Axley, and P.C. Kearney. 1971b. The chemistry and phytotoxicity of arsenic in soils. I. Contaminated field soils. Soil Sci. Soc. Am. Proc. 35:938–943.

Woolson, E.A., J.H. Axley, and P.C. Kearney. 1973. The chemistry and phytotoxicity of As in soils. II. Effects of time and phosphorous. Soil Sci. Soc. Am. Proc. 37:254–259.

Woolson, E.A., and A.R. Isensee. 1981. Soil residue accumulation from three applied arsenic sources. Weed Sci. 29:17–21.

Wu, C.H., N. Buehring, J.M. Davidson, and P.W. Santelmann. 1975. Napropamide adsorption, desorption, and movement in soils. Weed Sci. 23:454–457.

Yaron, B., and S. Saltzman. 1978. Soil-parathion surface interactions. Res. Rev. 69:1–34.

Yaron, B., A.R. Swoboda, and G.W. Thomas. 1967. Aldrin adsorption by soils and clays. J. Agric. Food Chem. 15:671–675.

11 Bioavailability of Sorbed Naphthalene to Bacteria: Influence of Contaminant Aging and Soil Organic Carbon Content[1]

William F. Guerin and Stephen A. Boyd
Michigan State University
East Lansing, Michigan

A major gap in our understanding of non-ionic organic contaminant (NOC) fate and behavior in soils is the inability to predict the influence that sorption has on biodegradation. From studies in other systems and with other (primarily ionic) compounds, it is known that the mechanism of sorption, the duration of soil-contaminant contact, the physicochemical properties of both the sorbate and the sorbent, and the specific characteristics of the degrading organisms are important determinants of desorption kinetics and sorbed compound bioavailability. These studies have shown that sorption may increase, decrease, or have no effect on the degradation of organic compounds (van Loosdrecht et al., 1990).

Few studies have examined the influence of NOC sorption on biodegradation, despite the fact that this is the most problematic group of environmental contaminants. Studies by Ogram et al. (1985) and Steen et al. (1980) have indicated that sorbed pesticides are unavailable to microbial degraders and that desorption into the soil solution is a prerequisite for biodegradation. The kinetics of NOC desorption may thus play an important role in determining the rates and extents of NOC biodegradation. Models developed to describe the desorption kinetics of NOCs from soils include the radial (aggregate) diffusion model (Wu & Gschwend, 1986), the two compartment mass transfer model (Karickhoff, 1980), and the intraorganic matter diffusion model (Brusseau et al., 1991).

In this chapter, we describe the application of a simple kinetic method to assess the bioavailability of soil-sorbed naphthalene by comparing the mineralization kinetics of two bacterial species in soil-free and soil-containing

[1] This work was supported by grant DE-FG02-89ER60809 from the U.S. Department of Energy, Subsurface Science Program, and the Michigan Agricultural Experiment Station.

slurries. Our results show that: (i) sorbed naphthalene is differentially available to individual bacterial species, (ii) availability is sorbent-specific, and (iii) availability is decreased upon prolonged naphthalene-soil contact.

MATERIALS AND METHODS

Soil Characterization

Four surface soils (Table 11–1) from southern Michigan were dried, ground to pass a 2-mm sieve, and analyzed for organic carbon (OC) content (Huffman Labs, Golden, CO). Soil texture was determined by the hydrometer method (Grigal, 1973), and cation exchange capacity was measured as the uptake of Mg by barium-saturated soils (Rhoades, 1982). Sorption isotherms were constructed by measuring the uptake of ^{14}C-naphthalene by soils from naphthalene solutions [in phosphate-buffered saline (PBS): 8.5 g of NaCl, 0.6 g of Na_2HPO_4, 0.3 g of KH_2PO_4 per liter of distilled water, pH 7.0] after overnight equilibration. Initial naphthalene concentrations were varied from approximately half-saturation (solubility = 31.69 μg mL^{-1}, May et al., 1978) to about 40 ng mL^{-1}. Soil slurries (in 25-mL glass centrifuge tubes) were centrifuged (12 100 × g, 20 min), the radioactivity remaining in the supernatants measured by liquid scintillation spectrometry, and the amount of sorbed naphthalene determined by difference. Sorptive equilibrium was complete after overnight equilibration, yielding naphthalene partition coefficients that remained constant for at least 30 d of additional soil-naphthalene contact.

Organisms and Growth Conditions

The two bacterial species used in these studies were *Pseudomonas putida* (ATCC 17484), and a gram-negative isolate from petroleum-contaminated soil, designated NP-Alk. Organisms were grown to early stationary phase in minimal medium [per liter of distilled water: 3.0 g of $(NH_4)_2HPO_4$, 2.0 g of NaCl, 1.2 g of KH_2PO_4, 3 mg of $MgSO_4$, 1 mL of trace element solution (Shelton & Tiedje, 1984), 1 mL of vitamin solution (Wolin et al., 1963), 1 mL of ferric quinate (Blakemore et al., 1979), pH 7.0] containing naphthalene (200 μg mL^{-1}) as sole C and energy source, at room temperature with

Table 11–1. Physicochemical properties of the soils used in bioavailability experiments.

Soil series	Order	% Organic C	Texture	CEC†	K_p‡	K_{oc}§	Log K_{oc}
Capac	Alfisol	3.46	Loam	12.2	12.51	362	2.56
Colwood	Mollisol	5.36	Loam	21.6	28.93	540	2.73
Oshtemo	Alfisol	0.76	Sandy loam	5.6	4.56	600	2.78
Schoolcraft	Mollisol	1.84	Loam	8.0	8.07	440	2.64

† Cation exchange capacity (cmol$_c$ kg^{-1}).
‡ Naphthalene partition coefficient (L kg^{-1}).
§ Organic C-normalized partition coefficient (K_p fraction organic C^{-1}).

shaking (150 rpm). Cells were harvested by centrifugation (12 100 × g, 10 min) and resuspended in PBS to a density of 10^9 cells mL^{-1} before use in bioavailability assays.

Bioavailability Assays

Sterilized (autoclaved) soils were pre-equilibrated with ^{14}C-naphthalene solutions (25–300 mg of soil mL^{-1}) of known activity and initial concentrations (measured by absorbance at 277 nm) of 40 to 100 ng mL^{-1} for periods up to 1 yr in 155 mL serum vials sealed with teflon-faced septa (or in 25 mL glass tubes with teflon-lined caps, the contents of which were transferred to serum vials at the start of experiments). Slurries and soil-free controls (75 mL total volume) were inoculated with a washed suspension of early stationary phase cells (750 μL) to give a final cell density of approximately 10^7 cells mL^{-1}. At precisely timed intervals, duplicate aliquots of the suspensions plus headspaces were withdrawn by syringe and transferred to sealed tubes containing 1 mL of 2*N* HCl. Evolved $^{14}CO_2$ was trapped on KOH-saturated filter paper contained in a plastic cup suspended from the tube stopper. After overnight degassing, the filter papers were placed in scintillation vials containing 7.5 mL of scintillation cocktail along with 5 mL 95% ethanol used to rinse the plastic cup. Samples were counted, counts were converted to disintegrations per minute by external standards quench correction, and the percent naphthalene mineralized was calculated on the basis of the total initial activity (determined at the time of slurry preparation).

Percent mineralization (P) was plotted as a function of incubation time (t) and data were fit to a first order production model:

$$P = P_{max}[1 - e^{-kt}]$$

to derive estimates of the first-order mineralization rate constant, k, and the maximal percent mineralization, P_{max}, by nonlinear regression analysis (SYSTAT). Initial mineralization rates (ng mL^{-1} min^{-1}) were calculated by normalizing k (min^{-1}) to P_{max} (dimensionless) and the initial naphthalene concentration (ng mL^{-1}). P_{max} was treated as an unknown parameter in nonlinear regression analyses to provide estimates of the extent of mineralization in soil slurries.

Theory

At substrate concentrations below the half-saturation constant, K_m, and with nongrowing cells, mineralization rates are linearly proportional to substrate concentration (Larson, 1980). *Assuming that sorbed naphthalene is unavailable to microbial degraders*, increasing the soil/solution ratio in slurries, and concomitantly decreasing the equilibrium aqueous phase concentration of naphthalene, should result in linear decreases in the initial mineralization rate. Plots of initial mineralization rates (y axis) in slurries vs. the equilibrium aqueous phase naphthalene concentrations (x axis) should, therefore, result in linear plots in which data points fall on a theoretical line connecting a soil-free control rate with the origin. Deviations above the line indicate that cells experience a higher naphthalene concentration than that which is present in the aqueous phase, whereas points below the line indicate that cells experience, on average, a lower naphthalene concentration. Similarly, if sorbed naphthalene is unavailable and if desorption is slow relative to biodegradation, plots of P_{max} vs. the equilibrium aqueous phase naphthalene concentration should result in linear plots in which data points fall on a theoretical line connecting a soil-free P_{max} with the origin. Upward deviations from this line indicate that organisms have access to sorbed material or that desorption is rapid relative to degradation over the time course of the assays.

RESULTS

Sorption isotherms for naphthalene on the four soils used in this study were all highly linear and gave partition coefficients (K_ps) ranging from 4.6 to 28.9 (Table 11-1). When K_ps were normalized to the fractional OC contents of the soils, the resulting log K_{oc} values converged on the literature value for naphthalene of 2.74 (Abdul et al., 1987). Knowing the partition coefficients for the soils used here, allowed calculation of the equilibrium aqueous phase naphthalene concentrations at given soil/solution ratios in bioavailability assays.

Both organisms used in this study brought about the complete and coordinate mineralization of naphthalene as sole C and energy source at concentrations up to 200 mg L^{-1}. Analysis of culture supernatant fluids by HPLC and colorimetric methods (Guerin & Jones, 1988) showed that naphthalene metabolites did not accumulate during degradation. This enabled the use of radiorespirometry to serve as an index of degradation in bioavailability assays.

Naphthalene mineralization was measured over a range of initial naphthalene concentrations (20–555 ng mL^{-1}) and initial mineralization rates were calculated by nonlinear regression analysis of mineralization curves. Rate data were fitted to the Michaelis-Menten equation by nonlinear regressions to obtain estimates of K_m and V_{max} (the maximal mineralization rate). Values of K_m for strains 17484 and NP-Alk were 211 and 91 ng mL^{-1}, respectively. These values agreed well with those obtained by linear regression of transformed data (Lineweaver-Burke plots) and set the upper limits for initial naphthalene concentrations used in bioavailability assays. To ensure linearity between initial mineralization rates and naphthalene concentrations, the latter was kept at values below 0.5 (K_m).

Bioavailability assays with NP-Alk were conducted in Capac soil slurries in which the initial naphthalene concentration was 37.9 ng mL^{-1} and the soil/solution ratio varied from 0 to 200 mg mL^{-1}. The equilibrium aqueous phase naphthalene concentrations ranged from 37.9 to 11.4 ng mL^{-1} in the soil slurries. Figure 11-1 shows the mineralization time courses for NP-Alk. Progressive decreases in both the rate and extent of mineraliza-

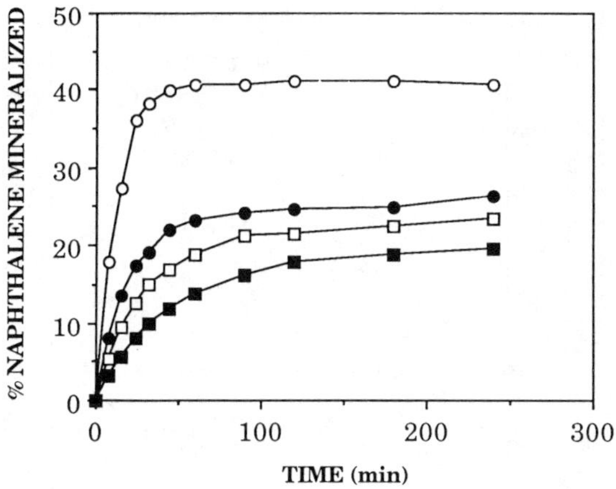

Fig. 11-1. Time courses for mineralization of ^{14}C-naphthalene sorbed to Capac soil by NP-Alk (1.06×10^7 cells mL^{-1}). Data for soil-free control (open circles), and Capac soil slurries at 50 (solid circles), 100 (open boxes) and 200 (solid boxes) mg soil (dry weight) mL^{-1} are shown. Slurries were equilibrated for 2 d prior to inoculation. Initial naphthalene concentration was 37.9 ng mL^{-1}.

tion were observed as the soil concentration increased and the dissolved naphthalene concentration decreased. Initial mineralization rates were calculated from mineralization data and plotted as a function of the equilibrium aqueous phase naphthalene concentration. Figure 11-2a shows that for NP-Alk, all points fell near, but slightly below the theoretical line, indicating that sorbed naphthalene was unavailable to this organism. Sorption also limited the extent of naphthalene mineralization by NP-Alk as seen in the progress curves of Fig. 11-1 and the P_{max} data in Fig. 11-2b. Minimal desorption of bound naphthalene was evident from the P_{max} values, which fell slightly above the theoretical line at high sorbent levels.

For *P. putida* strain 17484, mineralization was rapid and extensive at all sorbent levels tested. To eliminate the possibility that cell division or metabolic stimulation by alternate C sources present in the slurries was responsible for the higher-than-predicted rates of naphthalene mineralization, bioavailability experiments were conducted in which chloramphenicol (34 μg mL^{-1}) was added to the soil slurries prior to inoculation. We had previously determined that chloramphenicol at this concentration completely inhibited cell division without affecting naphthalene mineralization by preinduced cells. Figure 11-3 shows the naphthalene mineralization time courses in slurries containing Capac and Colwood soils (0, 77, and 192 mg soil mL^{-1}) at an initial naphthalene concentration of 96 ng mL^{-1}. As in experiments without chloramphenicol, mineralization was rapid and extensive in these systems.

Initial mineralization rates calculated using the data in Fig. 11-3 (and data for slurries containing 38.5 mg soil mL^{-1} but omitted from Fig. 11-3 for clarity) are plotted in Fig. 11-4a as a function of the equilibrium aqueous phase naphthalene concentration. All points fell above the theoretical line indicating that a fraction of the sorbed naphthalene was directly available to this organism to support higher-than-predicted rates of mineraliza-

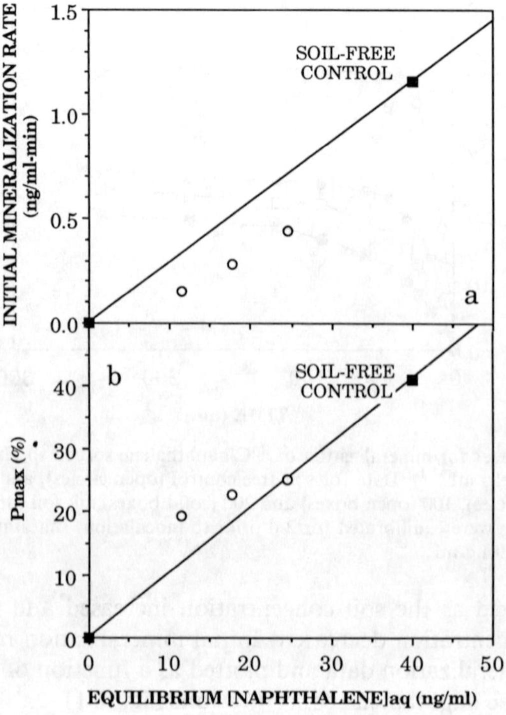

Fig. 11-2. Plots of initial mineralization rates (a) and P_{max}s (b) as a function of the equilibrium aqueous phase naphthalene concentrations for NP-Alk derived from the data in Fig. 11-1. Theoretical lines connecting the soil-free control rate and P_{max} to the origins define the conditions in which sorbed naphthalene is unavailable to the degradative bacteria or desorption is slow relative to the time course of the assay.

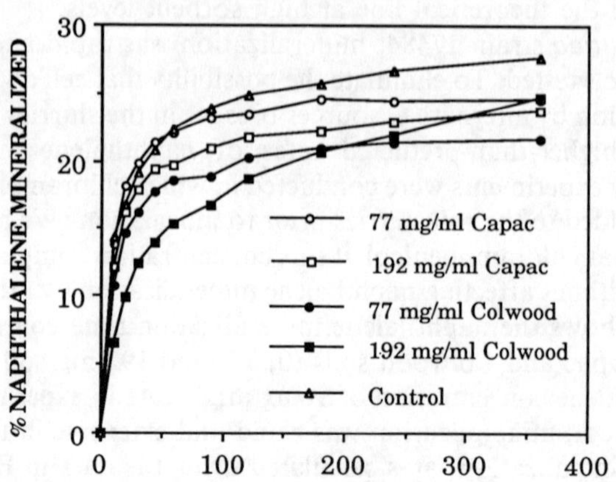

Fig. 11-3. Time courses for mineralization of ^{14}C-naphthalene in a soil-free control and soil slurries containing Capac and Colwood soils by *P. putida* strain 17484 in the presence of chloramphenicol (34 μg mL^{-1}). Soil concentrations are shown in the legend. Slurries were equilibrated for 8 d prior to inoculation. Initial naphthalene concentration was 96 ng mL^{-1}.

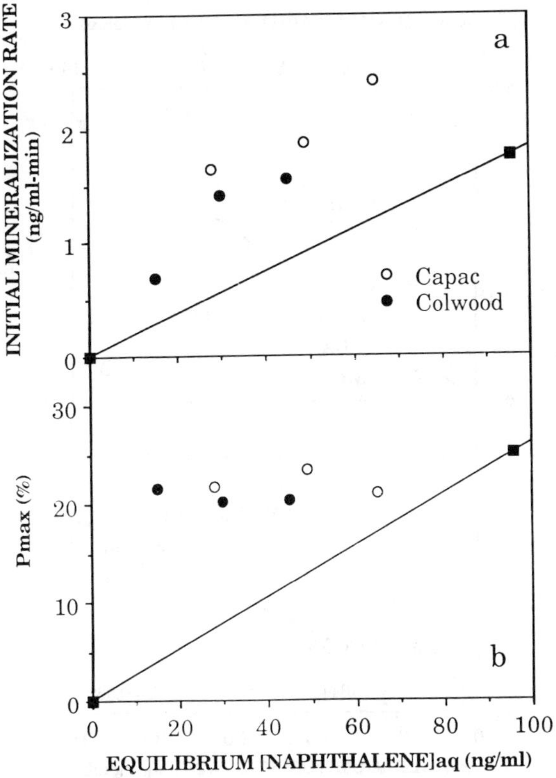

Fig. 11-4. Plots of initial mineralization rates (a) and P_{max}s (b) as a function of the equilibrium aqueous phase naphthalene concentration from the mineralization curves shown in Fig. 11-3. Also shown are data for slurries (38.5 mg soil mL^{-1}) omitted from Fig. 11-3 for clarity. All points fall above the theoretical lines indicating direct access of sorbed naphthalene to this organism or a rapid desorption of naphthalene over the time course of the assays.

tion. Sorption had little influence on the extent of naphthalene mineralization by strain 17484 over the 4-h time course of the assay as shown in the progress curves in Fig. 11-3 and the P_{max} data of Fig. 11-4b. This indicates that strain 17484, by using a fraction of the sorbed naphthalene, is able to promote the desorption of the remainder of the sorbed naphthalene.

To further illustrate the differences between the two organisms, bioavailability assays were conducted in slurries containing an identical mass of added naphthalene and equal amounts of the four soils listed in Table 11-1. Fig. 11-5 shows the relationship between the initial mineralization rates and the equilibrium aqueous phase naphthalene concentrations for NP-Alk and 17484. For NP-Alk (Fig. 11-5a), all rates fell near the theoretical line, indicating that sorbed naphthalene was equally unavailable regardless of the specific sorbent used. For strain 17484 (Fig. 11-5b), initial mineralization rates in soil slurries were highly correlated with the equilibrium aqueous phase naphthalene concentration (r^2 = 0.962) but all fell above the theoretical line. Furthermore, deviations from the theoretical rates increased with de-

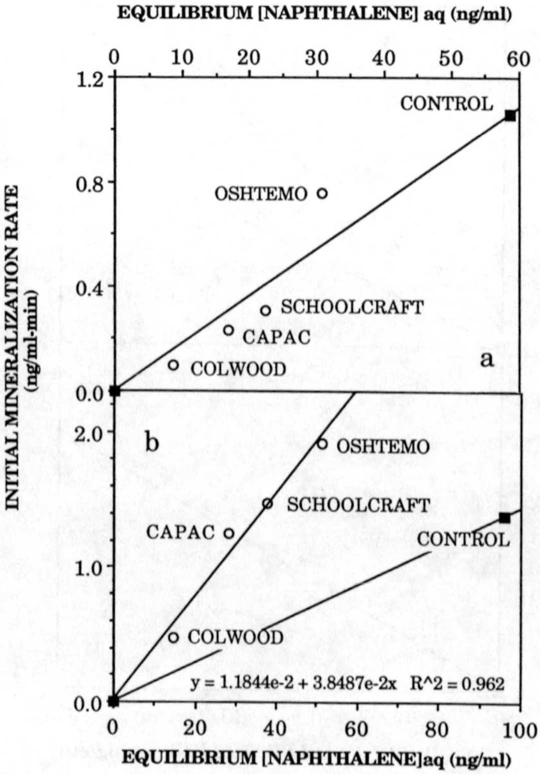

Fig. 11–5. Relationship between initial mineralization rates and the equilibrium aqueous phase naphthalene concentration for NP-Alk (a) and *P. putida* 17484 (b) in soil-free controls and slurries containing 200 mg (dry weight) mL $^{-1}$ of each of the four soils listed in Table 11–1. In (a), initial naphthalene concentration = 58.7 ng mL $^{-1}$; equilibration time = 5 d. In (b), initial naphthelene concentration = 96 ng mL $^{-1}$; equilibration time = 14 d. In (b), the linear regression line and equation for soil slurry data are shown in addition to the theoretical line.

creasing OC content of the soil, indicating differential, sorbent-dependent availability of the sorbed naphthalene.

According to the two-compartment model of NOC sorption kinetics of Karickhoff (1980), the distribution of sorbed NOCs between labile and nonlabile phases depends on the partition coefficient of the system. With increasing OC content of the sorbent (or decreasing solubility, S, of the NOC) and with increasing equilibration times, the ratio of nonlabile to labile sorbed NOC increases (Fig. 11–6). A reasonable interpretation of the bioavailability data presented for strain 17484 is that this organism has direct access to labile-sorbed naphthalene, the utilization of which promotes the desorption and mineralization of nonlabile sorbed naphthalene. To further test this hypothesis, aging experiments were conducted in which naphthalene was equilibrated in Colwood soil slurries (200 mg mL $^{-1}$) for periods up to 1 yr. At selected times, duplicate control (soil-free) and soil slurry vials were inoculated with strain 17484 and mineralization kinetics were measured. The

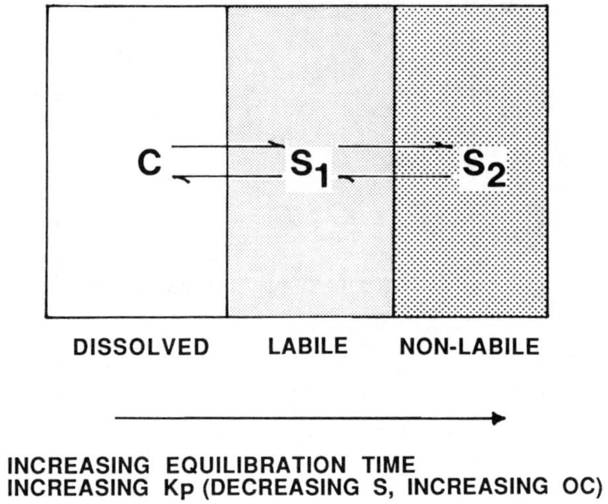

Fig. 11-6. Schematic representation of the two-compartment model of NOC sorption kinetics proposed by Karickhoff (1980).

ratios of initial mineralization rates and P_{max}s in slurries to corresponding values in controls were calculated. Figure 11-7a shows that as equilibration times increased, initial mineralization rates in slurries relative to rates in controls progressively decreased. At short equilibration times, measured rate ratios exceeded the theoretical rate ratio (assuming that only dissolved naphthalene was available) in accordance with previous observations. After several months, however, the measured rate ratio approached the theoretical rate ratio. Throughout the study, the ratio of slurry P_{max} values to control P_{max} values remained fairly constant at approximately unity (Fig. 11-7b), indicating that sorption did not limit the extent of naphthalene mineralization even when the sorbed naphthalene was predominantly in a nonlabile state. Values of P_{max} in controls were constant (within normal limits due to variation in inoculum density, culture status, etc.) over the course of aging experiments, and were equal in aged controls and fresh controls prepared on the days of the assays. Adsorptive or volatilization losses were therefore deemed to be minor.

DISCUSSION

Soil-sorbed NOCs may act as a long-term source of aquifer contaminants thwarting cleanup efforts using pump and treat or other technologies. Bioremediation offers an attractive alternative if it can be demonstrated that sorbed contaminants are accessible to potential microbial degraders. An understanding of the fundamental interactions between sorption and biodegradation is clearly needed. Despite considerable research efforts to elucidate the individual processes of NOC sorption and biodegradation, few studies have attempted to study the interactions of these processes. The

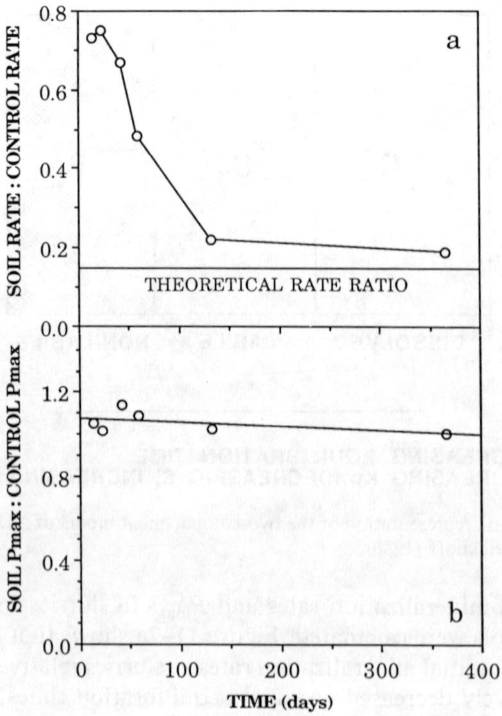

Fig. 11-7. Influence of naphthalene-slurry equilibration time on the initial rates (a) and extents of (P_{max}s) (b) naphthalene mineralization by strain 17484 in Colwood soil slurries (200 mg mL^{-1}) relative to rates and extents in aged controls. In (a), the horizontal line represents the theoretical rate ratio assuming only dissolved naphthalene is available to the organism. Initial naphthalene concentration was 99 ng mL^{-1}.

simple kinetic approach described here provides a powerful means of evaluating the effect of sorption on the bioavailability of sorbed NOCs.

The question of whether sorbed naphthalene is available to microbial degraders was found here to be strictly dependent on the organism in question. For NP-Alk, sorbed naphthalene was completely unavailable for degradation. Initial mineralization rates for this organism were consistently, though not considerably, lower than predicted rates. Extensive attachment of NP-Alk to the soils (W.F. Guerin and S.A. Boyd, 1993, unpublished data) may have occluded membrane-transport sites for dissolved naphthalene uptake. Depletion of naphthalene in solution resulted in slow desorption rates (and mineralization) of bound naphthalene during the time course of the incubations.

From experiments with 17484, a completely different picture emerged. This organism was able to immediately use both solution phase and sorbed naphthalene as evidenced by initial mineralization rates that consistently exceeded rates predicted from the dissolved naphthalene concentrations in slurries. P_{max} values in bioavailability experiments with strain 17484 were equal to soil-free control values, indicating that sorbed naphthalene was extensively desorbed from the soil over the time course of the assays.

Using nonlinear regression, the first-order model provided excellent fits to the data for NP-Alk. Large residual sums of squares in analysis of the 17484 data were an indication that non-equilibrium conditions existed in experiments with this organism, that is, that non-instantaneous but fairly rapid desorption of bound naphthalene occurred during the time course of the experiments. Since submission of this chapter, we have developed a three-parameter model that provides significantly better fits to mineralization data for strain 17484 (Guerin & Boyd, 1992), and which leads to the same general conclusions regarding the availability of sorbed naphthalene to this organism.

The two-compartment model of NOC sorption kinetics proposed by Karickhoff (1980) predicts that as the equilibration time or the partition coefficient of a sorbate-sorbent system increases (e.g., due to decreased solubility of the sorbate or increased OC content of the sorbent), the distribution of sorbate between labile and nonlabile sorbed phases shifts toward the latter. Since the kinetics of mass transfer between the labile and nonlabile phases are much slower than the kinetics of mass transfer between the labile-sorbed phase and the aqueous phase, naphthalene in a nonlabile phase would be expected to be less readily available to potential microbial degraders. Here, we observed that naphthalene bound to high OC soils was less available to strain 17484 than that bound to low OC soils. We also observed that upon prolonged aging (up to 1 yr), naphthalene mineralization rates in Colwood soil slurries approached theoretical rates predicted on the basis of the equilibrium aqueous phase concentrations. In both instances (higher OC of sorbent or prolonged aging), mineralization rates decreased as more of the naphthalene was present in a nonlabile state.

The distinction between labile and nonlabile sorbed NOCs is a conceptual one based on empirical evidence of desorption hysteresis (Karickhoff, 1980). Similar evidence led to the formulation of physical models in which organic contaminant desorption from the interior of organo-mineral aggregates was retarded by sorption-desorption interactions along tortuous diffusion paths (Wu & Gschwend, 1986). Soil crushing was found to hasten the desorption process (Wu & Gschwend, 1986) indicating that the extent of desorptive non-equilibrium was related to the diffusion path length. Contaminant diffusion through a polymer-like soil organic matrix may also lead to retarded diffusion (Brusseau et al., 1991). The labile sorbed phase of Karickhoff (1980) may thus represent material sorbed in surface localized (and microbially accessible) sites, while the nonlabile sorbed phase may represent material bound more deeply within the soil particle or organic matter matrix.

Consistent with these models is the conclusion that strain 17484 has direct access to naphthalene present in a surface localized or labile-sorbed phase. By depleting the naphthalene in this phase, desorptive diffusion of internally bound or nonlabile sorbed naphthalene from soils is facilitated.

The basis for the strikingly different results obtained for the two organisms in this study is currently under investigation in our laboratory. Both organisms are motile, gram-negative rods, but differ in their propensities to attach to soils and in their chemotactic behaviors. These results suggest that generalizations regarding the bioavailability of soil-bound NOCs are inappropriate. A variety of physicochemical and microbiological factors may strongly influence the bioavailability of soil-sorbed NOCs.

REFERENCES

Abdul, A.S., T.L. Gibson, and D.N. Rai. 1987. Statistical correlations for predicting the partition coefficient for nonpolar organic contaminants between aquifer organic carbon and water. Hazard. Waste Hazard. Mater. 4:211–222.

Blakemore, R.P., D. Maratea, and R.S. Wolfe. 1979. Isolation and pure culture of a freshwater magnetotactic spirillum in chemically defined medium. J. Bacteriol. 140:720–729.

Brusseau, M.L., R.E. Jessup, and P.S.C. Rao. 1991. Nonequilibrium sorption of organic chemicals: elucidation of rate-limiting processes. Environ. Sci. Technol. 25:134–142.

Grigal, D.F. 1973. Note on the hydrometer method of particle-size analysis. Minnesota Forestry Res. Notes. No. 245. Univ. of Minnesota, St. Paul.

Guerin, W.F., and S.A. Boyd. 1992. Differential bioavailability of soil-sorbed naphthalene to two bacterial species. Appl. Environ. Microbiol. 58:1142–1152.

Guerin, W.F., and G.E. Jones. 1988. Two-stage mineralization of phenanthrene by estuarine enrichment cultures. Appl. Environ. Microbiol. 54:929–936.

Karickhoff, S.W. 1980. Sorption kinetics of hydrophobic pollutants in natural sediments. p. 193–205. *In* R.A. Baker (ed.) Contaminants and sediments. Vol. 2. Ann Arbor Sci. Publ., Ann Arbor, MI.

Larson, R.J. 1980. Environmental extrapolation of biotransformation data. Role of biodegradation kinetics in predicting environmental fate. p. 67–86. *In* A.Q. Maki, et al. (ed.) Biotransformation and fate of chemicals in the aquatic environment. Am. Soc. Microbiol., Washington, DC.

May, W.E., S.P. Wasik, and D.H. Freeman. 1978. Determination of the solubility behavior of some polycyclic aromatic hydrocarbons in water. Anal. Chem. 50:997–1000.

Ogram, A.V., R.E. Jessup, L.T. Ou, and P.S.C. Rao. 1985. Effects of sorption on biological degradation rates of (2,4-dichlorophenoxy) acetic acid in soils. Appl. Environ. Microbiol. 49:582–587.

Rhoades, J.D. 1982. Cation exchange capacity. p. 149–157. *In* A.L. Page (ed.) Methods of soil analysis. Part 2. 2nd ed. Agron. Monogr. 9. ASA and SSSA, Madison, WI.

Shelton, D.R., and J.M. Tiedje. 1984. Isolation and partial characterization of bacteria in an anaerobic consortium that mineralizes 3-chlorobenzoic acid. Appl. Environ. Microbiol. 48:840–848.

Steen, W.C., D.F. Paris, and G.L. Baughman. 1980. Effects of sediment sorption on microbial degradation of toxic substances. p. 477–482. *In* R.A. Baker (ed.) Contaminants and sediments. Vol. 1. Ann Arbor Sci. Publ., Ann Arbor, MI.

van Loosdrecht, M.C.M., J. Lyklema, W. Norde, and A.J.B. Zehnder. 1990. Influence of interfaces on microbial activity. Microbiol. Rev. 54:75–87.

Wolin, E.A., M.J. Wolin, and R.S. Wolfe. 1963. Formation of methane by bacterial extracts. J. Biol. Chem. 238:2882–2886.

Wu, S.-C., and P.M. Gschwend. 1986. Sorption kinetics of hydrophobic organic compounds to natural sediments and soils. Environ. Sci. Technol. 20:717–725.

12 Utility of Sorption and Degradation Parameters from the Literature for Site-Specific Pesticide Impact Assessments

Richard E. Green, R. C. Schneider, and Robert T. Gavenda

University of Hawaii
Honolulu, Hawaii

Carl J. Miles

Toxikon Environmental Sciences
Jupiter, Florida

Assessments of the impact on groundwater quality of pesticide use in agriculture, on golf courses, parks, and roadsides require data on pesticide behavior in soils. Sorption partition coefficients and half-lives (or the associated first-order rate coefficients) are most frequently needed. It is common practice to use data from the literature, usually mean values calculated from a database (e.g., Rao & Davidson, 1980) or best estimates selected by experienced personnel, such as are provided by Wauchope et al. (1992). It is widely recognized that mean values of such parameters obtained from diverse soil environments are usually associated with high standard deviation (SD), as illustrated by the tabulated data by Rao and Davidson (1980) and more recently by Oshiro et al. (1993). Coefficients of variation (CV) exceeding 100% are not unusual for both sorption coefficients (K_d or K_{oc}) and degradation half-life ($t_{1/2}$). The parameter K_{oc} is derived by K_d/f_{oc}, where K_d is the linear partition coefficient for sorption from solution and f_{oc} is the fraction of organic C (OC) in the soil.

The potential for pesticides to leach to groundwater has been assessed with simple empirical leaching indices (e.g., Gustafson, 1989), simplified models that incorporate principal processes without dynamic simulation (e.g., Rao et al., 1985; Jury et al., 1987), and dynamic models of chemical transport, such as CMLS, PRZM, GLEAMS, or LEACHMP (Pennell et al., 1990). All of these tools are finding increasing use in a variety of practical assessments of pesticide leaching. All are more appropriate for ranking of pesticides for given situations or evaluation of the relative impact of alternative management practices rather than providing precise predictions for a given pesticide in site-specific situations. There is, however, an increasing demand for site-specific evaluations to aid regulators and land use planners. Despite the limitations of models in describing actual processes in real-world situations, the urgent need for quantitative impact assessments presses a variety of models into uses that are questionable.

Pesticide-leaching assessments based on leaching indices or models are subject to uncertainty from three sources: model error, parameter error, and input data error (Loague & Green, 1991). In this chapter, we will be focusing on the error in sorption and degradation parameters that are usually taken from the published literature and often used for site-specific assessments. The impact of uncertainty in parameters and input data has been addressed by Loague et al. (1989, 1990), using the attenuation factor (AF) of Rao et al. (1985) and soil and climatic input data for Hawaii. This work demonstrated that the SD of AF (a measure of uncertainty in AF) was, in some cases, more than an order of magnitude larger than AF itself (Loague et al., 1990). First-order uncertainty analysis revealed that the major sources of uncertainty were OC content of the soil, water recharge rate, and the two chemical property parameters, K_{oc} and $t_{1/2}$. A principal benefit derived from the characterization of components of uncertainty in a pesticide-leaching assessment is the potential for reducing uncertainty by obtaining more precise estimates of those model parameters and input data having the greatest uncertainty. Reliable estimates of soil and climatic data are clearly important for site-specific assessments. Improvement in the accuracy and precision of soil OC data can be achieved, in some cases, by additional sampling (Yost et al., 1993), but it is more difficult to improve water recharge estimates due to high temporal and spatial variation of precipitation and evaporation. Pesticide property data from the literature may be both inaccurate and imprecise for site-specific applications because the pesticide-soil interactions for an individual soil are not well represented by the population of soils included in studies from which published data originated. In addition, experimental conditions for published studies may be inappropriate for site-specific applications. These deficiencies in published chemical parameters can be circumvented if it is possible to either measure or otherwise obtain better estimates of sorption and degradation parameters for the area or site of interest.

The objective of this work was to evaluate the appropriateness of sorption and degradation parameters from the literature for four locations in Hawaii. Four field sites, having different soils, were selected for field-leaching measurements and characterization of key model inputs. Sorption and degradation were measured for five pesticides in surface and subsoil samples from each site; the measured data were then compared with data obtained from the literature by Oshiro et al. (1993). Measured data were also evaluated for conformity to standard linearity assumptions required in the use of K_{oc} as a sorption parameter and $t_{1/2}$ as a degradation parameter for environmental modeling.

METHODOLOGY

Pesticides Used

The pesticides selected are all important for pest control in Hawaii: three herbicides with a long history of use in sugarcane (*Saccharum officinarum* L.) production [hexazinone (3-cyclohexyl-6-(dimethylamino)-1-methyl-1,3,5-triazine-2,4-(1H,3H)-dione), atrazine (2-chloro-4-ethylamino-6-isopropylamino-*s*-triazine), and ametryn (2-ethylamino-4-isopropylamino-6-methyl-thio-*s*-triazine)]; a nematicide used in pineapple [*Ananas comosus* (L.) Merr.] production [fenamiphos (ethyl 3-methyl-4-(methylthio)phenyl 1-methylethylphosphoramidate)]; and an insecticide that is widely used in agriculture and turf for soil and plant insect control and also for ant and ground termite control in urban areas [chlorpyrifos (0,0-diethyl 0-(3,5,6-trichloro-2-pyridyl)-phosphorothioate)]. These pesticides have a wide range of solubilities, sorption characteristics, and degradation rates as summarized by Oshiro et al. (1993), providing a good set of chemicals with which to evaluate the adequacy of literature values for environmental assessments. Atrazine and ametryn are weak bases that are capable of protonation to ionic form at low pH; the pK_b values are 12.3 for atrazine and 10.1 for ametryn (Wauchope et al., 1992).

Soils, Sampling, and Preparation of Samples

Soil samples from depth increments of 0 to 20 cm and 60 to 80 cm were taken at six sites in each of the four experimental locations, three on the island of Oahu and one (Hilo series) on the island of Hawaii. The soil series and great group names of the four soils are presented in Table 12–1. Detailed

Table 12–1. Characteristics of soils used for sorption and degradation measurements.

Soil series	Great group†	Depth	Organic C Mean	Organic C SD	pH
		cm	— % —		
Molokai silty clay loam	Torrox	0–20	1.30	0.12	6.7
		60–80	0.55	0.07	6.8
Wahiawa silty clay	Eutrustox	0–20	1.63	0.08	6.4
		60–80	0.76	0.25	5.7
Kaneohe silty clay variant	Tropohumult	0–20	2.88	0.17	5.3
		60–80	1.51	0.35	5.3
Hilo silty clay loam	Hydrandept	0–20	6.22	0.38	5.3
		60–80	3.19	0.25	5.6

† Recently revised but not yet published by Soil Conservation Service as follows: Molokai: Eutrotorrox; Kaneohe variant: Acrudox; Hilo: Hydrudand.

soil profile descriptions for the Oahu soils are given by Foote et al. (1972); the Hilo soil was described by Sato et al. (1973). All four sites have a history of cultivation. The Hilo soil location was in grass cover for about 15 yr prior to tillage of the soil for the pesticide leaching study associated with the work reported here. All of the soils have deep, well-drained profiles.

Soil samples from 0 to 20 cm and 60 to 80 cm at each location were used to represent Ap (surface soil) and B2 (subsoil) horizons, respectively. The samples were mixed well and passed through a 4-mm sieve at field water content or after controlled drying at 22 °C if the samples from the field were too wet for sieving. Soil pH was measured in a 1:1 soil/distilled water mixture. Soil OC content was determined by combustion and thermal conductivity using a LECO WR-112 Carbon Determinator.[1]

Three of the locations had a history of pesticide use within 2 yr of sampling time. All samples were analyzed for residues of the five pesticides of interest. Soil samples (50 g, moist) were extracted twice with ethyl acetate using a wrist action shaker; the extracts were dried with anhydrous sodium sulfate, rotoevaporated to dryness, and rewetted with acetone. Soil extracts were analyzed using a Hewlett Packard 5890A-GC with nitrogen-phosphorus detector (NPD), split injection (1:20) and DB-1701 (J. & W. Scientific) capillary column (7 m × 0.25 mm, film thickness 25 μm). Analysis conditions used to separate the five pesticides and the two fenamiphos metabolites (fenamiphos sulfoxide and fenamiphos sulfone) were: injector temperature, 270 °C, a temperature program, 170 °C for 3 min, a rate of 40 °C/min to 240 °C, hold time 2 min. Analysis for the five pesticides revealed trace amounts (<15 μg/kg) of atrazine and ametryn in the surface layer of the Molokai soil, 6 to 56 μg/kg chlorpyrifos in the surface layer of the Wahiawa soil, and in the surface layer of the Kaneohe soil, 20 to 61 μg/kg ametryn and 13 to 50 μg/kg chlorpyrifos. The subsoil samples from all sites were free of residual pesticides.

Sorption Measurements

Sorption of the pesticides from aqueous solution was determined at ambient laboratory temperature (22 ± 2 °C) in a batch suspension with a solution/soil ratio of 5:1 and 4-h equilibration (Lee et al., 1986). Two different sets of measurements were made: a sorption isotherm determination with four samples from each location-depth combination, three solution concentrations (0.15, 1.0, and 2.0 μg/mL for chlorpyrifos and 1, 5, and 25 μg/mL for the other chemicals), and duplicate laboratory equilibrations; the second set, for determination of K_{oc} at a single concentration (1.0 μg/mL) for each pesticide, employed six samples for each location-depth combination and duplicate laboratory equilibrations. The results from the second set were used for comparison with literature K_{oc} values because the larger number of samples provided a better estimate of the SD for each pesticide-soil combi-

[1] Mention of trade names is solely for the information of the reader and does not constitute any endorsement by the Univ. of Hawaii over comparable products that are not named.

nation. Isotherm measurements from the first set allowed an evaluation of the assumption of sorption linearity that is inherent in the computation of K_{oc} (Green & Karickhoff, 1990).

Analysis for the pesticides in solution was conducted by radioassay for ^{14}C with radiolabeled atrazine, fenamiphos, and chlorpyrifos and by GC for hexazinone and ametryn. Radiolabeled pesticide solutions (activity, 4×10^4 dpm/mL) were supplemented with analytical pesticide standards to bring the concentrations up to the desired values. Aqueous samples were diluted with Optifluor and counted in a Packard Tri-Carb 4430 Liquid Scintillation Counter using an external standard and quench correction. For the unlabeled pesticides, ametryn and hexazinone, aqueous solutions were extracted three times with methylene chloride (Miles et al., 1990). Extracts were analyzed by GC-NPD as described earlier for the soil extracts.

Degradation Measurements

The same soil samples that were used for sorption measurements, as described above, were prepared for degradation experiments within 8 to 12 d after field sampling. Pesticide degradation measurements were conducted as described by Lee (1987), with only minor modification. Soil water contents were adjusted to 33 kPa tension prior to incubation with pesticides. Moist soils equivalent to 16 g oven-dry (OD) soil (with the exception of the Hilo soil, for which 20 g moist soil were used) were placed on aluminum foil and dosed individually with 1 mL of an aqueous solution containing all five pesticides. For hexazinone, ametryn, and fenamiphos the application corresponded to 5 μg/g OD soil for surface soils and 0.5 μg/g for subsoils, with the exception of the Hilo soil (4 μg/g moist surface soil and 0.4 μg/g moist subsoil). Applications to the Hilo soil were not based on OD soil quantities because the soil is naturally highly hydrated; the total amount applied per incubation tube was comparable to the other soils. Atrazine and chlorpyrifos applications to all soils were lower due to their restricted aqueous solubilities; for these two chemicals, saturated solutions were applied in the same volume as the other pesticides, and the application quantity was inferred from the zero-time extraction: 2.5 μg/g for atrazine and 3.5 μg/g for chlorpyrifos in surface soil, and 0.5 and 0.35 μg/g, respectively, in subsoil. After mixing, the treated soils were transferred to culture tubes (2×15 cm) and capped with a permeable membrane cap (Kimble) that permits exchange of oxygen and carbon dioxide while restricting loss of water vapor. Treated samples were analyzed by solvent extraction and gas chromatography at 0, 7, 21, and 42 d after incubation at 25 °C in a constant temperature incubator. Replication included the six replicates of field samples plus an additional duplication of two samples for each sampling depth to provide an estimate of laboratory measurement error, or a total of 16 units for each soil at each time. The results were fit by a first-order kinetic equation, using the extracted quantity of a pesticide at zero time to establish the initial concentration in the soil. Background levels of residual pesticides that existed in some samples (discussed earlier) were subtracted.

RESULTS

Soil Properties

The OC content and pH of the four soils are given in Table 12–1. Organic carbon is the principal soil component affecting sorption of nonpolar organic chemicals, and pH is known to affect sorption of ionizable chemicals, such as the s-triazine herbicides (Green & Karickhoff, 1990). pH also affects hydrolysis of pesticides such as fenamiphos and chlorpyrifos. The soils are arranged in the table from lowest OC to highest OC, with a range of 1.3 to 6.2% in the surface soils and a range of 0.55 to 3.2% in the subsoils. In general, the surface soil OC is about twice that of the subsoil. The SD for OC provides a measure of the variability existing in the sampled field plot area (about 25 m^2). Corresponding CV values range from 6 to 33%, with a mean of 13%. This magnitude of variability falls in the low variation to medium variation grouping of Warrick and Nielsen (1980) for soil physical properties having relevance to water and chemical movement in soils.

Mean soil pH values were slightly higher for the Torrox and Eutrustox soils than for the Tropohumult and Hydrandept. As anticipated, the lower pH values occur in soils that are in higher rainfall areas.

Measured Sorption and Degradation

Sorption

Although sorption was measured at only three concentrations (0.15, 1.0, and 2.0 μg/mL for chlorpyrifos and 1, 5, and 25 μg/mL for the other chemicals), the spread in concentration was sufficient to provide a basis for evaluating the linearity of isotherms. Also, the concentration range was appropriate considering normal expected field applications of 2 to 10 kg/ha. With the exception of chlorpyrifos, pesticide sorption was relatively linear on all soils. Measured K_d values and associated coefficients of determination for linear regression (r^2) for each soil-pesticide combination are shown in Table 12–2. Some low r^2 values for subsoils resulted from low sorption, so that isotherm linearity in these cases is difficult to establish. A typically good fit for surface soils is exemplified by atrazine sorption on the Eutrustox soil in Fig. 12–1 ($r^2 = 0.99$), while a poor fit is indicated by chlorpyrifos sorption on the Eutrustox in the lower graph of Fig. 12–1 ($r^2 = 0.85$).

The other assumption inherent in the application of the K_{oc} approach is a linear relationship between K_d and OC. Conformity of our measured data to linearity is indicated by a plot of K_d vs. fraction OC for all soils with each pesticide. The best fit, based on high r^2 and low standard error of regression, was obtained for atrazine (Fig. 12–2, upper graph: $r^2 = 0.86$) and the poorest for hexazinone (lower graph: $r^2 = 0.43$). The r^2 values for ametryn, fenamiphos, and chlorpyrifos are 0.55, 0.60, and 0.86, respectively.

The results discussed above should be kept in mind in the evaluation of subsequent sections where measured sorption results are compared with

Table 12-2. Sorption partition coefficient (K_d)† and coefficient of determination for linear regression (r^2) for five pesticides on four Hawaii soils.

Soil	Hexazinone		Atrazine		Ametryn		Fenamiphos		Chlorpyrifos	
	K_d	r^2	K_d	r^2	K_d	r^2	K_d	r^2	K_d	r^2
Surface soil (0–20 cm)										
Torrox	0.54	0.90	1.50	0.98	1.33	0.73	2.84	0.96	9.43	0.84
Eustrustox	0.42	0.64	1.12	0.99	1.35	0.94	1.78	0.98	10.4	0.85
Tropohumult	0.66	0.85	1.34	0.99	2.42	0.94	1.82	0.98	19.0	0.92
Hydrandept	2.31	0.99	3.08	0.99	5.12	0.99	5.19	0.99	49.1	0.82
Subsoil (60–80 cm)										
Torrox	0.17	0.73	0.44	0.99	0.08	0.32	0.49	0.97	0.67	−0.47
Eustustox	0.40	0.87	0.67	0.99	0.46	0.86	0.76	0.97	2.86	0.40
Tropohumult	0.82	0.84	0.70	0.99	1.22	0.88	0.84	0.98	4.31	0.78
Hydrandept	2.90	0.97	2.74	0.99	3.31	0.98	3.46	0.99	10.9	0.92

† K_d derived from equation $S = K_dC$, in which S is sorbed concentration (μg/g) and C is solution concentration (μg/mL). Initial solution concentrations were 1, 5, and 25 μg/mL, except for chlorpyrifos (0.2, 1, 2 μg/mL).

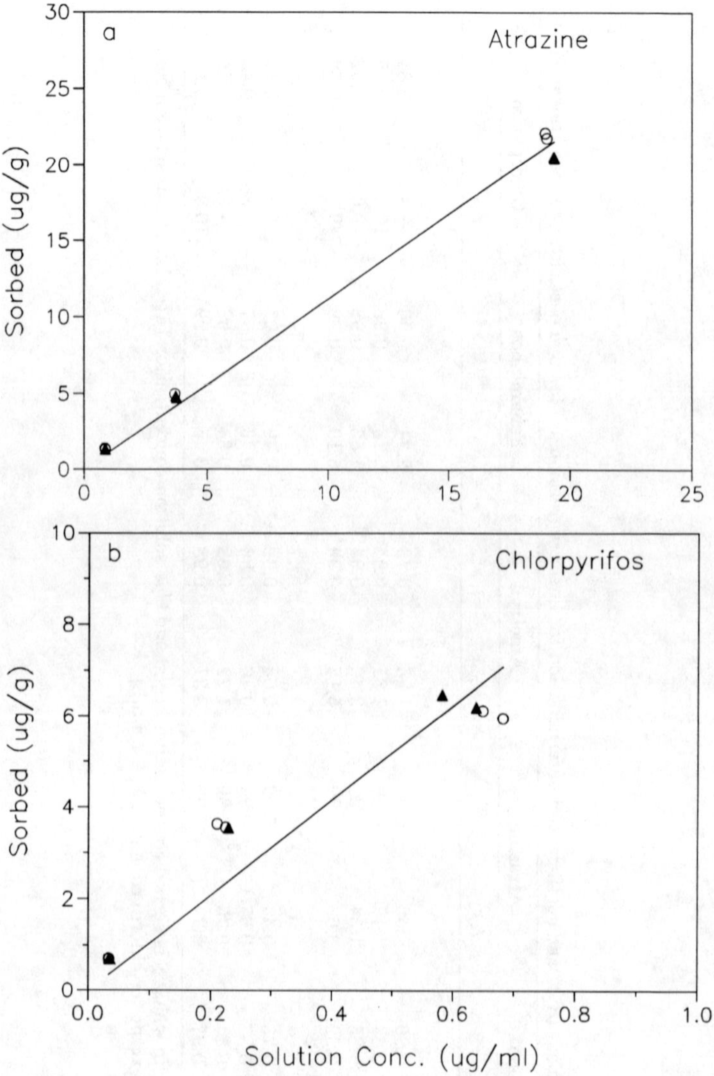

Fig. 12-1. Examples of linear sorption isotherms having reasonably good fit to the measured data (atrazine on Tropohumult soil, 0-20 cm) and poor fit (chlorpyrifos on the same soil). Different symbols represent two different sites within a field plot.

values from the literature. Chlorpyrifos and hexazinone do not meet the requirements for calculating K_{oc} values as well as do atrazine, ametryn, and fenamiphos.

Degradation

Calculation of half-lives or first-order rate coefficients for pesticide degradation is based on the assumption that the relationship between log concentration and time is linear. Visual examination of degradation plots and

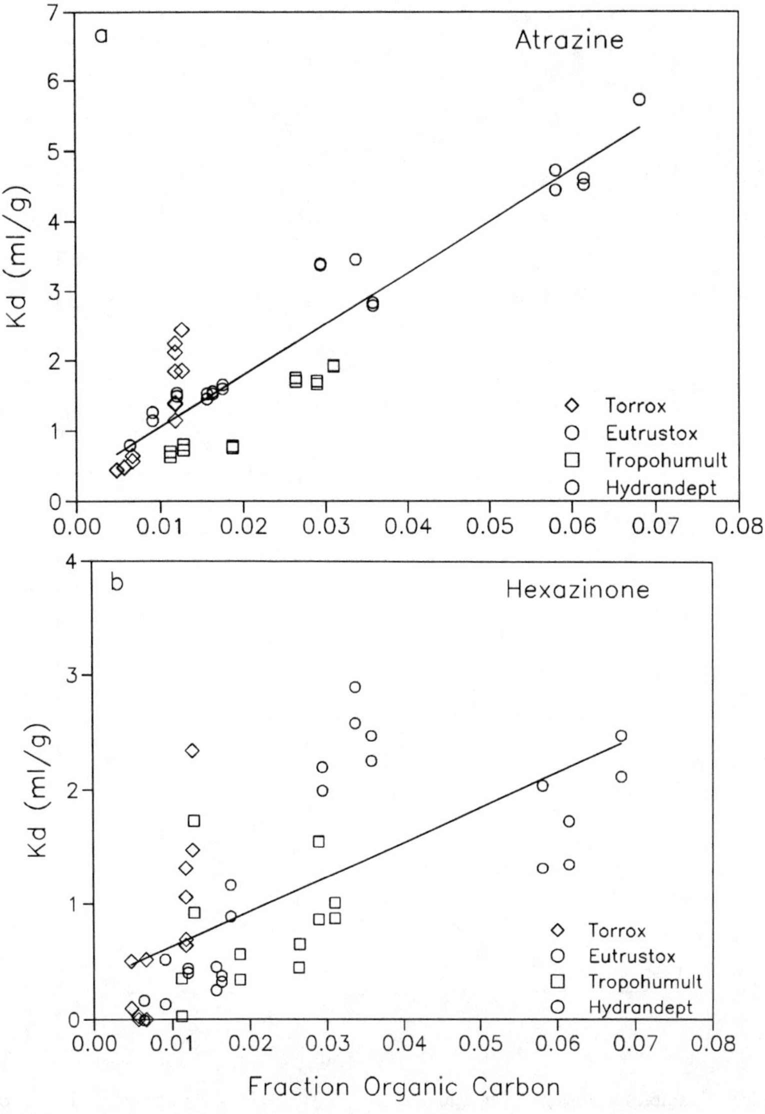

Fig. 12-2. Linear plot of K_d vs. f_{oc} for four Hawaii soils: examples of relatively good fit (atrazine, $r^2 = 0.86$) and poor fit (hexazinone, $r^2 = 0.43$). Multiple symbols for a given soil represent replicate field soil samples.

corresponding r^2 values suggested reasonable conformity to first-order kinetics. Examples of good and poor fits of the linear plot to the data are shown in Fig. 12-3; the upper graph for chlorpyrifos degradation in the Hydrandept surface soil ($r^2 = 0.97$) is typical of a reasonably good fit, while the lower graph for chlorpyrifos in the Hydrandept subsoil ($r^2 = 0.76$) illustrates a poor fit. Values of r^2 for all pesticide-soil combinations are given in Table 12-3, along with the intercept b for the linear equation $\ln(C/C_0)$

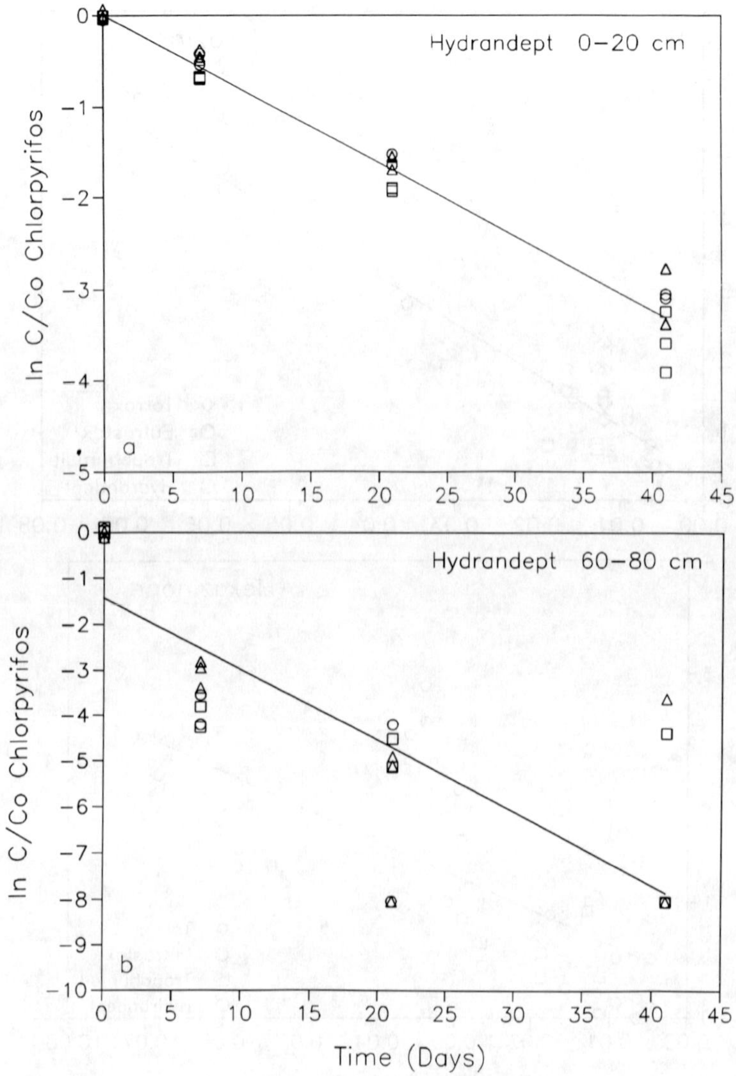

Fig. 12-3. Examples of linear fit for equation $\ln(C/C_0) = -kt + b$ to pesticide degradation data for Hawaii soils: (a) relatively good fit to first-order kinetics for chlorpyrifos in Hydrandept surface soil ($r^2 = 0.97$, intercept $b = 0$) and (b) poor fit for chlorpyrifos in the Hydrandept subsoil ($r^2 = 0.76$, intercept $b = -1.54$). Symbols represent replicate soil samples from four field sites.

$= -kt + b$, in which C/C_0 represents the fraction of chemical remaining at time t, k is the first-order rate coefficient, and b is the intercept. If b for a given pesticide-soil combination is a statistically significant non-zero number, the assumption of first-order kinetics can be questioned (Lee, 1987). Several of the fits represented in Table 12–3 have significant intercepts, including that shown in the lower graph of Fig. 12–3. Degradation was better represented by first-order kinetics in surface soil than in subsoil.

Table 12–3. Pesticide transformation/degradation on four Hawaii soils: degree of conformity to first-order kinetics as indicated by the coefficient of determination (r^2) and nonzero intercept.†

	Pesticide									
	Atrazine		Ametryn		Hexazinone		Chlorpyrifos		Fenamiphos	
Soil	r^2	Intercept	r^2	Intercept	r^2	Intercept	r^2	Intercept	r^2	Intercept
Torrox										
Surface	0.88	0	0.88	0	0.90	−0.07	0.88	−0.25	0.93	−0.34
Subsurface	0.55	0	0.73	0	0.71	0	0.84	−0.55	0.62	0
Eustrustox										
Surface	0.92	0.09	0.90	−0.12	0.90	0	0.96	0	0.87	−1.06
Subsurface	0.60	0	0.74	0	0.55	0	0.89	0	0.88	−1.02
Tropohumult										
Surface	0.96	0	0.96	0	0.78	0	0.98	0	0.53	−2.06
Subsurface	0.66	0	0.59	0	0.22	0	0.83	−0.65	0.52	−1.37
Hydrandept										
Surface	0.78	0	0.90	−0.38	0.90	0	0.97	0	0.57	−1.93
Subsurface	0.33	0	0.80	0	0.22	0	0.76	−1.54	0.77	−1.22

† Intercepts that are statistically different from zero on the ln (C/C_0) scale are shown; see examples for chlorpyrifos in Fig. 12–3.

Comparison of Measured and Literature Data

Sorption

Sorption (K_{oc}) data for the two depth intervals (0–20 cm and 60–80 cm) of the four soils are shown in Table 12–4. Means and SD for K_{oc} values from the literature (Oshiro et al., 1993) are compared with values collected in this study. The overall means and SD were determined from four soil means, thus providing a measure of the deviation of the four soil means relative to the overall mean for each chemical. For the surface soils, several observations are noteworthy:

1. Mean measured K_{oc} values for all soils except the Torrox are generally lower than literature mean values.
2. K_{oc} values for the Torrox are generally higher than for the other soils. This is especially apparent for ametryn; the Torrox value is too high to be included within the interval for literature values, while measured values for the other three soils are too low to fit the literature values. The high K_{oc} values on the Torrox soil may be the result of an accumulation of charcoal from the burning of sugarcane before harvest (a common practice in Hawaii); this response to charcoal additions was demonstrated previously for ametryn and atrazine (Yamane & Green, 1972).
3. With the exception of chlorpyrifos, the interval given by the mean ± 1 SD for the overall measured values overlaps that of the literature values.
4. Only 55% (11 out of 20) of the individual mean measured values for specific soil-pesticide combinations fall within the mean ± 1 SD interval for literature values.
5. The ranking of K_{oc} values from high to low values is the same for literature values and measured values: chlorpyrifos > ametryn > fenamiphos > atrazine > hexazinone.
6. The CV for literature values (not shown in the table) are on the order of 50%, while for measured values on individual soils they are considerably lower (generally <30%).

Subsoil K_{oc} values follow the same general pattern as surface soils, although the overall means are lower.

Degradation

Half-lives of literature and measured values for surface soils and subsoils are shown in Table 12–5. We note the following:

1. Measured half-lives for surface soils are generally about one-third the literature values.
2. Subsoil half-lives are not always longer than those for surface soils; on the contrary, fenamiphos and chlorpyrifos half-lives are shorter for subsoils in most cases.

Table 12-4. K_{oc} (mL/g) means and standard deviations (in parentheses) for five pesticides: comparison of literature values with measured values for four Hawaii soils.

	Hexazinone		Atrazine		Ametryn		Fenamiphos		Chlorpyrifos	
Literature†:	33	(18)	129	(84)	334	(122)	327	(249)	7240	(3471)
Measured—Surface soil (0–20 cm)										
Overall‡	49	(35)	96	(38)	257	(182)	179	(116)	922	(256)
Torrox	101	(44)	150	(36)	530	(156)	344	(44)	1010	(75)
Eustrustox	34	(20)	94	(2)	170	(36)	179	(10)	765	(61)
Tropohumult	31	(12)	62	(3)	156	(21)	94	(10)	673	(187)
Hydrandept	29	(6)	79	(5)	171	(29)	100	(5)	1241	(143)
Measured—Subsoil (60–80 cm)										
Overall	49	(19)	92	(30)	170	(86)	119	(58)	522	(317)
Torrox	32	(18)	89	(5)	235	(263)	88	(20)	259	(33)
Eustustox	33	(16)	127	(6)	254	(58)	202	(57)	939	(266)
Tropohumult	48	(49)	54	(10)	87	(59)	73	(10)	292	(54)
Hydrandept	73	(8)	99	(17)	106	(16)	114	(22)	600	(130)

† From Oshiro et al. (1993).

‡ Overall means and standard deviations are calculated from individual soil means.

Table 12-5. Degradation half-lives (days) and standard deviations (in parentheses) for five pesticides: comparison of literature values with measured values for four Hawaii soils.

	Hexazinone		Atrazine		Ametryn		Fenamiphos		Chlorpyrifos	
Literature†:	145	(57)	53	(26)	44	(22)	29	(15)	52	(31)
Measured—Surface soil (0–20 cm)										
Overall‡	44	(12)	18	(13)	15	(10)	8	(2)	16	(5)
Torrox	56	(19)§	14	(5)	12	(5)	11	(3)	17	(7)
Eustrustox	38	(13)	37	(11)	29	(10)	7	(3)	21	(5)
Tropohumult	52	(29)	16	(3)	12	(2)	9	(9)	15	(2)
Hydrandept	30	(10)	8	(4)	6	(6)	6	(6)	9	(1)
Measured—Subsoil (60–80 cm)										
Overall	206	(153)	38	(31)	25	(21)	7	(3)	7	(3)
Torrox	56	(19)	11	(10)	7	(5)	9	(7)	7	(3)
Eustustox	92	(87)	39	(32)	37	(22)	6	(2)	11	(4)
Tropohumult	346	(676)	81	(60)	50	(43)	10	(10)	6	(3)
Hydrandept	330	(622)	21	(31)	9	(5)	4	(2)	3	(2)

† From Oshiro et al. (1993).
‡ Overall means and standard deviations are calculated from individual soil means.
§ Standard deviation for each soil-pesticide combination is derived from regression standard error.

3. For surface soils, of the 20 pesticide-soil combinations, only 15% of the mean measured half-lives fall within the mean ± 1 SD interval for literature values. For subsoils, 25% of measured values correspond to literature values.
4. The ranking of chemicals with respect to half-life is the same for literature values and for overall measured values for the surface soil: hexazinone > atrazine > chlorpyrifos > ametryn > fenamiphos. Ametryn and chlorpyrifos switched positions for subsoils.

DISCUSSION

It is not surprising that published sorption and degradation parameters are highly variable because the soils or sediments used and the experimental conditions employed often differed between studies. Degradation half-lives are especially sensitive to substrate differences as well as to temperature and moisture conditions during incubations. Even so, Rao and Davidson (1980) found that the CV for published half-life data was <100% for most pesticides. Thus, they concluded that disappearance rates could be estimated within a factor of 2 to 4 for most pesticides using presently available data. The data summarized by Oshiro et al. (1993), which we have compared with measured data in this study, demonstrate a similar level of variation as those of Rao and Davidson (1980), probably because the database used by both summaries was similar. It is of interest that the half-lives that we measured for each pesticide-soil combination, that is for six sampling sites at a single location in Hawaii, have similar CV values as the estimates from published data, which encompass a wide range of soils and conditions. However, the overall mean measured half-life values for the Hawaii surface soils were about one-third the mean values from the literature. Thus, it would be risky to use literature values of pesticide degradation for a site-specific assessment of pesticide leaching in Hawaii.

Measured K_{oc} values for a given chemical obtained from different studies might be expected to be less sensitive to soil and environmental conditions than degradation half-life measurements. Several assumptions, however, are inherent in the computation and use of K_{oc} values (Green & Karickhoff, 1990): (i) equilibrium exists; (ii) isotherms are linear and singular; (iii) sorption occurs on OC only and not on clay minerals; (iv) all OC in soils has similar sorption characteristics (Green & Karickhoff, 1990). Violation of these assumptions results in errors that may be either positive or negative (Rao & Davidson, 1980; Green & Karickhoff, 1990). Chlorpyrifos and hexazinone sorption measurements did not conform well to some assumptions required for K_{oc} computation, and these two chemicals did have a relatively high SD for the individual sites. Additionally, the literature estimate of sorption for chlorpyrifos was on the order of 10 times greater than what was measured on Hawaii soils. This may be the result of chlorpyrifos sorption on smectite clays in addition to sorption on OC (Mingelgrin & Gerstl, 1983) in some of the published studies from which chlorpyrifos data were obtained; the Hawaii soils used in this study did not contain smectite clays. In summary, literature estimates of sorption were satisfactory for about half of the pesticide-soil combinations.

The overall ranking of pesticides with respect to both sorption and degradation was consistent between literature and measured estimates. This is encouraging in that literature estimates are often used to compare the environmental fate of various pesticides. The correct ranking of relative sorption and degradation behavior with published data, however, does not qualify such data for site-specific assessments of pesticide leaching.

ACKNOWLEDGMENT

This research was supported in part by the State of Hawaii Department of Agriculture. Stephanie Whalen, Hawaii Sugar Planters' Experiment Station, assisted in the selection of analytical methods and arrangements for field sites. George Ulep and Tony Pham assisted with field sampling and laboratory analysis.

REFERENCES

Foote, D.E., E.L. Hill, S. Nakamura, and F. Stephens. 1972. Soil Survey of the Islands of Kauai, Oahu, Maui, Molokai, and Lanai, State of Hawaii. Soil Conservation Service, USDA, and Univ. Hawaii Agric. Exp. St., Honolulu.

Green, R.E., and S.W. Karickhoff. 1990. Sorption estimates for modeling. p. 80–101. *In* H.H. Cheng (ed.) Pesticides in the soil environment: Processes, impacts and modeling. SSSA Book Ser. 2. SSSA, Madison, WI.

Gustafson, D. 1989. Ground water ubiquity score: A simple method for assessing pesticide leachability. J. Environ. Tox. Chem. 8:339–357.

Jury, W.A., D.D. Focht, and W.J. Farmer. 1987. Evaluation of pesticide groundwater pollution potential from stnadard indices of soil-chemical adsorption and biodegradation. J. Environ. Qual. 16:422–428.

Lee, C.-C. 1987. Sorption and degradation parameters for modeling nematicide fate in soil. Ph.D. diss. Univ. Hawaii, Honolulu.

Lee, C.-C., R.E. Green, and W.J. Apt. 1986. Transformation and adsorption of fenamiphos, f. sulfoxide and f. sulfone in Molokai soil and simulated movement with irrigation. J. Contam. Hydrol. 1:211–225.

Loague, K.M., and R.E. Green. 1991. Statistical and graphical methods for evaluating solute transport models: Overview and application. J. Contam. Hydrol. 7:51–73.

Loague, K.M., R.E. Green, T.W. Giambelluca, T.C. Liang, and R.S. Yost. 1990. Impact of uncertainty in soil, climatic, and chemical information in a pesticide leaching assessment. J. Contam. Hydrol. 5:171–194.

Loague, K.M., R.S. Yost, R.E. Green, and T.C. Liang. 1989. Uncertainty in a pesticide leaching assessment for Hawaii. J. Contam. Hydrol. 4:139–161.

Miles, C.J., K. Yanagihara, S. Ogata, G. van de Verg, and R. Boesch. 1990. Soil and water contamination at pesticide mixing and loading sites on Oahu, Hawaii. Bull. Environ. Contam. Toxicol. 44:955–962.

Mingelgrin, U., and Z. Gerstl. 1983. Reevaluation of partitioning as a mechanism of nonionic chemicals adsorption in soils. J. Environ. Qual. 12:1–11.

Oshiro, W.C., C.J. Miles, and R.E. Green. 1993. Physical-chemical properties of selected pesticides for environmental fate modeling. Hawaii Inst. Trop. Agric. and Human Resour. Res. Ser. 069.

Pennell, K.D., A.G. Hornsby, R.E. Jessup, and P.S.C. Rao. 1990. Evaluation of five simulation models for predicting aldicarb and bromide behavior under field conditions. Water Resour. Res. 26:2679–2693.

Rao, P.S.C., and J.M. Davidson. 1980. Estimation of pesticide retention and transformation parameters required in nonpoint source pollution models. p. 23–67. *In* M.R. Overcash and J.M. Davidson (ed.) Environmental impact of nonpoint source pollution. Ann Arbor Sci. Publ., Ann Arbor, MI.

Rao, P.S.C., A.G. Hornsby, and R.E. Jessup. 1985. Indices for ranking the potential for pesticide contamination of groundwater. Proc. Soil Crop Sci. Soc. Fla. 44:1–8.

Sato, H.H., W. Ikeda, R. Paeth, R. Smythe, and M.Jr. Takehiro. 1973. Soil survey of the island of Hawaii, State of Hawaii. Soil Conservation Service, USDA, and Univ. Hawaii Agric. Exp. Stn., Honolulu.

Warrick, A.W., and D.R. Nielsen. 1980. Spatial variability of soil physical properties in the field. p. 319–355. *In* D. Hillel (ed.) Applications of soil physics. Academic Press, New York.

Wauchope, R.D., T.M. Buttler, A.G. Hornsby, P.W.M. Augustijn Beckers, and J.P. Burt. 1992. The SCS/ARS/CES pesticide properties database for environmental decision-making. Rev. Environ. Contam. Toxicol. 123:1–164.

Yamane, V.K., and R.E. Green. 1972. Adsorption of ametryne and atrazine in Oxisol, montmorillonite, and charcoal in relation to pH and solubility effects. Soil Sci. Soc. Am. Proc. 36:58–64.

Yost, R.S., K. Loague, and R.E. Green. 1993. Improvement of soils data for pesticide leaching assessments: Sampling strategies based on geostatistics or soil taxonomy. Geoderma 57:247–262.

13 Enhanced Biodegradation of Soil Pesticides: Interactions Between Physicochemical Processes and Microbial Ecology

Allan S. Felsot

Illinois Natural History Survey
Champaign, Illinois

Daniel R. Shelton

USDA-ARS Pesticide Degradation Laboratory
Beltsville, Maryland

Microbial adaptation for the metabolism of extrinsic substrates through enzyme induction has long been studied by biochemists, but its occurrence among soil microorganisms that degrade pesticides had remained relatively obscure until the 1970s. The phenomenon was first observed in soil perfusion columns (Audus, 1949). The degradation of 2,4-D (see Table 13–1 for chemical name) was preceded by a lag period before it rapidly disappeared from the perfusate. Upon subsequent passage of perfusate containing 2,4-D, the lag period shortened, and 2,4-D was immediately degraded. Audus (1964) hypothesized that microbial adaptation to 2,4-D had occurred, perhaps via enzyme induction. Kirkland and Fryer (1966) later observed that prior treatment of soil in the field with phenoxyacetic herbicides enhanced the rate of loss of subsequent applications.

During the 1960s the general public was awakened to the problems of chemical persistence and biomagnification associated with the chlorinated hydrocarbon insecticides (Carson, 1962; Woodwell, 1967), but an evolution in use of pesticides was already occurring. Biodegradable and less persistent organophosphorus insecticides were replacing the very recalcitrant organochlorines. To be effective, however, soil-applied pesticides must be bioactive for several months. Thus, as a result of changes in soil pesticides use, the significance of microbial adaptation in agroecosystems was realized. In rice (*Oryza sativa*) paddy soils, Sethunathan (1971) noted that diazinon rapidly degraded after several previous applications. More importantly, control of the brown planthopper [*Nilaparvata lugens* (Stal)] failed. These observations

 Sorption and Degradation of Pesticides and Organic Chemicals in Soil. SSSA Special Publication no. 32.

Table 13–1. Pesticides known to condition soil for enhanced biodegradation.†.

Pesticide	Class
2,4-D	H
2,4-dichlorophenoxyacetic acid	
Alachlor	H
2-chloro-*N*-(2,6-diethylphenyl)-*N*-(methoxymethyl)acetamide	
Aldicarb	I
(2-methyl-2-(methylthio)propionaldehyde) O-methylcarbamoyl oxime	
Bendiocarb	I
2,2-dimethyl-1,3-benzodioxol-4-yl methylcarbamate	
Benfuracarb	I
ethyl *N*-[2,3-dihydro-2,2-dimethylbenzofuran-7-yloxycarbonyl(methyl) aminothio]-*N*-isopropyl-β-alaninate	
Benomyl	
methyl 1-(butylcarbamoyl)benzimidazol-2-ylcarbamate	
Butylate	H
S-ethyl bis(2-methylpropyl) carbamothioate	
Carbaryl	I
1-naphthyl methylcarbamate	
Carbofuran	I
2,3-dihydro-2,2-dimethyl-7-benzofuranyl methylcarbamate	
Carbosulfan	I
2,3-dihydro-2,2-dimethyl-7-benzofuran-7-yl (dibutylaminothio) methylcarbamate	
Chlorpropham	H
1-methylethyl(3-chlorophenyl)carbamate	
Cloethocarb	I
2-(2-chloro-1-methyoxyethoxy)phenyl methylcarbamate	
Dalapon	H
2,2-dichloropropanoic acid	
Diallate	H
S-(2,3-dichloro-2-propenyl) bis(1-methylethyl)carbamothioate	
Diazinon	I
O,O-diethyl *O*-(2-isopropyl 6-methyl-4-pyrimidinyl) phosphorothioate	
Diphenamid	H
N,N-dimethyl-2,2-diphenylacetamide	
Disulfoton sulfone	I
O,O-diethyl *S*-[2-(ethylsulfonyl)ethyl] phosphorodithioate	
Diuron	H
N'-(3,4-dichlorophenyl)-*N,N*-dimethylurea	
Endothall	H
7-oxabicyclo[2.2.1]heptane-2,3-dicarboxylic acid	
EPTC	H
S-ethyl dipropylcarbamothioate	
Ethoprop	I
O-ethyl *S,S*-dipropyl phosphorodithioate	
Fenamiphos	I
ethyl 4-(methylthio)-m-tolyl isopropylphosphoramidate	
Fensulfothion	I
O,O-diethyl *O*-[4-(methylsulfinyl)phenyl] phosphorothioate	
Fonofos	I
O-ethyl *S*-phenyl ethylphosphonodithioate	
Furathiocarb	I
butyl 2,3-dihydro-2,2-dimethylbenzofuran-7-yl *N,N*-dimethyl-*N,N*-thiodicarbamate	

(continued on next page)

Table 13-1. Continued.

Pesticide	Class
Iprodione 3-(3,5-dichlorophenyl)-*N*-(1-methylethyl)-2,4-dioxo-1-imidazolidinecarboxamide	F
Isofenphos 1-methylethyl 2-[[ethoxy[(1-methylethyl)amino]phosphinothioyl]oxy] benzoate	I
Isopropcarb 2-isopropylphenyl methylcarbamate	I
Linuron *N'*-(3,4-dichlorophenyl)-*N*-methoxy-*N*-methylurea	H
MCPA 4-chloro-2-methylphenoxyacetic acid	H
Mephosfolan 2-(diethoxyphosphinylimino)-4-methyl-1,3-dithiolane	I
Metalaxyl methyl *N*-(2-methoxyacetyl)-*N*-(2,6-xylyl)-DL-alaninate	F
Methomyl methyl *N*-[[methylaminocarbonyl]oxy]ethanimidothioate	I
Napropamide *N,N*-diethyl-2-(1-naphthalenyloxy)propanamide	H
Oxamyl methyl 2-(dimethylamino)-*N*-[[(methylamino)carbonyl]oxy]-2-oxoethanimidothioate	I
Parathion *O,O*-diethyl *O*-(4-nitrophenyl) phosphorothioate	I
Phorate sulfone *O,O*-diethyl *S*-[ethylsulfonyl)methyl] phosphorodithioate	I
Pronamide 3,5-dichloro-*N*-(1-dimethyl-2-propynyl)benzamide	H
Propham 1-methylethyl phenylcarbamate	H
Propoxur 2-isopropoxyphenyl methylcarbamate	I
Pyrazon 5-amino-4-chloro-2-phenyl-3(2H)-pyridazinone	H
Terbufos sulfoxide *S*-[[(1,1-dimethylethyl)sulfinyl]methyl] *O,O*-diethyl phosphorodithioate	I
Triallate *S*-(2,3,3-trichloro-2-propenyl) bis(1-methylethyl)carbamothioate	H
Trimethacarb 3,4,5- & 2,3,5-trimethylphenyl methylcarbamate	I
Vernolate *S*-propyl dipropylcarbamothioate	H
Vinclozolin 3-(3,5-dichlorophenyl)-5-ethenyl-5-methyl-2,4-oxazolidinedione	F

† F = fungicide; H = herbicide; I = Insecticide.

were the first to show the practical effect of microbial adaptations in soil, but the phenomenon continued to remain obscure until EPTC and carbofuran, used widely in the midwestern Corn Belt, did not work well after years of continuous use (Felsot, 1991).

Today, microbial adaptation for pesticide metabolism is commonly known as enhanced biodegradation (Kaufman & Edwards, 1983). Enhanced biodegradation can be defined as the accelerated degradation of a pesticide

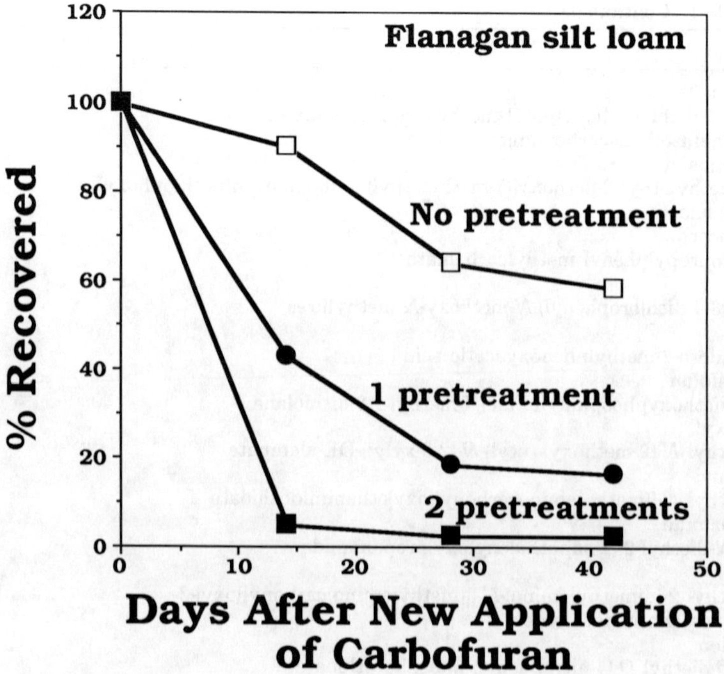

Fig. 13-1. Degradation of technical carbofuran in Flanagan silt loam conditioned by one or two previous applications of granular formulated carbofuran (Furadan 15G) (Felsot, 1986).

after repeated application to soil or water (Fig. 13-1). Development of the phenomenon is thought to require at least a microbial population or community capable of using the pesticide as a C or nutrient source (Fig. 13-2). Metabolism of the pesticide may occur by inductive or constitutive enzyme systems, but microbial populations have been hypothesized to proliferate in either case. Expression of enhanced biodegradation through increased microbial populations or metabolic activity would depend on favorable environmental conditions.

Enhanced biodegradation is the consequence of natural processes of metabolic adaptation, but its development is a double-edged sword. It has adversely affected control of soil pests across the world, but metabolic adaptations in general are also major mechanisms for degrading, detoxifying, or assimilating pollution. Many types of pesticides have been reported capable of conditioning soil for enhanced biodegradation (Table 13-1), and the phenomenon has been studied from ecological, biochemical, and genetic perspectives (Felsot, 1989). This chapter extends previous work (e.g., Racke & Coats, 1990) by reviewing some of the idiosyncracies of enhanced biodegradation of pesticides and exploring the interactions between microbial ecology and physicochemical processes that seem to affect the development of the phenomenon.

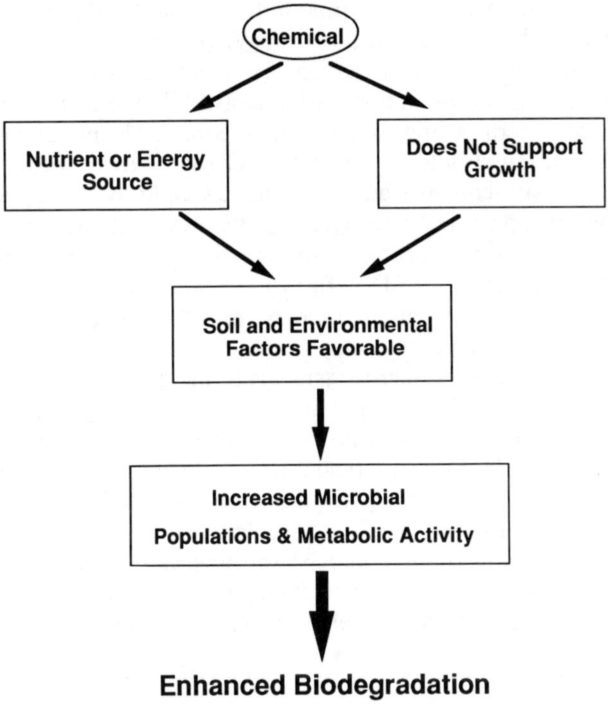

Fig. 13-2. Conceptual model of two pathways leading to enhanced biodegradation of pesticides. If the chemical serves as an energy or nutrient source and environmental conditions are favorable, degradation can be enhanced and the effect is comparatively stable in the absence of further pesticide inputs. If microbial populations are stimulated by addition of extraneous nutrients or physical alteration of the environment, then degradation can also be enhanced, but the effect only lasts as long as favorable conditions are present.

Ecological Perspective of Enhanced Biodegradation

Microorganisms exhibit two ecological strategies towards substrate assimilation or metabolism: mineralization and cometabolism (Alexander, 1981). In the mineralization strategy, the absorbed substrate is broken down into smaller molecules, which are further metabolized via energy-generating pathways. In this case, the biomass of the population increases at the expense of the substrate. The rate of change in substrate concentration in the medium increases coincidentally with the expanding microbial population. As the substrate concentration in the medium declines to a low level, growth of the microbial population reaches a plateau at a higher cell density. An excellent illustration of this hypothesis is the rapid disappearance in water of paranitrophenol (PNP) (a hydrolytic metabolite of parathion) coincident with increased numbers of PNP-mineralizing bacteria (Spain et al., 1984).

Enhanced biodegradation of compounds like 2,4-D and parathion seems best explained by the mineralization strategy. If enhanced biodegradation is due to microorganisms that derive energy or nutrients from metabolism of the pesticide, then microbial population density should increase when the pesticide is added to soil, water, or a culture medium. The metabolic capacity for biodegradation should be much greater than before the addition of the pesticide, and subsequently added pesticide should then disappear from the medium at an augmented rate. The population density of specific 2,4-D degrading microorganisms was measured in soil treated twice in succession with 2,4-D (Fournier et al., 1981). After the second treatment, the lag period decreased coincidentally with an increase in the population of herbicide-specific degraders. The lag period may simply be the time required for small populations of mineralizing microorganisms to increase to the cell densities necessary to cause detectable changes in the substrate (Wiggins et al., 1987).

In contrast to mineralization, no net energy is derived during cometabolism (Alexander, 1981). Many pesticides are incidentally transformed in soil by metabolic pathways containing enzymes with broad substrate specificity; the net result is a degradation product that is recalcitrant to further degradation or is bound to the organic matrix. In some cases, however, transformation products produced by one species may be used as a C source by other species (Alexander, 1979). Nevertheless, when the parent pesticide cannot

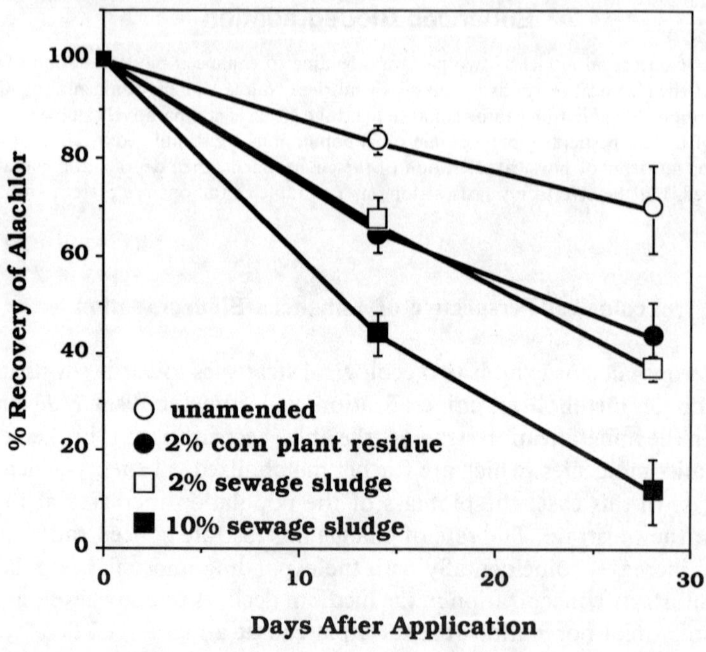

Fig. 13-3. Loss of alachlor from a mixture of Ipava silt loam (fine, montmorillonitic, mesic Aquic Argiudolls) and Sable silty clay loam (fine-silty, mixed, mesic Typic Haplaquolls) after amendment with 2% w/w ground corn plant residue or 2% w/w and 10% w/w digested municipal sewage sludge (Dzantor & Felsot, 1992). Initial concentration of alachlor was 100 mg kg^{-1}, and the soil was incubated in the dark at 25 °C.

be used as a growth substrate, the microbial population transforming it does not expand, and the rate of change of pesticide concentration in the medium remains relatively constant.

Under certain conditions, however, cometabolism can lead to a temporary augmentation of biodegradation rate (Fig. 13–2). For example, adding organic amendments like plant residue or sewage sludge to soil can enhance the degradation of cometabolically biodegradable compounds like alachlor (Felsot & Dzantor, 1990; Dzantor & Felsot, 1992) (Fig. 13–3). Thus, physical or chemical conditions favorable to microbial activity can result in the appearance of enhanced biodegradation. Under these circumstances biodegradation would seem rapid only as long as favorable conditions existed, but the capability to mineralize pesticides would afford a much longer term enhancement of biodegradation.

Several studies have shown that enhanced biodegradation resulting solely from previous pesticide applications depends on the adequacy of certain physical requirements like soil moisture and soil type (Chapman et al., 1986a,b; Dzantor & Felsot, 1990). These observations imply that physicochemical interactions of the pesticide with the soil can influence the expression of microbial adaptation in the same manner that conditions favorable to general microbial activity can influence degradation rate of cometabolized compounds.

Idiosyncrasies of Enhanced Biodegradation

Development of Enhanced Biodegradation May or May Not Be Associated with Changes in Population Dynamics

Perhaps the most controversial characteristic of enhanced biodegradation is the inability to definitively conclude from available data that populations of specific pesticide degraders do increase in response to utilization of the pesticide as a C or nutrient source. The most probable number (MPN) method in conjunction with monitoring either the disappearance of pesticide as the sole C source or the appearance of products has been the experimental technique most commonly used to monitor the size of pesticide-degrading populations. Moorman (1990) has reviewed some of the studies on dynamics of pesticide-degrader populations. In summary, studies on 2,4-D (Fournier et al., 1981), parathion (Sudhakar-Barik et al., 1979; Spain et al., 1980), isofenphos (Racke & Coats, 1987), and carbofuran (Hendry & Richardson, 1988; Dzantor & Felsot, 1989) have shown significantly larger populations of pesticide degraders after preincubation of soil with the pesticide in question. Although populations of EPTC-degrading bacteria and actinomycetes were as equally abundant in EPTC-conditioned soil as in nonconditioned soil, butylate- and vernolate-degrading microorganisms were more abundant in a vernolate- and butylate-pretreated soil than in the same soil without carbamothioate herbicide treatment (Mueller et al., 1989).

Measurable changes in numbers of specific pesticide-degrading microorganisms after pretreatment of soil with a mineralizable pesticide may depend upon initial application rate. When Moorman (1988) exposed soils to high rates of EPTC (60 vs. 6 mg kg^{-1}), he observed an increase in EPTC degraders. Population numbers have also varied in response to low and high 2,4-D concentrations (Fournier et al., 1981).

Recently, several studies have questioned whether microbial adaptation resulting in enhanced biodegradation results from increases in microbial numbers or from increases in activity of enzymes capable of degrading the pesticide. Using kinetic models to analyze MPN analyses associated with carbofuran mineralization (Merica & Alexander, 1990), Scow et al. (1990) did not detect significant changes in carbofuran-degrading populations in soils preexposed to 0.1 or 50 mg kg^{-1} carbofuran. The kinetic models best fitting the data were not consistent with utilization of carbofuran as a growth substrate. In the specific soil studied (Kendaia clay loam), populations of carbofuran degraders seemed high ($>10^8$ cells kg^{-1} soil) at the start of the experiments (Merica & Alexander, 1990), but soils with an initially low abundance of specific degraders may have responded with population increases. For example, in studies that reported increases in numbers of carbofuran degraders, initial population densities were on the order of 10^6 cells kg^{-1} soil (Hendry & Richardson, 1988; Dzantor & Felsot, 1989).

The MPN methods may be too insensitive to resolve small changes in population density. If enzymatic activity per cell is assumed to be constant, then a doubling of degradation rate would be associated with only a doubling of pesticide-degrading populations. This change is simply too small for hypothesis testing when the 95% confidence interval of the MPN method may span almost an order of magnitude.

Although bacteria that can grow on carbofuran as a sole C or N source have been isolated (Karns et al., 1986; Chaudry & Ali, 1988), Scow et al. (1990) concluded that carbofuran was cometabolized in soil with a concomitant increase in metabolic activity per cell upon preexposure to carbofuran rather than a substantial increase in numbers of carbofuran degraders. In support of this hypothesis, Dzantor and Felsot (1989) found several bacterial isolates growing on agar-containing carbofuran as sole C source, but tests in liquid culture showed bacterial biomass increased only if soil extract (i.e., other C sources) was present with the carbofuran.

The controversy surrounding population dynamics of pesticide degraders and its relationship to enhanced biodegradation raises the issue of how the phenomenon can be maintained from year to year under field conditions. If enhanced biodegradation is solely a phenomenon of enzyme induction or augmentation of physiological activity, then such responses must be maintained in the absence of substrate (or at least in the presence of trace concentrations). Clearly, the amount of pesticide normally added to soil can only support small increases in biomass, thus induction of increased metabolic activity must play some role in expression of enhanced biodegradation (Turco & Konopka, 1990). A contributing factor could be the spread of catabolic plasmids following preexposure of soil to mineralizable substrates (Waid, 1972; Spain et al., 1980). A role for increases in numbers of specific degraders is also feasible if consideration is given to the comparatively high concentrations in the top inch of soil, where most of the pesticide resides after application. The somewhat confusing picture of the biochemical ecology underlying enhanced pesticide biodegradation warrants further investigations.

Enhanced Biodegradation Can Develop after a Single Pesticide Application

From the viewpoint of crop protection, the number of prior treatments required for development of enhanced biodegradation is important. In replicated field plots, isofenphos degraded more rapidly following the second year of treatment than during the previous year (Abou-Assaf et al., 1986). Laboratory studies have confirmed that enhanced biodegradation of isofenphos can occur following a single pretreatment (Chapman et al., 1986c). Similar observations have been made for EPTC (Obrigawitch et al., 1982). Using a series of laboratory incubations of aqueous soil extracts, Chapman and Harris (1990) reported the rapid development of enhanced biodegradation within weeks to months after a single application of various insecticides to soil.

Enhanced Biodegradation of Some Pesticides Can be Transferred from One Soil to Another

Earlier studies on enhanced biodegradation of diazinon and carbofuran showed that soil water derived from soil pretreated with either pesticide could be added to previously untreated soil or water to stimulate rapid biodegradation of either compound (Sethunathan et al., 1971; Sethunathan & Pathak, 1972; Felsot et al., 1981). Inoculation of soil untreated with thiocarbamate herbicides with 1% w/w soil with prior EPTC treatments caused enhanced biodegradation of EPTC (Dick et al., 1990). These studies support the contention that adapted organisms can be transported from one field to another on machinery carrying soil residues (Read, 1986).

For some compounds, microorganisms responsible for enhanced biodegradation cannot be easily transferred to untreated areas under field conditions. For example, fonofos degraded as slowly in soil collected from untreated interrow areas of a fonofos-treated plot as it did in soil from a plot without prior fonofos treatments (Table 13-2; Felsot, 1990). In contrast, carbofuran degraded as rapidly in soil collected from untreated interrow areas as in soil collected from adjacent rows with prior carbofuran treatments (Table 13-2). For both insecticide treatments, there was an equal probability of soil being transferred from the 18-cm treated band to the interrow areas during planting and tillage operations, yet only carbofuran degradation was enhanced in soil from the untreated interrows.

Initial Concentration of the Pesticide May Influence the Development of Enhanced Biodegradation

The concentration of the initial pesticide treatment can influence the development of enhanced biodegradation. Low parts per billion ($\mu g\ kg^{-1}$) of carbofuran resulted in enhanced biodegradation (Hendry & Richardson,

Table 13-2. Recovery of carbofuran and fonofos incubated for 21 d in soil previously treated in the field for 1 to 5 yr (Felsot, 1990).†

	Percentage of applied recovered at 21 d	
Pretreatment history	Carbofuran	Fonofos
No insecticide	37.8 ± 9.0	42.6 ± 12.7
One yr	0.0 ± 0.0	3.1 ± 0.4
Five yr	0.2 ± 0.3	3.0 ± 0.3
Five yr (off-band)	1.1 ± 0.7	44.4 ± 3.5

† Treatments were randomized within each of four replicate blocks in a 2-acre field plot at the Univ. of Illinois NW Illinois Agricultural Research and Demonstration Center near Monmouth, IL (Felsot & Tollefson, 1990). The soil was a Muscatine silt loam; the slope was about 1%. Soil was collected from within an 18-cm insecticide-treated band on either side of the seed furrow. The off-band (interrow) soil was collected from the space between corn rows.

1988), but soil type may significantly influence the effect of concentration. For example, comparatively greater concentrations of carbofuran were required to condition a muck soil for enhanced biodegradation than a sandy soil (Chapman et al., 1986a). Even though a pesticide may have become subject to enhanced biodegradation, high doses in soil could still inhibit its degradation (e.g., aldicarb, Read, 1987).

Cross-Conditioning by Related Compounds and Metabolites Is Common

Some compounds shown to undergo enhanced biodegradation can condition soil for rapid degradation of related chemicals even though the soil had not been treated previously with the related compound. This phenomenon, known as cross-adaptation or -conditioning, has been shown for phenoxyacetates (Audus, 1964), thiocarbamate herbicides (Wilson, 1984), and methylcarbamate insecticides (Harris et al., 1984; Dzantor & Felsot, 1989). Thus far, cross-conditioning seems to occur between analogous compounds within a pesticide class, although one report suggested cross-conditioning between fensulfothion and carbofuran (Read, 1983).

Hydrolytic metabolites of several pesticides can condition soil for enhanced biodegradation of the parent compound. Under aerobic conditions, dichorophenol, paranitrophenol, and salicylic acid conditioned soil for the rapid degradation of 2,4-D, parathion, and isofenphos, respectively (Somasundaram et al., 1989); the hydrolysis products of carbofuran and diazinon did not cause microbial adaptation. Under anaerobic conditions (flooded soil), however, carbofuran phenol and 1-napthol caused enhanced biodegradation of the parent pesticides, carbofuran and carbaryl, respectively (Rajagopal et al., 1986). The phenolic metabolites accumulated in the flooded soil and were not used as energy sources, so the mechanism of microbial adaptation for metabolism of the parent compounds remained unknown.

One utilitarian aspect of cross-conditioning phenomena is an opportunity to study metabolic pathways involved in microbial adaptation. This aspect is suggested by observation that the thiocarbamate EPTC conditions soil for rapid degradation of the analogs vernolate and butylate (Wilson, 1984); however, substitution of a cyclohexane ring for the alkyl side chain seems to inhibit enhanced biodegradation. Thus, in the case of thiocarbamate degradation, either a specific enzyme system is involved in metabolism or alternatively, altering the physicochemical properties by changing substituents reduces bioavailability.

Persistence of Enhanced Biodegradation Varies Among Pesticides

Enhanced biodegradation of phenoxyacetates and methylcarbamates seems persistent under field conditions in the absence of further inputs of pesticide. For example, 5 yr after application of MCPA to an experimental field in Oxford, UK, degradation of freshly added MCPA was still significantly faster in previously treated soil than in untreated soil (Fryer et al., 1980). Carbofuran degradation remained significantly enhanced in soil collected yearly from 1981 to 1984 from a field in Illinois where carbofuran use had ceased after 1979 (A.S. Felsot, 1992, unpublished data).

The longevity of enhanced biodegradation is further exemplified by Chapman and Harris (1990) who have characterized enhanced biodegradation by calculating the ratio of pseudo-first-order degradation rate constant in the pesticide-retreated soil to the rate constant in the untreated control soil. In this scheme, the ratio was called the E factor; any $E >> 1$ indicated that enhanced biodegradation was occurring. A comparison of E factors among insecticides showed enhanced biodegradation of fonofos and the oxidative metabolites of terbufos {*S*-[[(1,1-dimethylethyl)thio]methyl] *O,O*-diethyl phosphorodithioate} persisted for about 1 yr in contrast to the greater than 3 yr for carbofuran and isofenphos. Recently, Ou (1991) showed that enhanced biodegradation of fenamiphos had disappeared in soil collected 3 or more years after a field was first treated.

EPTC is also subject to a comparatively quick reversion to greater persistence within a year after termination of use (Harvey, 1987). In the case of EPTC degradation, the degradative enzyme is believed to reside on a bacterial plasmid (Tam et al., 1987), and instability of enhanced biodegradation may be due to spontaneous loss of the plasmid in the absence of pesticide (Lee, 1984; Tam et al., 1987).

Instability of enhanced biodegradation offers the opportunity to manage the phemonenon for long-term maintenance of efficacious pest control. After annual rotations with alachlor or cyanazine, weed control by EPTC was superior to weed control after continuous use of EPTC (Harvey, 1990). In banana (*musa* x *paradisiaca* L.) plantations, where two to four applications of nematicides are made within a year, alternations of fenamiphos with other organophosphorus nematicides greatly improved persistence and efficacy of fenamiphos (Anderson & Lafuerza, 1990). Annual rotations of fonofos with terbufos or carbofuran in field plots significantly increased recovery of fonofos residues (Fig. 13–4) (Felsot & Tollefson, 1990; Felsot, 1991).

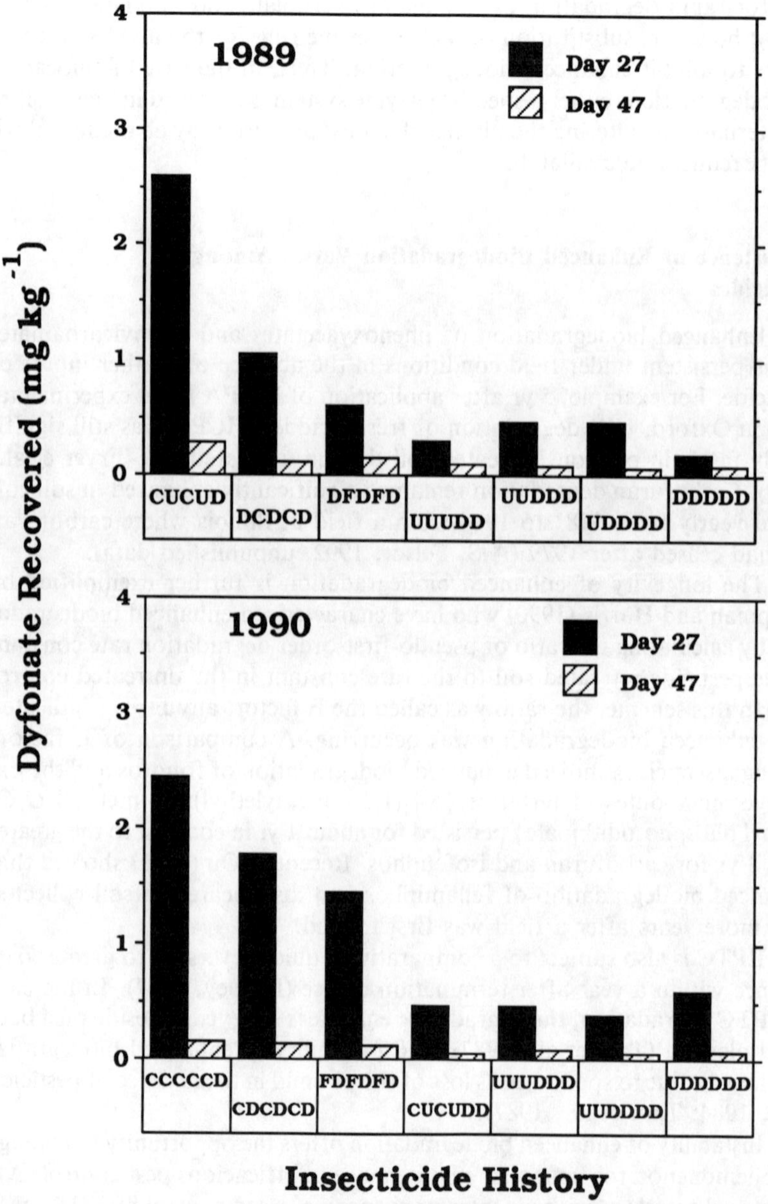

Fig. 13-4. Effect of annual pesticide rotations on recovery of fonofos 27 and 47 d following an application of Dyfonate 20G (1.12 kg ha^{-1}) to field plots at the Univ. of Illinois NW Illinois Agricultural Research and Demonstration Center during 1989 and 1990 (Felsot, 1990). Combinations of letters under each bar represent successive pesticide treatments; C = Counter 15G (terbufos), D = Dyfonate 20G (fonofos), F = Furadan 15G (carbofuran), and U = untreated (described in Felsot and Tollefson, 1990). Thus, CCCCCD would represent 5 yr of treatment by Counter 15G prior to a 1990 treatment of Dyfonate 20G; similarly, CDCDCD would represent alternate year treatments with Dyfonate 20G starting with Counter 15G in 1985.

Enhanced Biodegradation May Be Accompanied by a Change in Metabolic Pathways

Characterization of metabolic pathways following the development of enhanced biodegradation has lagged behind ecologically oriented studies. Studies with parathion, isofenphos, and EPTC, however, indicate that a shift in importance of one metabolic pathway over another may occur as a result of microbial adaptation. Retreatment of flooded soil with parathion resulted in enhanced biodegradation and a shift away from formation of aminoparathion (reduction) to formation of *p*-nitrophenol (hydrolysis) (Sudhakar-Barik et al., 1979). In isofenphos-retreated soils, oxidation to the oxon was significantly decreased compared to transformations resulting in CO_2 evolution and bound residues (Racke & Coats, 1987). This shift in metabolism implied a greater propensity for hydrolysis than for oxidation. Finally, enhanced biodegradation of EPTC in retreated soils was explained by increased hydroxylation of the alpha-propyl C as the major route of degradation (Dick et al., 1990). In nonconditioned soil, sulfoxidation was the major degradative pathway.

Interactions Between Physicochemical Processes and Microbial Ecology

Appropriate Experimental Systems for Studying Enhanced Biodegradation

If one is to study the relationships between physicochemical processes and microbial ecology of enhanced pesticide biodegradation, then the correct experimental design must be used to ensure that observations of increased degradation rates are due solely to prior chemical exposure and not inherent soil factors. Four approaches have been taken to study the development, occurrence, and ecology of enhanced biodegradation.

1. Comparing degradation rate in soils collected from fields separated in space but having well-defined pesticide-use histories.
2. Comparing degradation rate in soils collected from adjacent areas, for example, an untreated fence row around a treated field.
3. Comparing degradation rate in soil collected from a single, untreated source divided into two batches that are either left untreated as a control or conditioned with a pesticide.
4. Comparing degradation rate in a field with treated and untreated plots that have been replicated and randomized among one another.

Enhanced biodegradation of carbofuran was first observed in soils that were collected from disparate fields with known histories of pesticide use (Felsot et al., 1981, 1982). Subsequent studies compared degradation rates in soil collected from a level field with a long history of carbofuran use and in soil collected from a surrounding ditch (Felsot, 1986; Fig. 13-5). Although carbofuran degradation was much slower in soil from the surrounding ditch or in soils collected from fields without prior use of carbofuran, these approaches were inadequate to definitively prove the existence of enhanced bio-

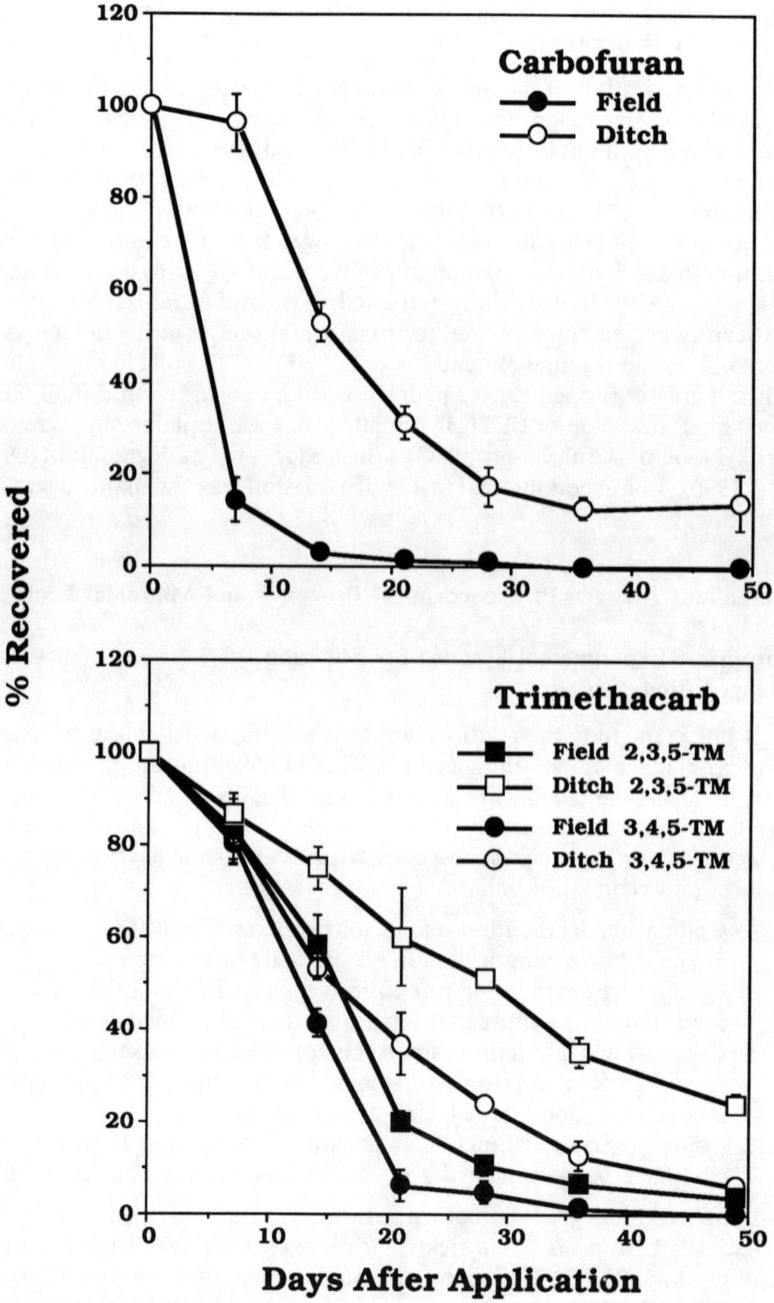

Fig. 13-5. Degradation of carbofuran and trimethacarb (1:1 mixture of 2,3,5-trimethylphenyl methylcarbamate and 3,4,5-trimethylphenyl methylcarbamate) in soil (Andres-Reddick-Symmerton association) collected from a corn field treated repeatedly with carbofuran and in soil collected from a surrounding, sod-covered ditch (Felsot, 1986). Application rate was 10 mg kg^{-1}, and the soils were incubated at 25 °C in the dark.

degradation because they lacked appropriate controls. For example, the observed rates of carbofuran and trimethacarb degradation depicted in Fig. 13–5 could have been due to inherent differences in microbial ecology that existed independently of chemical conditioning and resulted from differences in vegetative cover.

The best laboratory system for studying the ecology of enhanced biodegradation would be the one in which soil was collected from a single location. In the laboratory, the soil would be divided into batches that are then treated with pesticides; one batch would always be left untreated as a control. Such an approach was successfully used to definitively prove that prior treatment of soil with carbofuran was caused by microbial adaptations as opposed to differences in original soil ecology (Harris et al., 1984). Furthermore, this approach has been used to study population changes that occurred upon conditioning (Dzantor & Felsot, 1989; Merica & Alexander, 1990) or to study the effects of altered physical conditions like soil moisture (Chapman et al., 1986b).

The interest in the significance of enhanced biodegradation on pest control necessitated the development of experiments in the field. One approach that proved successful was the use of microplots; soil was removed from a single untreated site and placed at an experiment station in small bordered plots (Harris et al., 1988). The plots were large enough to repeatedly sample soils over many years after successive pesticide treatments. Furthermore, a crop was planted to encourage a rhizosphere normally associated with the pesticide. The importance of the rhizosphere in enhancing biodegradation rates has not been well studied and has only recently been recognized (Hsu & Bartha, 1979; Walton & Anderson, 1990).

An approach taken on a larger scale involved delineating a field into replicated treatments that were randomized among each other in block fashion (Harvey et al., 1986; Felsot & Tollefson, 1990). At an agricultural experiment station in Illinois, commitment of land over time allowed the establishment of pesticide-use histories ranging from 0 to 5 yr that were studied simultaneously (Felsot & Tollefson, 1990). Degradation was studied in the field (Fig. 13–4) as well as in soil collected from the field (Table 13–2) and returned to the laboratory (Felsot, 1990). In the long-term field study, degradation of fonofos (Dyfonate) was enhanced by 1 to 4 yr of previous treatments. The greater recovery of fonofos from the insecticide rotational treatments (DCDCD in 1989, CDCDCD, FDFDFD in 1990) suggested that enhanced biodegradation of fonofos was not stable beyond 1 yr after application (Fig. 13–4).

Sorption and Enhanced Biodegradation

Sorption of pesticides by soil is generally thought to protect pesticides from biologidal degradation (Ogram et al., 1985). A role for sorption in influencing the expression of enhanced biodegradation is suggested by observations on differences in biodegradation between soil types and the general effects of moisture tension. For example, higher concentrations of carbofuran were required to condition a muck than a sandy loam (Chapman et al., 1986a); presumably, sorption was greater in the muck than in the sandy loam because of a higher organic C content (Felsot & Wilson, 1980).

Research with dry formulations also suggests direct interactions between sorption and enhanced biodegradation. For example, granular and controlled release formulations, which sorb pesticide on a clay or polymeric matrix, can slow the degradation rate in soils conditioned for enhanced biodegradation (Chapman et al., 1986b; Felsot & Tollefson, 1990). Addition of charcoal to soil conditioned for enhanced biodegradation significantly slowed the rate of carbofuran degradation (Turco & Konopka, 1990). These studies show that sorption controls bioavailability of pesticides to microbial populations and slows degradation rate. Such phenomena may be advantageous for improving crop protection if bioavailability to the target pests is not also significantly affected.

Kaufman et al. (1985) suggested that development of enhanced biodegradation somehow altered the sorption potential of soil. The desorption of diphenamid was greater in conditioned soils than in nonconditioned soils. When conditioned and nonconditioned soils were mixed in equal amounts, degradation was faster and percentage desorption was greater if diphenamid was added first to the conditioned soil, which reinforced the observation of differences in desorption potential caused by preexposure to the pesticide.

The hypothesis that sorption potential is affected by enhanced biodegradation was tested on two sets of paired soils, one of each set being conditioned by carbofuran (A.S. Felsot, 1992, unpublished data). In the first case, K_f (Freundlich adsorption coefficient) for carbofuran was lower on a conditioned soil ($K_f = 0.44$) than on unexposed soil ($K_f = 0.85$) collected from a ditch surrounding the field. The unexposed ditch soil, however had a different rhizosphere and higher organic C content (2.3 vs. 1.8%) than the soil collected from the field. In a more controlled study, soil was collected from an experimental field where plots treated for 5 yr with carbofuran were replicated and randomized among plots not receiving any pesticide during the same time (Felsot & Tollefson, 1990). K_f for carbofuran was lower in the soil retreated with carbofuran ($K_f = 0.80$) than in the nontreated control soil ($K_f = 1.02$). These observations tended to support the idea that sorption and thus bioavailability might have been affected by the development of enhanced biodegradation, especially if there was some increase in pesticide-degrading microbial populations associated with clay and organic surfaces that would otherwise be available for pesticide sorption.

As depicted in Fig. 13–2, enhanced biodegradation can occur following stimulation of microbial activity with nutrient amendments. Under these conditions, the effects of biostimulation may overcome the increased sorption that would be expected from high loadings of organic amendments. For example, alachlor degradation can be enhanced by amending soil with sewage sludge or corn (*Zea mays* L.) meal (Fig. 13–3) (Dzantor & Felsot, 1992), but the sorption coefficient of alachlor can also be significantly increased (Guo et al., 1993). In sandy soil columns under unsaturated flow for 30 d, mass balance calculations of alachlor in soil and leachate suggested enhanced bio-

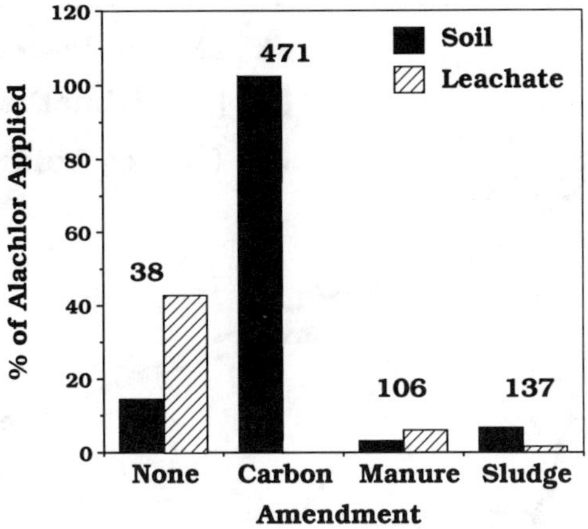

Fig. 13–6. Effect of C-rich amendments on mass balance of alachlor recovered from Plainfield sand soil columns and leachate under conditions of unsaturated flow (Guo et al., 1993). Bold numbers above each bar represent the K_{oc} experimentally determined for mixtures of the soil and the amendments.

degradation upon addition of manure and sludge, yet the K_{oc} (adsorption coefficient normalized for organic C content) had increased three-fold with the amendments (Fig. 13–6). Activated C, which has no nutrient value, increased K_{oc} by 10-fold and inhibited degradation. These results implied that during bioremediation of waste-contaminated soils, the potentially inhibitory effects of sorption on biodegradation could be overcome by use of organic amendments containing readily available nutrients.

Soil Type and Enhanced Biodegradation

Development of enhanced biodegradation can differ between soil types (Dzantor & Felsot, 1990), but it is unclear whether the differences are due to microbial ecology or physicochemical interactions. One prior application of trimethacarb to a Drummer-Catlin silty clay loam (fine-silty, mixed, mesic Typic Haplaquolls; fine-silty, mixed, mesic Typic Argiudolls, respectively) accelerated the degradation of a subsequent application (Fig. 13–7). However, repeated exposure of a Flanagan silt loam (fine, montmorillonitic, mesic Aquic Argiudolls) to trimethacarb failed to cause enhanced biodegradation. In contrast, enhanced biodegradation of carbofuran easily developed in both soils; furthermore, carbofuran cross-conditioned the Flanagan soil for trimethacarb degradation (Fig. 13–7). Such cross-conditioning by one compound for enhanced biodegradation of another compound without the second compound causing microbial adaptation itself was reported by Audus (1964); MCPA cross-conditioned soil for rapid biodegradation of 4-CPA, but 4-CPA could not condition soil for its own enhanced biodegradation. Although the specific reason for the lack of enhanced biodegradation of trimethacarb in the Flanagan soil was not investigated, distribution coefficients for trimethacarb were markedly higher in the Flanagan soil than

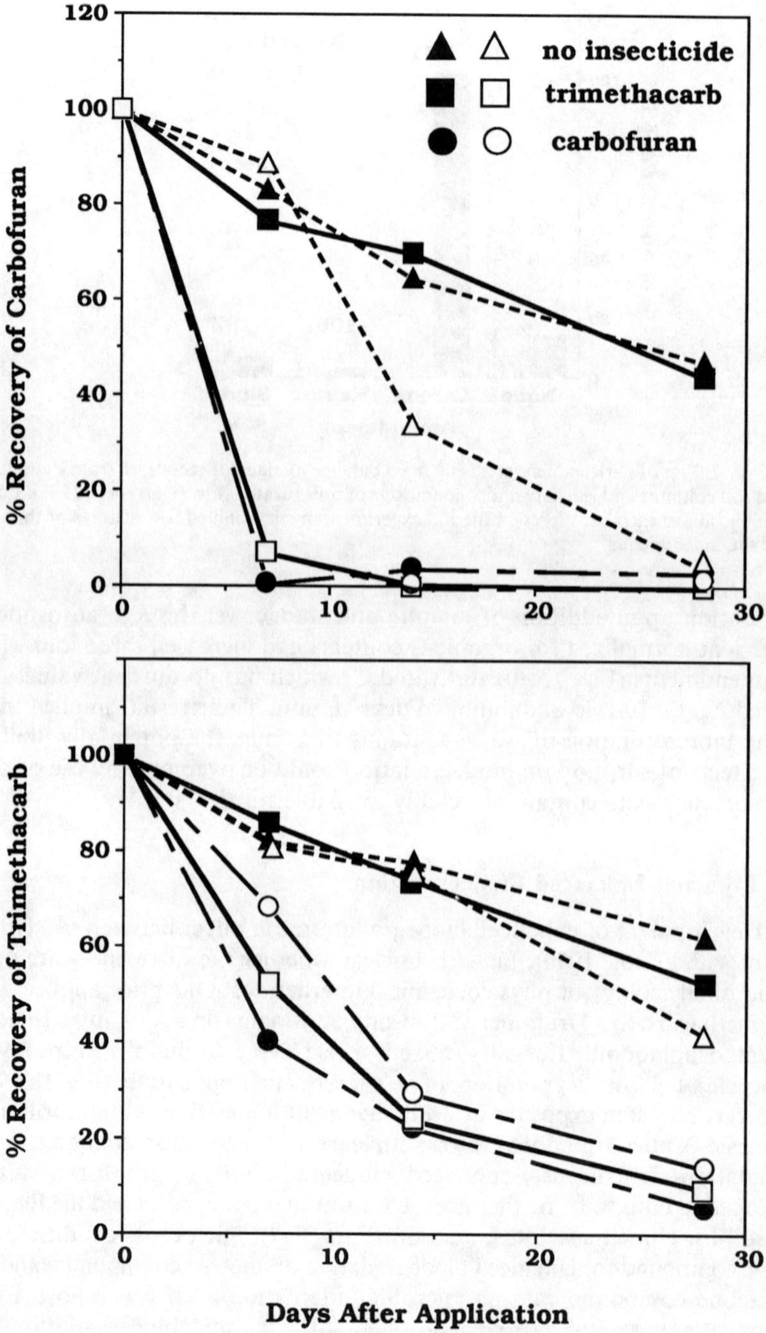

Fig. 13-7. Recovery of technical carbofuran and trimethcarb in a Flanagan silt loam (shaded symbols) and a Drummer-Catlin silty clay loam (open symbols) that had been pretreated with granular formulations of carbofuran (Furadan 15G, circles) or trimethacarb (Broot 15G, squares) (adapted from Dzantor & Felsot, 1990). The control was the no-insecticide treatment (triangles).

Table 13-3. Adsorption of carbofuran and trimethacarb (Dzantor & Felsot, 1990).

Insecticide	Distribution coefficient, K_d (L kg^{-1}) Drummer-Catlin SiCL†	Flanagan SiL†	K_d ratio Drummer:Flanagan
Carbofuran	0.71	0.83	1.17
3,4,5-TM‡	0.97	3.93	4.05
2,3,5-TM‡	1.12	2.91	2.60

† SiCL = silty clay loam, 2.0% organic C content; SiL = silt loam, 3.5% organic C content.
‡ Trimethacarb has two isomers, 3,4,5-trimethylphenyl methylcarbamate and 2,3,5-trimethylphenyl methylcarbamate.

in the Drummer-Catlin soil; distribution coefficients for carbofuran were similar in both soils (Table 13-3).

Effects of Soil Moisture

Low soil moistures inhibited the degradation of carbofuran (Ou et al., 1982) and the development of enhanced biodegradation (Chapman et al., 1986a). When soil moisture of a sandy loam was held to 11.3% w/w, granular formulations of carbofuran degraded at similar rates in conditioned and nonconditioned soils (Chapman et al., 1986b); however, when moisture was 27.0% w/w, granular carbofuran degraded faster in the conditioned soil than in the nonconditioned soil. Technical carbofuran was not affected by moisture content. These observations suggested that the effect of soil moisture was manifested through an effect on desorption rate from the granular carrier rather than an effect on the physiological status of the microorganisms.

In contrast to the observations of technical carbofuran in a sandy loam soil, low soil moistures (<15% w/w) in a Muscatine silt loam (22.5% moisture at 33 kPa) significantly slowed the degradation of technical carbofuran and fonofos even though the soil had been pretreated in the field for 5 yr with granular formulations (Fig. 13-8) (A.S. Felsot, 1991, unpublished data). At comparatively higher soil moistures (≥15-30% w/w), enhanced biodegradation was definitively expressed. These results suggested that biodegradation rate was influenced by the physiological status of the microorganisms in addition to physicochemical interactions with the soil.

Recently, Shelton and Parkin (1991) conducted a series of experiments that shed further light on the role of soil moisture and desorption in influencing biodegradation. Radiolabelled carbofuran was incubated in a conditioned soil held at five moisture tensions (40-1500 kPa). Periodically, carbofuran was measured as the percentage soluble, sorbed, and evolved as $^{14}CO_2$. Distribution coefficients (K_d) were calculated from the soluble and sorbed concentrations. Carbofuran degradation was significantly slower at 700 and 1500 kPa than at other moisture contents (Table 13-4). K_d, however, was smaller (i.e., more carbofuran in solution) in the drier soil than in the comparatively more moist soil. These results suggested that under relatively dry conditions carbofuran degradation slowed because microbial activity was adversely affected rather than because bioavailability (i.e., desorption) was rate limit-

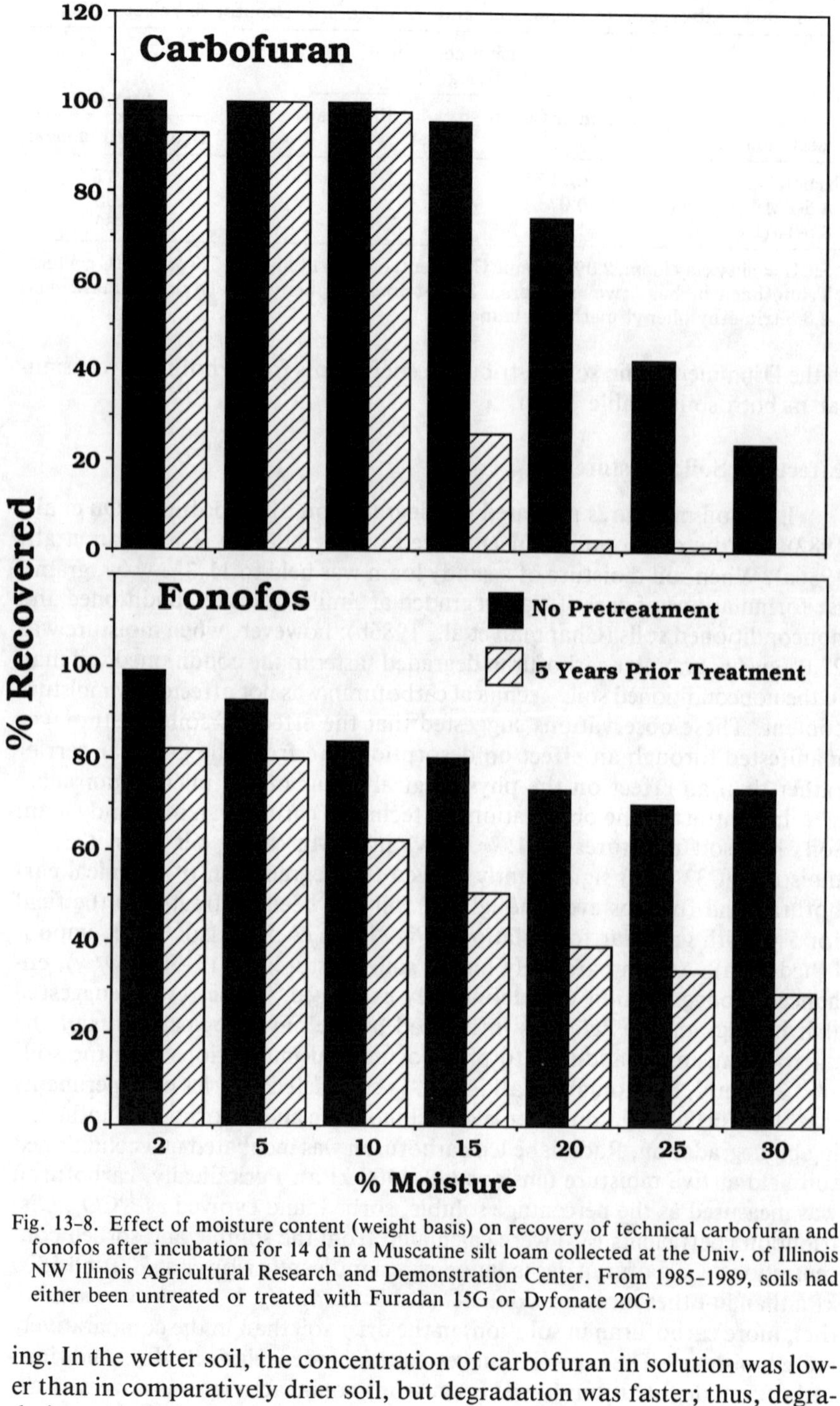

Fig. 13–8. Effect of moisture content (weight basis) on recovery of technical carbofuran and fonofos after incubation for 14 d in a Muscatine silt loam collected at the Univ. of Illinois NW Illinois Agricultural Research and Demonstration Center. From 1985–1989, soils had either been untreated or treated with Furadan 15G or Dyfonate 20G.

ing. In the wetter soil, the concentration of carbofuran in solution was lower than in comparatively drier soil, but degradation was faster; thus, degradation was limited by the desorption rate. Studies with less water-soluble

Table 13-4. Relationship among moisture tension, half-life, and apparent distribution coefficient in a Hatsboro silt loam treated for 3 yr in the field with carbofuran. (Derived from Shelton and Parkin, 1991.)

Moisture tension	DT-50%†	K_d‡
kPa	d	L kg^{-1}
40	6.4	0.15
80	6.6	0.14
160	6.9	0.14
340	10.2	0.13
700	16.2	0.11

† DT-50% is the time when 50% of the total carbofuran applied has degraded.
‡ K_d, the distribution coefficient, was determined at the day equivalent to the DT-50%. If no measurement was available on the day corresponding to the DT-50%, then the average of the K_d's on the days bracketing the DT-50% were used.

compounds (like organophosphorus insecticides) known to cause microbial adaptations are needed to clarify the relationship between desorption and microbial physiological ecology.

CONCLUSION

When repeatedly added to soil, a variety of insecticides, herbicides, and fungicides can cause microbial adaptations resulting in enhanced biodegradation. The mechanisms by which the biodegradation rate is accelerated remains elusive. Because some of the pesticides studied seemed to serve as energy sources for bacterial isolates, there has been an assumption that microbial numbers increase as a result of the pesticide application. Some studies have shown elevated numbers of specific pesticide-degraders following repeated applications of the pesticide in question. Other studies, however, have questioned the significance of the increase in numbers and concluded that there is an increase in specific enzyme activity, although direct measurements of these metabolic systems has been lacking. It has been documented that enhanced biodegradation can develop after as little as one prior application of a pesticide, and the phenomenon is persistent for at least 1 yr following initial treatment.

Development and expression of enhanced biodegradation is affected by physicochemical interactions of a pesticide with the soil in addition to microbial ecology. Biodegradation has been inhibited when sorption potential is increased. Furthermore, the development of enhanced biodegradation itself may alter slightly the sorption potential of the soil; some studies indicate a reduction in sorption upon conditioning of the soil. Soil moisture is critical to expression of enhanced biodegradation; not only does moisture affect the physiology of the pesticide-degrading microflora, but it can also affect the desorption rate of pesticides applied as granular or controlled release formulations.

Further studies of enhanced biodegradation are important for long-term maintenance of soil pesticide efficacy, but many lessons that are applicable to bioremediation of chemical waste can also be learned. Future studies need to focus on population ecology of specific pesticide degraders, enzyme systems involved in pesticide metabolism, and interactions between microbial ecology and physicochemical processes affecting the pesticide.

REFERENCES

Abou-Assaf, N., J.R. Coats, M.E. Gray, and J.J. Tollefson. 1986. Degradation of isofenphos in cornfields with conservation tillage practices. J. Environ. Sci. Health B21:425–446.

Alexander, M. 1979. Role of cometabolism. p. 67–75. *In* A.W. Bourquin and P.H. Pritchard (ed.) Microbial degradation of pollutants in marine environments. EPA 600/9-79-012. USEPA, Gulf Breeze, FL.

Alexander, M. 1981. Biodegradation of chemicals of environmental concern. Science 211:132–138.

Anderson, J.P.E., and A. Lafuerza. 1990. Microbiological aspects of accelerated pesticide degradation in soils. Poster 07B-23. 7th Int. Congr. Pesticide Chemistry, Book of Abstracts. Vol. III. Int. Union Pure Appl. Chem., Hamburg, Germany.

Audus, L.J. 1949. The biological detoxication of 2:4-dichlorphenoxyacetic acid in soil. Plant Soil 2:31–36.

Audus, L.J. 1964. Herbicide behaviour in the soil. II. Interactions with soil microorganisms. p. 163–206. *In* L.J. Audus (ed.) The physiology and biochemistry of herbicides. Academic Press, New York.

Carson, R. 1962. Silent spring. Fawcett Publ., Greenwich, CT.

Chapman, R.A., and Harris. C.R. 1990. Enhanced degradation of insecticides in soil. Factors influencing the development and effects of enhanced microbial activity. p. 82–96. *In* K.D. Racke and J.R. Coats (ed.) Enhanced biodegradation of pesticides in the environment. ACS Symp. Ser. 426. Am. Chem. Soc., Washington, DC.

Chapman, R.A., C.R. Harris, and C. Harris. 1986a. Observations on the effect of soil type, treatment intensity, insecticide formulation, temperature and moisture on the adaptation and subsequent activity of biological agents associated with carbofuran degradation in soil. J. Environ. Sci. Health B21:125–141.

Chapman, R.A., C.R. Harris, and C. Harris. 1986b. The effect of formulation and moisture level on the persistence of carbofuran in a soil containing biological systems adapted to its degradation. J. Environ. Sci. Health B21:57–66.

Chapman, R.A., C.R. Harris, P. Moy, and K. Henning. 1986c. Biodegradation of pesticides in soil: Rapid degradation of isofenphos in a clay loam after a previous treatment. J. Environ. Sci. Health B21:269–276.

Chaudhry, G.R., and A.N. Ali. 1988. Bacterial metabolism of carbofuran. Appl. Environ. Microbiol. 54:1414–1419.

Dick, W.A., R.O. Ankumah, G. McClung, and N. Abou-Assaf. 1990. Enhanced degradation of S-ethyl N,N-dipropylcarbamothioate in soil and by an isolated soil microorganism. p. 98–112. *In* K.D. Racke and J.R. Coats (ed.) Enhanced biodegradation of pesticides in the environment. ACS Symp. Ser. 42. Am. Chem. Soc., Washington, DC.

Dzantor, E.K., and A.S. Felsot. 1989. Effects of conditioning, cross-conditioning, and microbial growth on development of enhanced biodegradation of insecticides in soil. J. Environ. Sci. Health B24:569–597.

Dzantor, E.K., and A.S. Felsot. 1990. Soil differences in the biodegradation of carbofuran and trimethacarb following pretreatment with these insecticides. Bull. Environ. Contam. Toxicol. 45:531–537.

Dzantor, E.K., and A.S. Felsot. 1992. Landfarming and biostimulation for decontaminating herbicide wastes in soil. p. 46–67. *In* T.D. Ferguson (ed.) Proc. Int. Workshop on Research in Pesticide Treatment/Disposal/Waste Minimization. EPA-600/59-9/047. U.S. Gov. Print. Office, Washington, DC.

Felsot, A.S. 1986. Effect of conditioning soil for enhanced biodegradation of carbofuran on dissipation of other methyl carbamate insecticides. Poster 6B-08. Abstr. 6th Int. Congr. Pesticide Chem., Int'l. Union Pure Appl. Chem., Ottawa, ON, Canada.

Felsot, A.S. 1989. Enhanced biodegradation of insecticides in soil: Implications for agroecosystems. Annu. Rev. Entomol. 34:453–476.

Felsot, A.S. 1990. Effect of annual rotations of corn rootworm soil insecticides on their persistence and efficacy. Display Presentation no. 1174. 37th Annu. Meet. Entomol. Soc. Am., New Orleans.

Felsot, A.S. 1991. Enhanced biodegradation of soil insecticides in the USA—significance and management. p. 41–50. *In* A. Walker (ed.) Pesticides in soils and water: Current perspective. Monogr. no. 47. Brit. Crop Prot. Counc., Farnham, Surrey, UK.

Felsot, A.S., and E.K. Dzantor. 1990. Enhancing biodegradation for detoxification of herbicide waste in soil. p. 192–213. *In* K.D. Racke and J.R. Coats (ed.) Enhanced biodegradation of pesticides in the environment. ACS Symp. Ser. 426. Am. Chem. Soc., Washington, DC.

Felsot, A., J.V. Maddox, and W. Bruce. 1981. Enhanced microbial degradation of carbofuran in soils with histories of Furadan use. Bull. Environ. Contam. Toxicol. 26:781–788.

Felsot, A.S., and J.J. Tollefson. 1990. Evaluation of some methods for coping with enhanced biodegradation of soil insecticides. p. 249–268. *In* K.D. Racke and J.R. Coats (ed.) Enhanced biodegradation of pesticides in the environment. ACS Symp. Ser. 426. Am. Chem. Soc., Washington, DC.

Felsot, A., and J. Wilson. 1980. Adsorption of carbofuran and movement on soil thin layers. Bull. Environ. Contam. Toxicol. 24:778–782.

Felsot, A.S., J.G. Wilson, D.E. Kuhlman, and K.L. Steffey. 1982. Rapid dissipation of carbofuran as a limiting factor in corn rootworm (Coleoptera: Chrysomelidae) control in fields with histories of continuous carbofuran use. J. Econ. Entomol. 75:1098–1103.

Fournier, J.C., P. Codaccioni, G. Soulas, and C. Repiquet. 1981. Soil adaptation to 2,4-D degradation in relation to the application rates and the metabolic behaviour of the degrading microflora. Chemosphere 10:977–984.

Fryer, J.D., P.D. Smith, and R.J. Hance. 1980. Field experiments to investigate long-term effects of repeated applications of MCPA, tri-allate, simazine and linuron: II. Crop performance and residues 1969–78. Weed Res. 20:111–116.

Guo, L., T.J. Bicki, A.S. Felsot, and T.D. Hinesly. 1993. Sorption and movement of alachlor in soil modified by carbon-rich wastes. J. Environ. Qual. 22:186–194.

Harris, C.R., R.A. Chapman, C. Harris, and C.M. Tu. 1984. Biodegradation of pesticides in soil: Rapid induction of carbamate degrading factors after carbofuran treatment. J. Environ. Sci. Health B19:1–11.

Harris, C.R., R.A. Chapman, J.H. Tolman, P. Moy, K. Henning, and C. Harris. 1988. A comparison of the persistence in a clay loam of single and repeated annual applications of seven granular insecticides used for corn rootworm control. J. Environ. Sci. Health B23:1–32.

Harvey, R.G. 1987. Herbicide dissipation from soils with different herbicide use histories. Weed Sci. 35:583–589.

Harvey, R.G. 1990. Systems allowing continued use of carbamothioate herbicides despite enhanced biodegradation. p. 214–221. *In* K.D. Racke and J.R. Coats (ed.) Enhanced biodegradation of pesticides in the environment. Am. Chem. Soc. Symp. Ser. 426. Am. Chem. Soc., Washington, DC.

Harvey, R.G., G.R. McNevin, J.W. Albright, and M. Kozak. 1986. Wild proso millet (*Panicum milliaceum*) control with thiocarbamate herbicides on previously treated soils. Weed Sci. 34:773–780.

Hendry, K.M., and C.J. Richardson. 1988. Soil biodegradation of carbofuran and furathiocarb following soil pretreatment with these pesticides. Environ. Toxicol. Chem. 7:763–774.

Hsu, T.-S., and R. Bartha. 1979. Accelerated mineralization of two organophosphate insecticides in the rhizosphere. Appl. Environ. Microbiol. 37:36–41.

Karns, J.S., W.W. Mulbry, J.O. Nelson, and P.C. Kearney. 1986. Metabolism of carbofuran by a pure bacterial culture. Pestic. Biochem. Physiol. 25:211–217.

Kaufman, D.D., and D.F. Edwards. 1983. Pesticide/microbe interaction effects on persistence of pesticides in soil. IUPAC Pest. Chem. 3:177–182.

Kaufman, D.D., Y. Katan, D.F. Edwards, and E.G. Jordan. 1985. Microbial adaptation and metabolism of pesticides. p. 437–451. *In* J.L. Hilton (ed.) Agricultural chemicals of the future. BARC Symp. no. 8. Rowman and Allanheld Publishers, Totowa, NJ.

Kirkland, K., and J.D. Fryer. 1966. Pretreatment of soil with MCPA as a factor affecting persistence of a subsequent application. p. 616–621. *In* Proc. British Weed Control Conf., Brighton, U.K., 8th, Br. Weed Control Council, Croyden, U.K.

Lee, A. 1984. EPTC (S-ethyl N,N-dipropylthiocarbamate)-degrading microorganisms isolated from a soil previously exposed to EPTC. Soil Biol. Biochem. 16:529–531.

Merica, R.R., and M. Alexander. 1990. Populations and activity of carbofuran-degrading microorganisms in soil. Plant Soil 126:101–108.

Moorman, T.B. 1988. Populations of EPTC-degrading microorganisms in soils with accelerated rates of EPTC degradation. Weed Sci. 36:96–101.

Moorman, T.B. 1990. Adaptation of microorganisms in subsurface environments. p. 167–180. *In* K.D. Racke and J.R. Coats (ed.) Enhanced biodegradation of pesticides in the environment. ACS Symp. Ser. 426. Am. Chem. Soc., Washington, DC.

Mueller, J.G., H.D. Skipper, E.G. Lawrence, and E.L. Kline. 1989. Bacterial stimulation by carbamothioate herbicides. Weed Sci. 37:424–427.

Obrigawitch, T., R.G. Wilson, A.R. Martin, and F.W. Roeth. 1982. The influence of temperature, moisture, and prior EPTC application on the degradation of EPTC in soils. Weed Sci. 30:175–181.

Ogram, A.V., R.E. Jessup, L.T. Ou, and P.S.C. Rao. 1985. Effects of sorption on biological degradation rates of (2,4-dichlorophenoxy)acetic acid in soils. Appl. Environ. Microbiol. 49:582–587.

Ou, L.-T. 1991. Interactions of microorganisms and soil during fenamiphos degradation. Soil Sci. Soc. Am. J. 55:716–722.

Ou, L.-T., D.H. Gancarz, W.B. Wheeler, P.S.C. Rao, and J.M. Davidson. 1982. Influence of soil temperature and soil moisture on degradation and metabolism of carbofuran in soils. J. Environ. Qual. 11:293–298.

Racke, K.D., and J.R. Coats. 1987. Enhanced degradation of isofenphos by soil microorganisms. J. Agric. Food Chem. 5:94–99.

Racke, K.D., and J.R. Coats (ed.) 1990. Enhanced biodegradation of pesticides in the environment. ACS Symp. Ser. 426. Am. Chem. Soc., Washington, DC.

Rajagopal, B.S., S. Panda, and N. Sethunathan. 1986. Accelerated degradation of carbaryl and carbofuran in a flooded soil pretreated with hydrolysis products, 1-naphthol and carbofuran phenol. Bull. Environ. Contam. Toxicol. 36:827–832.

Read, D.C. 1983. Enhanced microbial degradation of carbofuran and fensulfothion after repeated applications to acid mineral soil. Agric. Ecosyst. Environ. 10:37–46.

Read, D.C. 1986. Influence of weather conditions and microorganisms on persistence of insecticides to control root maggots (Diptera: Anthomyiidae) in rutabagas. Agric. Ecosyst. Environ. 16:165–173.

Read, D.C. 1987. Greatly accelerated microbial degradation of aldicarb in retreated field soil, in flooded soil, and in water. J. Econ. Enotomol. 80:156–163.

Scow, K.M., R.R. Merica, and M. Alexander. 1990. Kinetic analysis of enhanced biodegradation of carbofuran. J. Agric. Food Chem. 38:908–912.

Sethunathan, N. 1971. Biodegradation of diazinon in paddy fields as a cause of its inefficiency for controlling brown planthoppers in rice fields. PANS 17:18–19.

Sethunathan, N., S. Caballa, and M.D. Pathak. 1971. Absorption and translocation of diazinon by rice plants from submerged soils and paddy water and the persistence of residues in plant tissues. J. Econ. Entomol. 64:571–576.

Sethunathan, N., and M.D. Pathak. 1972. Increased biological hydrolysis of diazinon after repeated application in rice paddies. J. Agric. Food Chem. 20:586–589.

Shelton, D.R., and T.B. Parkin. 1991. Effect of moisture on sorption and biodegradation of carbofuran in soil. J. Agric. Food Chem. 39:2063–2068.

Somasundaram, L., J.R. Coats, and K.D. Racke. 1989. Degradation of pesticides in soil as influenced by the presence of hydrolysis metabolites. J. Environ. Sci. Health B24:457–478.

Spain, J.C., P.H. Pritchard, and A.W. Bourquin. Effects of adaptation on biodegradation rates in sediment/water cores from estuarine and freshwater environments. Appl. Environ. Microbiol. 40:726–734.

Spain, J.C., P.A. Van Veld, C.A. Monti, P.H. Pritchard, and C.R. Cripe. 1984. Comparison of *p*-nitrophenol biodegradation in field and laboratory test systems. Appl. Environ. Microbiol. 48:944–950.

Sudhakar-Barik, P.A. Waid, C. Ramakrishna, and N. Sethunathan. 1979. A change in the degradation pathway of parathion after repeated applications to flooded soil. J. Agric. Food Chem. 27:1391–1392.

Tam, A.C., R.M. Behki, and S.U. Khan. 1987. Isolation and characterization of an *s*-ethyl-*N,N*-dipropylthiocarbamate-degrading *Arthrobacter* strain and evidence for plasmid-associated *s*-ethyl-*N,N*-dipropylthiocarbamate degradation. Appl. Environ. Microbiol. 53:1088–1093.

Turco, R.F., and A. Konopka. 1990. Biodegradation of carbofuran in enhanced and non-enhanced soils. Soil Biol. Biochem. 22:195–201.

Waid, J.S. 1972. The possible importance of transfer factors in the bacterial degradation of herbicides in natural ecosystems. Residue Rev. 44:65–71.

Walton, B.T., and T.A. Anderson. 1990. Microbial degradation of trichloroethylene in the rhizosphere. Potential application to biological remediation of waste sites. Appl. Environ. Microbiol. 56:1012–1016.

Wiggins, B.A., S.H. Jones, and M. Alexander. 1987. Explanations for the acclimation period preceding the mineralization of organic chemicals in aquatic environments. Appl. Environ. Microbiol. 53:791–796.

Wilson, R.G. 1984. Accelerated degradation of thiocarbamate herbicides in soil with prior thiocarbamate herbicide exposure. Weed Sci. 32:264–268.

Woodwell, G.M. 1967. Toxic substances and ecological cycles. Sci. Am. 216:24–31.

14 Enhanced Biodegradation of the Nematicide Fenamiphos in Soil

L.-T. Ou, J. E. Thomas, and D. W. Dickson

University of Florida
Gainesville, Florida

Fenamiphos [ethyl-3-methyl-4-(methylthio)phenyl (1-methylethyl) phosphoramidate] is a broad spectrum, nonvolatile, organophosphorus nematicide. This chemical is rapidly oxidized in soil to fenamiphos sulfoxide (FSO), which in turn is slowly oxidized to fenamiphos sulfone (FSO_2) (Ou & Rao, 1986). Both FSO and FSO_2 have nematicidal activity and toxicity similar to fenamiphos (Waggoner & Khasawinah, 1974). Ou (1991) reported enhanced degradation of fenamiphos in soil collected from an experimental site planted in potato (*Solanum tuberosum* L.) 2 yr after a single field application of the chemical. Enhanced degradation has also been reported for several other organophosphorus insecticides (Sethunathan, 1973; Barik et al., 1979; Read, 1983), and other pesticides (Racke & Coats, 1990).

The objective of this study was to determine the mineralization rate and half-life of fenamiphos in soil collected from a turfgrass site. This site had a history of repeated applications of fenamiphos, annually or biannually, over a period of more than 15 yr.

MATERIALS AND METHODS

Soil

Soil samples were taken from the 15th hole of the Gainesville Golf and Country Club, Gainesville, FL on 4 Feb., 22 Apr., and 13 May 1991. This turfgrass site had been treated with fenamiphos annually or biannually since early 1970 at a rate of 11.3 to 16.8 kg ha^{-1}. Since 1986, this site was treated with fenamiphos annually either in April or May. On 22 Apr. 1991, after soil samples were collected, this site was treated with fenamiphos at 11.2 kg ha^{-1}. Soil samples were collected from three locations (Fig. 14-1), from the center of the fairway (three depths, 0–15, 15–30, and 30–60 cm deep), and the putting green (0–10 cm deep), and from the rough area (0–15 cm and 15–30 cm deep) located on the southside of the fairway. This area has a 0.5

 Sorption and Degradation of Pesticides and Organic Chemicals in Soil. SSSA Special Publication no. 32.

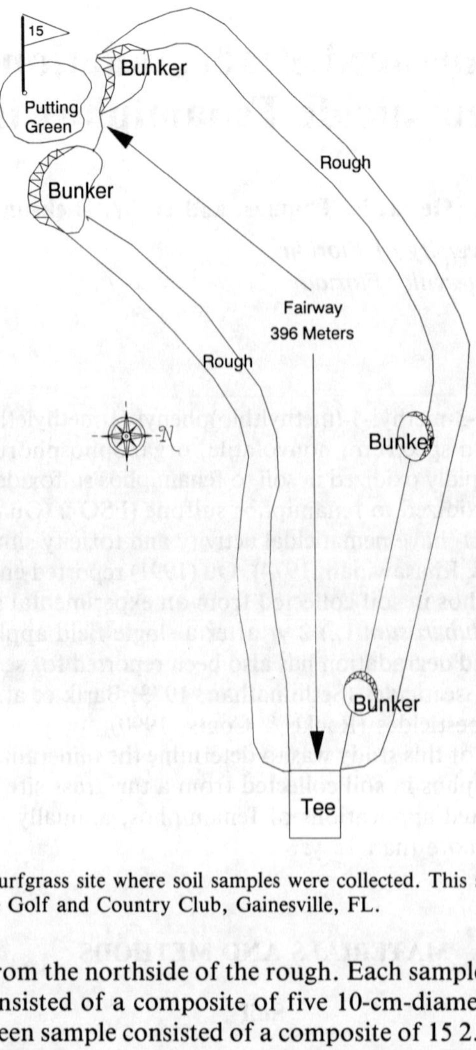

Fig. 14–1. The turfgrass site where soil samples were collected. This site is the 15th hole of the Gainesville Golf and Country Club, Gainesville, FL.

m elevation from the northside of the rough. Each sample from the fairway and rough consisted of a composite of five 10-cm-diameter cores, whereas the putting green sample consisted of a composite of 15 2.5-cm-diam. cores. All soil samples were placed in plastic bags and stored in the dark at 4 °C.

Soil collected from this site was classified as Blichton sand (a loamy, siliceous, hyperthermic Arenic Plinthic Paleaquults). Selected soil properties are shown in Table 14–1.

Chemicals

Analytical grade fenamiphos (99.1% purity) was provided by the USEPA (Research Triangle Park, NC). Technical grade fenamiphos (95% purity) and ring-1-labeled ^{14}C-fenamiphos (specific activity 9.51×10^8 Bq mmol^{-1}) were provided by Mobay Chemical Corp. (Kansas City, MO). All other chemicals were either analytical grade, pesticide grade, or the highest grade commercially available.

Table 14-1. Properties of the soil samples collected from the turfgrass site before and after the annual application of fenamiphos.

Soil	Depth	pH	Soil water content	Organic C
	cm		g kg^{-1}	
		Before annual application		
Fairway	0-15	6.8	190	23.1
Fairway	15-30	7.0	138	5.2
Fairway	30-60	7.1	117	6.1
Putting green	0-10	6.3	166	29.6
Rough	0-15	5.6	149	21.6
Rough	15-30	6.1	91	7.5
		After annual application		
Putting green	0-10	6.6	100	23.5
Rough	0-15	4.5	199	25.3
Rough	15-30	5.4	194	12.5

Soil Treatment

Similar procedures to those reported by Ou (1991) were used for the determination of mineralization of ^{14}C-fenamiphos in soil. Radio-labeled fenamiphos and analytical grade fenamiphos were added to 100 g of soil (oven-dry basis) at the rate of 37 kBq and 400 μg, respectively. After mixing, each flask was tightly closed with a rubber stopper under which a small stainless steel vial containing KOH (5 mol L^{-1}) was hung using a stainless steel wire. At 3 d, 7 d, and weekly thereafter, all vials were removed and replaced by new vials containing fresh KOH. At the same time, the weights of the flasks were checked, and deionized water was added to compensate for any water loss. Trapped $^{14}CO_2$ in the KOH traps was determined by liquid scintillation counting.

For determination of the disappearance of ^{14}C-activity in soil, 150 g of soil in 1 L glass bottles were treated with ^{14}C-fenamiphos and analytical grade fenamiphos at rates of 270 kBq and 600 μg, respectively. After mixing, the bottles were incubated in the dark at 25 °C. At 1- or 2-wk intervals, 10 g of soil were removed from the bottles for solvent extraction and placed in 50-mL culture tubes equipped with teflon-lined screw caps.

For demonstration of enhanced degradation of fenamiphos in the laboratory, surface soil (500 g) collected from the fairway before the annual application of fenamiphos was placed in a 1-L Erlenmeyer flask and treated with technical grade fenamiphos at a rate of 10 μg g^{-1}. The flask was incubated in the dark at 23 ± 1 °C. After 4 wk, the sample was treated with fenamiphos at the same rate.

Soil Extraction

Soil samples (10 g) in culture tubes were extracted three times with 25 mL of an equal volume of acetone and methanol. The soil extracts were filtered under vacuum using Whatman no. 41 filters. ^{14}C-activity in the extracts was quantified by liquid scintillation counting.

RESULTS AND DISCUSSION

Apparently, after repeated annual or biannual applications of fenamiphos for more than 15 yr, degradation of the chemical in soil at the site of the Gainesville Golf and Country Club was enhanced. Mineralization of fenamiphos in soil samples collected from the fairway and the putting green before the annual fenamiphos application was rapid (Fig. 14–2). In contrast, mineralization of fenamiphos in soil samples collected from the rough was much lower. The rough had never been treated with fenamiphos. Mineralization rates in both the surface samples collected from the fairway and the putting green before the annual fenamiphos application were similar. Mineralization rates of fenamiphos in soils collected from the fairway and the rough decreased with soil depth.

Although the rough had never been treated with fenamiphos, mineralization rates in soil samples collected from this area were greater than the rates observed in surface soils of Cecil sandy loam (clayey, kaolinitic, thermic Typic Kanhapludults), Arredondo fine sand (loamy, siliceous, hyperthermic Grossarenic Paleudults), and Ellzey fine sand (sandy, siliceous, hyperthermic Arenic Ochraqualfs) (Ou, 1991; Ou & Rao, 1986). These surface soils

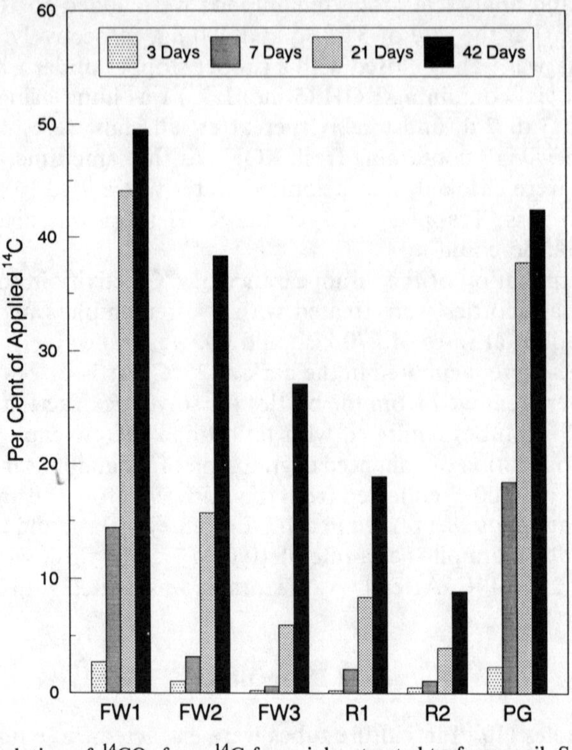

Fig. 14–2. Evolution of $^{14}CO_2$ from ^{14}C-fenamiphos treated turfgrass soil. Soil samples were collected before the annual fenamiphos application. FW1 = fairway, 0 to 15 cm deep. FW2 = fairway, 15 to 30 cm deep. FW3 = fairway, 30 to 60 cm deep; R1 = Rough, 0 to 15 cm deep; R2 = rough, 15 to 30 cm deep; PG = putting green, 0 to 10 cm deep.

were collected from sites that had never been exposed to the nematicide, and 4 to 10% of the applied ^{14}C-fenamiphos was mineralized in 42 d of incubation. In contrast, 20% of the applied ^{14}C-fenamiphos was mineralized in 42 d in the surface soil of the rough collected before annual application. Since the rough is adjacent to the fairway, it is likely that from time to time small amounts of fenamiphos may drift into the rough from the fairway through rainfall, irrigation, wind, and human activities. Furthermore, fenamiphos-degrading microorganisms could also be transmitted from the fairway into the rough through the same means. As a result, the rate of mineralization was slow in soil samples collected from sites that have no or little chance to be contaminated with fenamiphos.

Mineralization in the surface soil sample collected from the putting green 23 d after the annual fenamiphos application was higher during the first 7 d of incubation (Fig. 14–3) than for the corresponding sample collected from collected after the annual fenamiphos application. It was found that 2.3 and 16.3% of the applied ^{14}C-fenamiphos in soil samples collected before and after the annual application were mineralized, respectively. After 7 d, the mineralization pattern reversed between the two samples in which mineralization rates were higher in the soil collected before the annual application than in the soil collected after the annual application. The plausible scenario for rapid initial mineralization of fenamiphos in the soil collected after the annual application would be the presence of higher fenamiphos-degrading microbial population density in this soil stimulated by the field treatment of fenamiphos.

Mineralization rates of fenamiphos in soil samples collected from the rough after the annual fenamiphos application (Fig. 14–3) were similar to the corresponding soil samples collected before the annual application. Apparently, little fenamiphos had drifted into the rough from the fairway during the application of fenamiphos to the fairway.

Following one laboratory treatment of fenamiphos to the surface, soil sample of the fairway collected before the annual application mineralization of fenamiphos was slightly enhanced after 7 d of incubation (Fig. 14–4). Mineralization in the surface sample receiving two laboratory treatments, however, displayed greater enhancement in the first 7 d of incubation. After 7 d, mineralization in this sample was lower than that in the control and one laboratory treatment sample.

First-order rate constants (k_1) and half-life values ($t_{1/2}$) for disappearance of ^{14}C-activity in soil collected from the putting green, fairway and rough before and after the annual application of fenamiphos are shown in Table 14–2. These values were estimated based on the ^{14}C-activity in the solvent extracts. Ou (1991) found that the majority of the ^{14}C-activity in the solvent extracts was associated with the three toxic chemicals, fenamiphos, FSO, and FSO_2. He used the term total-toxic-residue (TTR) to represent the three chemicals. In this regard, the values of k_1 and $t_{1/2}$ for ^{14}C-activity disappearance would be close to the corresponding values for TTR. Similar to the mineralization results, k_1 values in soil collected from the fairway and the putting green were larger than those in soil collected from the rough.

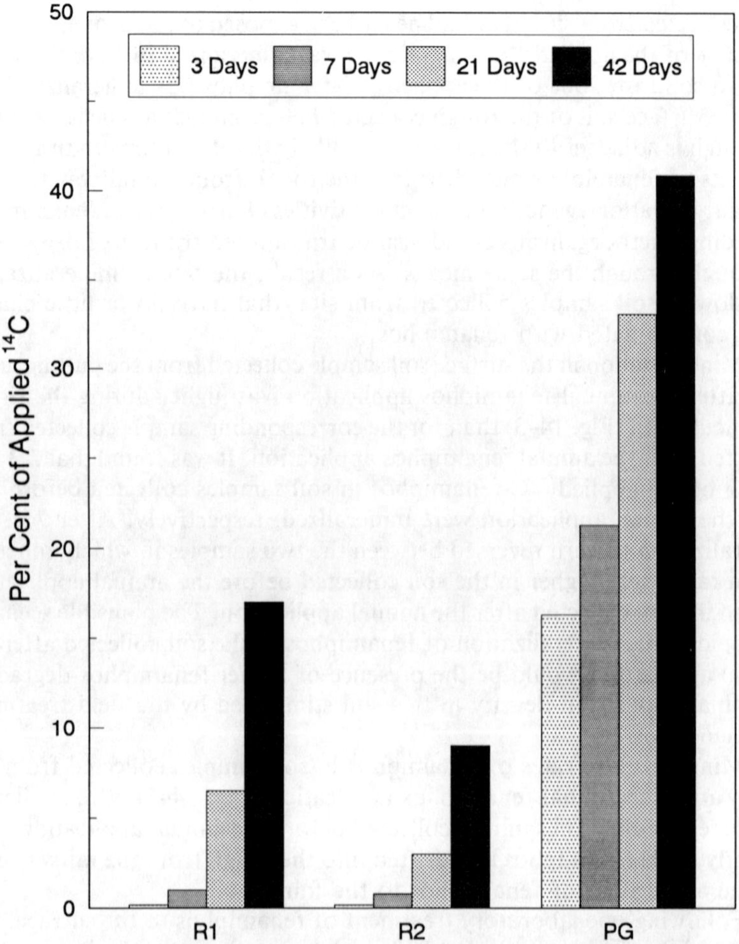

Fig. 14–3. Evolution of $^{14}CO_2$ from ^{14}C-fenamiphos treated turfgrass soil. Soil samples were collected after the annual fenamiphos application. R1 = rough, 0 to 15 cm deep; R2 = rough, 15 to 30 cm deep; PG = putting green, 0 to 10 cm deep.

Also $t_{1/2}$ values for fenamiphos in putting green and fairway soils were smaller than those from the rough. Half-life for the putting green surface soil collected after the annual application was three times shorter than in the soil collected before field application, 0.9 vs. 2.8 d. Ou (1991) reported that fenamiphos degradation in soil was enhanced after a single field application of fenamiphos at a rate of 4.48 kg ha^{-1} to an experimental site planted in potato. After one field application, $t_{1/2}$ for TTR was shortened from 131 to 22 d. In this study, after more than 15 yr of annual or biannual applications of fenamiphos to the turfgrass site, $t_{1/2}$ for TTR in surface soil was shortened to 1 to 3 d depending upon the time following the annual fenamiphos application.

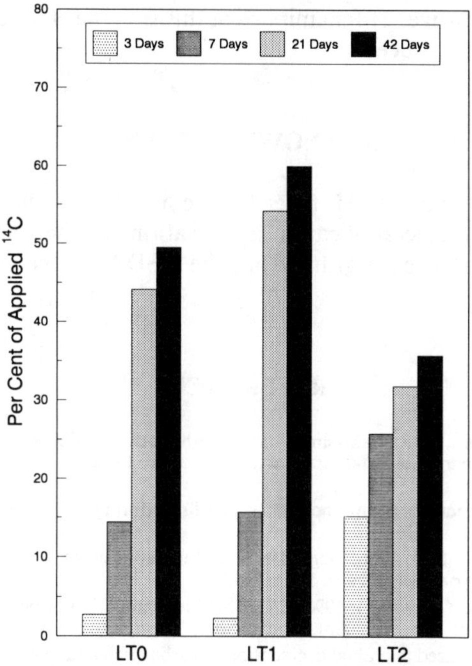

Fig. 14-4. Evolution of $^{14}CO_2$ from ^{14}C-fenamiphos treated fairway surface (0–15 cm deep) soil collected before the annual fenamiphos application. LT0 = no laboratory treatment; LT1 = one laboratory treatment; LT2 = two laboratory treatments.

CONCLUSION

Fenamiphos in soil collected from the Gainesville Golf and Country Club was rapidly degraded. The rapid degradation of fenamiphos was most likely due to an increase in fenamiphos-degrading microbes promoted by the repeated applications of fenamiphos to the site for more than 15 yr. Judging

Table 14-2. First-order rate constants (K_1) and half-lives ($t_{1/2}$) for disappearance of extractable ^{14}C-activity in turfgrass soil collected before and after the annual application of fenamiphos.

Soil	Depth	k_1	$t_{1/2}$	r^2
	cm	d^{-1}	d	
		Before annual application		
Fairway	0–15	2.477×10^{-1}	2.9	0.929
Fairway	15–30	1.279×10^{-1}	5.4	0.973
Putting green	0–10	2.487×10^{-1}	2.8	0.990
Rough	0–15	4.910×10^{-2}	14.1	0.942
Rough	15–30	5.160×10^{-2}	13.4	0.935
		After annual application		
Putting green	0–10	7.88×10^{-1}	0.9	0.999

from the short half-lives of fenamiphos at this site, the efficacy of the nematicide will be much reduced.

ACKNOWLEDGMENT

The authors thank T. Hoffner for technical assistance, and J. Hyden for providing pesticide-application information for the turfgrass site. This study was supported by a grant from the USDA Water Quality Research Program.

REFERENCES

Barik, S., P.A. Washid, C. Ramakrishna, and N. Sethunathan. 1979. A change in the degradation pathway of parathion after repeated applications to flooded soils. J. Agric. Food Chem. 27:1391–1392.

Ou, L.-T. 1991. Interactions of microorganisms and soil during fenamiphos degradation. Soil Sci. Soc. Am. J. 55:716–722.

Ou, L.-T., and P.S.C. Rao. 1986. Degradation and metabolism of oxamyl and phenamiphos in soils. J. Environ. Sci. Health B21:25–40.

Racke, K.D., and J.R. Coats (ed.). 1990. Enhanced biodegradation of pesticides in the environment. ACS Symp. Ser. 426. Am. Chem. Soc., Washington, DC.

Read, A.V. 1983. Enhanced microbial degradation of carbofuran and fensulfothion after repeated applications to acid mineral soils. Agric. Ecosyst. Environ. 10:37–46.

Sethunathan, N. 1973. Degradation of parathion in flooded acid soils. J. Agric. Food Chem. 21:602–604.

Waggoner, T.B., and A.M. Khasawinah. 1974. New aspects of organophosphorus pesticides. VII. Metabolism, biochemical, and biological aspects of nemacur and related phosphoamidate compounds. Residue Rev. 53:79–97.